velocity of electromagnetic waves in free space Co = $3 \times 10^8 \, m s^{-1}$

for all waves:
velocity = wavelength (λ) $\times$ frequency (f)
for e.m. waves Co = λf

u.v.: ultraviolet UHF: ultra high frequency
VHF: very high frequency

ertzian waves

o | medium radio | long radio | very long radio

UHF | VHF

ar

wavelength (λ) metres

10^0(1) 10^2 10^4 10^6 10^8

frequency (f) hertz

10^8 10^6 10^4 10^2 10^0(1)

DETROIT
PUBLIC
LIBRARY

TERESA RICKARDS

BARNES & NOBLE THESAURUS OF **PHYSICS**

Edited by DR R. C. DENNEY
and STEPHEN FOSTER

BARNES & NOBLE BOOKS
A DIVISION OF HARPER & ROW, PUBLISHERS
New York, Cambridge, Philadelphia, San Francisco
London, Mexico City, São Paulo, Sydney

BLA Publishing Limited and the author would like to thank Arthur Godman, Dr R. C. Denney, Stephen Foster and Rosie Vane-Wright for their helpful advice and assistance in the production of this book.

Library of Congress Cataloging in Publication Data

Rickards, Teresa
Barnes & Noble thesaurus of physics

Includes index.
1. Physics--Dictionaries. I. Title. II. Title: Thesaurus of Physics.
QC5.R5 1984 530'.03'21 83-47598
ISBN 0-06-015214-1
ISBN 0-06-463582-1 (pbk.)

This book was designed and produced by BLA Publishing Limited, Swan Court, London Road, East Grinstead, Sussex, England.

A member of the **Ling Kee Group**
LONDON·HONG KONG·TAIPEI·SINGAPORE·NEW YORK

Illustrations by Rosie Vane-Wright, Hayward & Martin, BLA Publishing Limited.
Phototypeset in Britain by Composing Operations Limited
Color origination by Planway Limited
Printed in Spain

Contents

How to use this book

This book combines the functions of a dictionary and a thesaurus: it will not only define a word for you, but it will also indicate other words related to the same topic, thus giving the reader easy access to one particular branch of the science. The emphasis of this work is on interconnections.

On pages 3 and 4 the contents pages list a number of broad groupings, sometimes with sub-groups, which may be used where reference to a particular theme is required. If, on the other hand, the reader wishes to refer to one particular word there is, at the back of the book, an alphabetical index in which approximately 1700 words are listed.

Looking up one particular word or phrase

Refer to the alphabetical index at the back of the book, then turn to the appropriate page. At the top of that page you will find the name of the general subject printed in bold type, and the specialised area in lighter type. For example, if you look up **meniscus**, you will find it listed on p.25, at the top of which page is **PROPERTIES OF MATTER**/SURFACE TENSION. If you were unsure of the meaning of the phrase, you may now not only read its definition, but also place it in context. Immediately after the word or phrase you will see in brackets (parentheses) the abbreviation indicating which part of speech it is: (*n*) indicates a noun, (*v*) a verb and (*adj*) an adjective. Then follows a definition of the word, expressed as far as is possible in language which is in common use. Where a related word is listed nearby, a simple system using arrows has been devised.

(↑) means that the related term may be found above or on the opposite page.
(↓) means that the related term may be found below or on the opposite page.

A page reference in brackets is given for any word which is linked to the topic but is to be found elsewhere in the book. You will soon appreciate the advantages of this scheme of cross-referencing. Let us take an example. On p.238 the entry **radioactive nucleus** is:

> **radioactive nucleus** an unstable nucleus (p.237) capable of emitting radioactive radiations (↓) on nuclear disintegration (↑).

To gain a broader understanding, the reader will look at the entry **radioactive radiations** below on the same page, at the entry **nuclear disintegration** above on the same page, and will also refer to the entry **unstable nucleus** on p.237.

Searching for associated words

As the reader will have observed, the particular organisation of this book greatly facilitates research into related words and ideas, and the extensive number of illustrations and diagrams assists in general comprehension.

Retrieving forgotten or unknown information

It would appear impossible to look up something one has forgotten or does not know, but this book makes it perfectly feasible. All that is required is a knowledge of the general area in which the word is likely to occur and the entries in that area will direct you to the appropriate word. If, for example, one wished to know more about **Young's experiment**, but had forgotten the term, it would be sufficient to know it was connected with **interference of waves**; the reader looking up **interference of waves** would be referred to **superposition of waves** which would then indicate **Young's experiment**, which is defined and/or further explained by means of a diagram.

Studying or reviewing a subject

Two methods of using this book will be helpful to the reader who wishes to know more about a topic, or who wishes to review knowledge of a topic.

(*i*) For a broader understanding of waves, for example, you would turn to the section dealing with this area and read through the different entries, following up the references which are given to guide you to related words.

(*ii*) If you have studied one particular branch of the science and you wish to review your knowledge, looking through a section on **magnetism**, by way of example, might refresh your memory or introduce an element which you had not previously realised was connected.

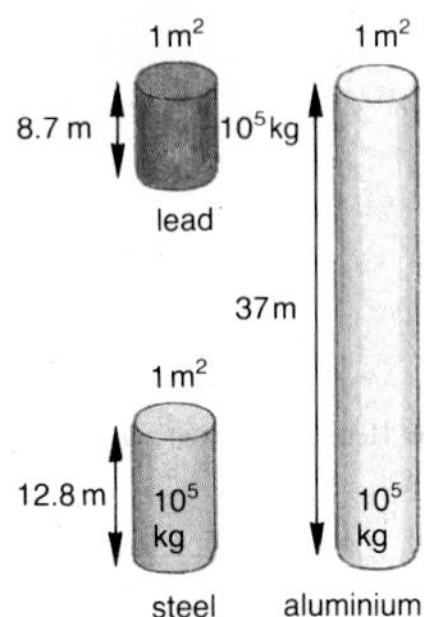

metal bars of equal masses

density = mass/volume

mass/density

lead:
volume = $8.7\,m^3$
density = $10^5/8.7$
$= 11.4 \times 10^3\,kg\,m^{-3}$

steel:
volume = $12.8\,m^{-3}$
density = $10^5/12.8$
$= 7.8 \times 10^3\,kg\,m^{-3}$

aluminium:
volume = $37\,m^3$
density = $10^5/37$
$= 2.7 \times 10^3\,kg\,m^{-3}$

atom (e.g. argon)

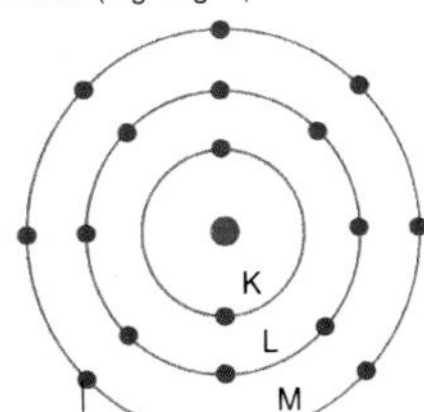

shell structure 2K, 8L, 8M electrons
nucleus has 18 protons and 18 neutrons

mass (*n*) the quantity of matter in a body (p.12) indicating its inertia (p.36); units of mass are gram (g), kilogram (kg). **massive** (*adj*).

density (*n*) mass (↑) per unit volume of a body or substance; ratio of mass/volume for a body or substance; units are gram/cm³ ($g\,cm^{-3}$), kilogram/m³ ($kg\,m^{-3}$). **dense** (*adj*).

specific gravity the density of a substance ÷ the density of water; it has no units.

particle (*n*) a very small body (p.12) which can be observed, e.g. Brownian movement (p.14), or studied indirectly, e.g. atom (↓), molecule.

fundamental particle a particle (↑) identified as a fundamental constituent of the atom (↓) and apparently indivisible, e.g. electron (↓), proton (p.8), neutron (p.8).

atom (*n*) the smallest particle (↑) of matter which can take part in a chemical reaction. **atomic** (*adj*).

element (*n*) substance with only one kind of atom (↑); 92 naturally occurring elements. **elementary** (*adj*).

nucleus (*n*) tiny, positively charged central part of an atom (↑) where most of its mass (↑) is concentrated; contains the fundamental particles (↑) protons (p.8) and neutrons (p.8); its diameter is of the order of 10^{-14}m. **nuclear** (*adj*).

electron (*n*) a very small fundamental particle (↑) of matter moving in orbit round the nucleus (↑) of an atom (↑); mass (↑) 9.11×10^{-31} kg, negative electric charge (p.165) 1.60×10^{-19} coulomb (C). **electronic** (*adj*).

electronic charge negative electric charge on an electron (↑), 1.60×10^{-19} coulomb (C); denoted by e.

specific charge electric charge (p.165) carried per unit mass (↑) by an electron (↑) or other charged particle; ratio of electronic charge (e)/electron mass (m); $e/m = 1.76 \times 10^{11}$ coulomb/kg ($C\,kg^{-1}$)

orbit (*n*) circular or elliptical path taken by a body or particle revolving round a centre of attraction, e.g. planets around the sun in the solar system (p.43), electrons around an atomic nucleus (↑). **orbit** (*v*).

energy level the energy state of a particular electron in orbit around the central nucleus (↑) of an atom; discrete energy levels (p.234).

orbital (*n*) path taken by an electron in orbit (↑) at a specific energy level (↑) around the central nucleus of an atom. **orbital** (*adj*).

nucleon (*n*) a particle in the nucleus of an atom, e.g. proton (p.8) neutron (p.8).

proton (*n*) a charged particle in the nucleus (p.7) of an atom; the nucleus of the commonest form of hydrogen atom; mass 1.67×10^{-27} kg or 1.007 a.m.u. (↓); positive charge 1.60×10^{-19} coulomb (C) is equal numerically to the negative electronic charge (p.7).
neutron (*n*) an uncharged particle in the nucleus (p.7) of an atom; mass 1.009 a.m.u. (↓).
nuclear forces short-range attractive forces between nucleons (p.7) holding an atomic nucleus (p.7) together against electrostatic repulsive forces (p.166) between protons (↑).
nucleon number (A) the number (A) of nucleons (p.7) in the nucleus of an atom of a particular element (p.7).
mass number alternative name for nucleon number (↑).
proton number (Z) the number (Z) of protons (↑) in the nucleus of an atom of a particular element (p.7).
atomic number alternative name for proton number (↑).
neutral atom describes the normal electrical state of an isolated un-ionized atom (↓) with equal numbers of positively charged protons (↑) in the nucleus and negatively charged electrons in orbit in energy levels (p.7) around the nucleus (p.7).
nuclide a symbolic way of representing the unique features of a particular atomic nucleus (p.7) which distinguish it from other nuclei, in the form of $^{A}_{Z}X$, where X = chemical symbol for the element (p.7) concerned, A = nucleon number (↑) and Z = proton number (↑), e.g. helium nucleus $^{4}_{2}He$ where A = 4 and Z = 2.
isotopes (*n*) two or more nuclides (↑) having the same atomic number (Z) (↑) but different nucleon numbers (A) (↑) so that they have identical electron configurations and chemical properties and cannot be separated by chemical means, e.g. uranium-238, $^{238}_{92}U$; uranium-235, $^{235}_{92}U$. **isotopic** (*adj*).
isotopes of hydrogen chemically identical nuclides (↑) of hydrogen gas, distinguished by name and chemical symbol: hydrogen H, deuterium D, tritium T; their compounds with oxygen give chemically identical forms of water distinguished by name and chemical formula: light water H_2O, heavy water D_2O, tritiated water T_2O; tritium is radioactive (p.239); hydrogen is the commonest naturally occuring isotope.
valence electrons electrons capable of participating in a chemical reaction.
revolve (*v*) to move in orbit (p.7) about a centre of attraction. **revolution** (*n*).

isotopes

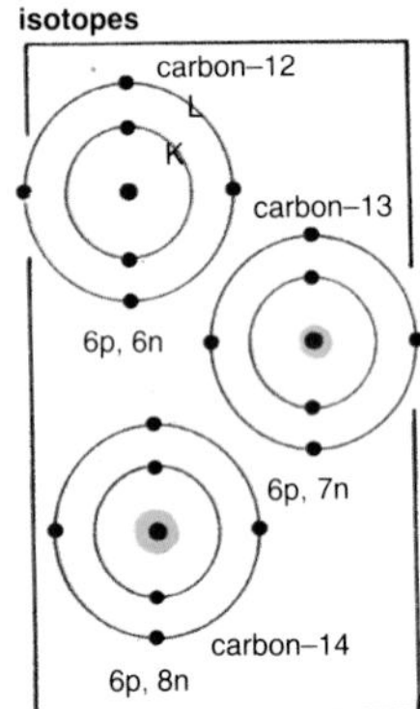

isotopes of carbon:
2K, 4L electrons
carbon–12 $^{12}_{6}C$; A = 12; Z = 6
carbon–13 $^{13}_{6}C$; A = 13; Z = 6
carbon–14 $^{14}_{6}C$; A = 14; Z = 6
p = number of protons in nucleus
n = number of neutrons in nucleus

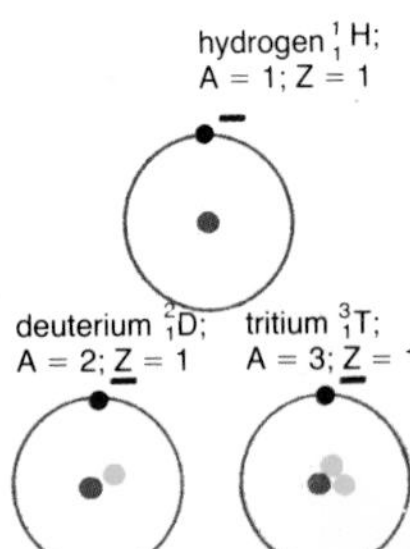

isotopes of hydrogen:
1 K electron

abundance (*n*) ratio of number of atoms of a particular isotope (↑) ÷ the total number of atoms of a particular element (p.7) present in a naturally occurring mixture of isotopes in an element (p.7) or compound; usually expressed as a percentage, e.g. natural uranium contains 0.71% of uranium-235 atoms. **abundant** (*adj*).

atomic mass unit (a.m.u.) mass of an atom of carbon-12 isotope (↑) ÷ 12.000; 1 a.m.u. = 1.660×10^{-27} kg.

relative atomic mass mass of 1 atom of an element relative to 1⁄12 of mass of carbon-12 isotope.

ionized atom atom from which one or more orbital electrons has been removed leaving the atom with a resultant positive charge. **ionize** (*v*).

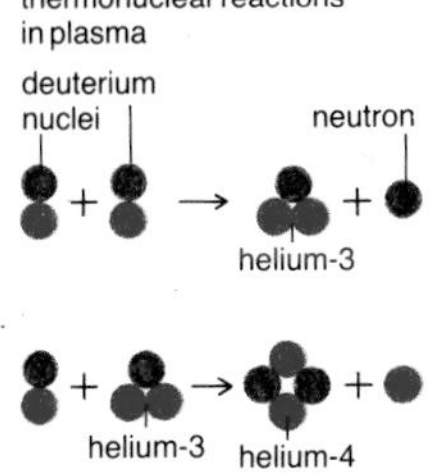

conduction electrons electrons (p.7) in an electrical conductor (p.155), e.g. metal, loosely bound to their parent atoms and released from outer electron shells by thermal energy (p.28) of vibration of atoms in the crystal lattice. Their movement in the direction of an applied electric field (p.165) transfers an electric charge (p.165) through the metal; represented as overlapping valence and conduction bands in energy band model for solids (p.15).

charge carriers particles carrying positive or negative electric charges which make up electric current (p.152) e.g. electrons (p.7) and positive ions (↓); holes (p.223) are positive charge carriers in semiconductors (p.222); the direction of the conventional current (p.150) is given by the resultant direction of flow of positive charge.

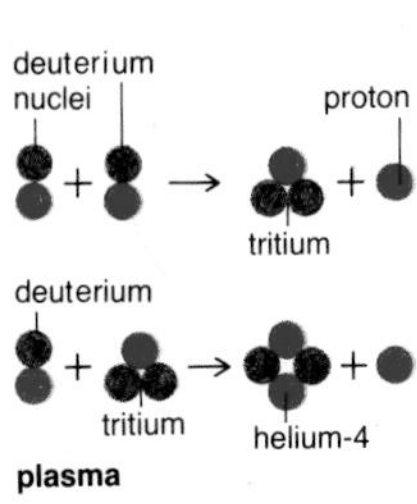

plasma

ion (*n*) an atom or group of atoms with positive or negative electric charge (p.165) found in electrolytes (p.10) and in the gas in a discharge tube and in a plasma (↓). **ionize** (*v*), **ionic** (*adj*).

plasma (*n*) a highly ionized substance at very high temperature, such as the material in the sun or in a torus; the atoms are nearly all fully ionized and the substance consists of electrons and atomic nuclei; it has been described as the fourth state of matter (p.14).

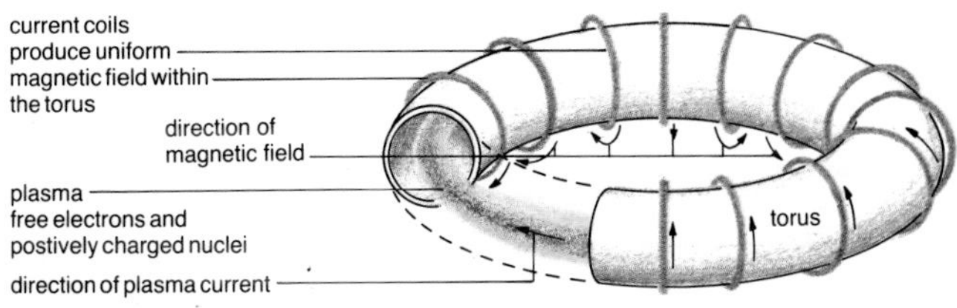

ionic dissociation separation of a neutral molecule into 2 oppositely charged ions (p.9) in solution in water, e.g. NaCl into Na^+ and Cl^-. **dissociate** (*v*), **dissociated** (*adj*).

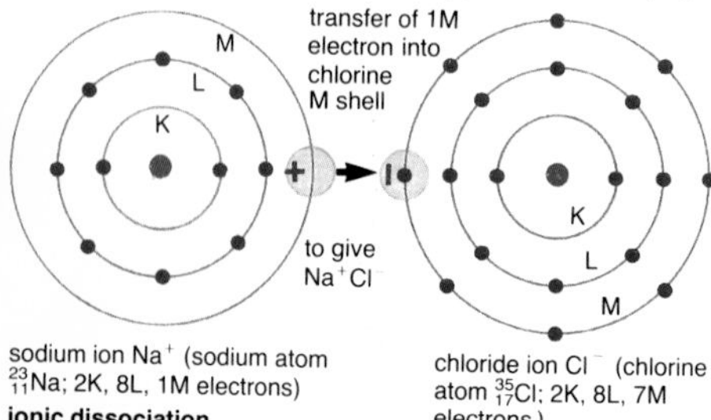

sodium ion Na^+ (sodium atom $^{23}_{11}Na$; 2K, 8L, 1M electrons)
chloride ion Cl^- (chlorine atom $^{35}_{17}Cl$; 2K, 8L, 7M electrons)
ionic dissociation

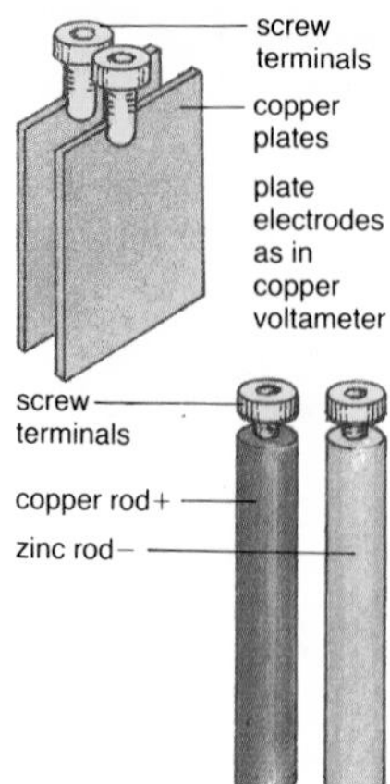

rod electrodes, poles or terminals in Leclanché cell
electrode

electrolyte (*n*) a solution which is electrically conducting because the solute molecules are dissociated (↑) into ions in water, e.g. sodium chloride, silver nitrate, copper sulphate. **electrolytic** (*adj*).

electrical conductivity the reciprocal (1/ρ) of electrical resistivity (p.158) of an electrolyte (↑) or other conductor; units are $ohm^{-1}cm^{-1}$.

electrode (*n*) the electrically conducting plate or rod through which electric current (p.152) enters or leaves a conducting medium.

cathode (*n*) the electrode (↑) maintained at negative electric potential (p.170) attracting positive ions (p.9) during electrolysis (↓); emitter of electrons (p.7) in vacuum discharge tube. **cathodic** (*adj*).

anode (*n*) the electrode (↑) maintained at positive electric potential (p.170), attracting negative ions (p.9) or electrons (p.7) during electrolysis (↓) or in a vacuum discharge tube. **anodic** (*adj*).

electrolysis (*n*) passage of an electric current (p.152) through an electrolyte (↑) causing chemical effects due to discharging ions at electrode (↑) surfaces. **electrolyse** (*v*), **electrolytic** (*adj*).

electrolytic cell 2 electrodes (↑) immersed in an electrolyte (↑) supplied with electric current (p.152) from a d.c. source (p.150); used to carry out electrolytic changes in industrial processes, e.g. purification of metals, electroplating of metals.

voltameter (*n*) 2 electrodes (↑) immersed in an electrolyte (↑) supplied with electric current (p.152) from a d.c. source (p.150) used to investigate electrochemical effects of current.

electrolysis
Hofmann voltameter for electrolysis of water

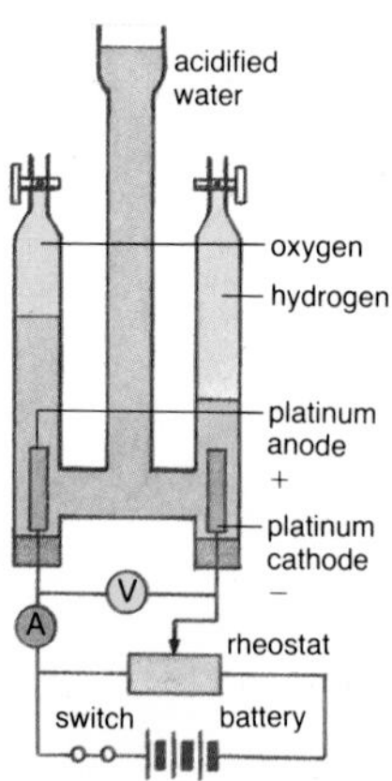

copper voltameter

copper plate cathode −
copper plate anode +
copper sulphate electrolyte
SO_4^{2-}
Cu^{2+}
switch
battery
A
rheostat

copper voltameter 2 copper plate electrodes (↑) immersed in copper sulphate electrolyte (↑); on passage of an electric current (p.152) copper is dissolved from the anode (↑), transferred through the electrolyte as ions (↑) and deposited as metal on the cathode (↑).

mole (*n*) quantity of substance containing the same number of elementary units as there are carbon atoms in 0.012 kilogram of the carbon-12 isotope (p.8); the elementary unit must be specified and can be an atom (p.7), a molecule, an ion (p.9) or an electron (p.7); the mole is a basic unit on the SI system, e.g. 1 mole of hydrogen has mass 1 gram; 1 mole of electrons has mass 5.486×10^{-4} gram, where 1 mole of electrons = electron mass $\times N_o$ and N_o = Avogadro number (↓); mole denoted by mol.

Avogadro constant the number of particles in 1 mole (↑); the approximate value is 6.02×10^{23}.

Avogadro number alternative name for Avogadro constant (↑).

Faraday's constant quantity of electric charge (p.165) carried by 1 mole (↑) of electrons; denoted by F; value is 9.65×10^4 coulomb/mole; $F = N_o e$ where e = electronic charge (p.7) and N_o = Avogadro Number (↑).

electrochemical equivalent (e.c.e.) mass of any ion liberated or deposited during electrolysis (↑) by the passage of 1 coulomb (p.165) of electric charge; denoted by z; mass liberated or deposited (kg) = zIt where I = electric current (ampere) (p.152) and t = time (s); units of z are kilogram/coulomb ($kg\,C^{-1}$) ($1A = 1C\,s^{-1}$).

electrolysis

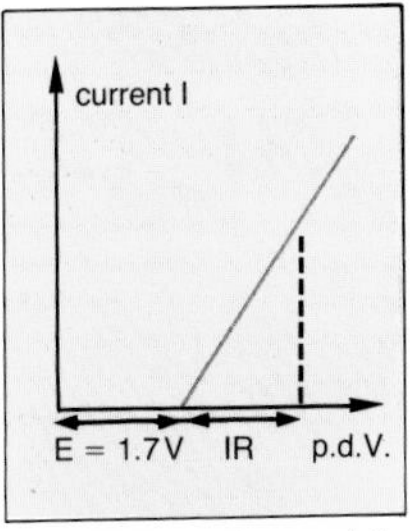

current/voltage characteristic of Hofmann water voltameter showing back e.m.f. E = 1.7 volt

Faraday's Laws of electrolysis when an electric current (p.152) is passed through an electrolytic cell (↑) or a voltameter (↑) the chemical effects occurring at the electrodes during electrolysis (↑) are summarized in two laws: 1. The mass of any substance liberated at or dissolved from the electrodes of a voltameter during electrolysis is directly proportional to the total quantity of electric charge (p.165) passed through the electrolyte; mass liberated (m) = zIt (kg) where z = electrochemical equivalent (↑). 2. The masses of different substances liberated at or dissolved from the electrodes of a voltameter during electrolysis by the same quantity of electric charge are in the ratio of their relative ionic masses divided by the number of charges carried by the respective ions.

dimension (*n*) linear measurement taken on a solid body (↓), e.g. length, breadth or height.
dimensions of a physical quantity a way of expressing a physical quantity in terms of powers of fundamental quantities such as mass M, length L and time T, e.g. area = L^2, velocity = LT^{-1}, force = MLT^{-2}; electrical and magnetic quantities require current I as an additional dimension, e.g. charge = IT, voltage = $ML^2T^{-3}I^{-1}$.
dimensional analysis a technique of using the dimensions of physical quantities (↑) to investigate the validity of an equation relating different physical quantities.
body (*n*) sample of solid matter (p.15) with a clearly recognizable shape, e.g. metal cube.
rigid body a body (↑) whose dimensions (↑) and shape are not easily altered by deforming forces (p.19), e.g. an iron bar resists bending or twisting.
lamina (*n*) plane solid plate of negligible thickness. **laminar** (*adj*).
crystal (*n*) sample of a solid substance (p.15) with a regular shape having its surface bounded by plane facets of characteristic shape with characteristic angles between them, e.g. diamond, quartz, calcite; a crystal is built up from millions of identical crystal unit cells (↓). **crystallization** (*n*), **crystallize** (*v*), **crystalline** (*adj*).
crystal unit cell simplest possible arrangement of chemically bonded atoms or molecules showing the characteristic internal structure of a crystalline solid (↓).
single crystal a crystal (↑) in which the crystal unit cell (↑) is repeated thousands of times in any direction without irregularity.
crystal lattice the continuous regular arrangement of atoms or molecules within a crystal or crystalline solid.
crystalline structure possessed by any material or substance with an ordered internal structure formed by repetition of the crystal unit cell (↑) in 3-dimensional space; many complex compounds have an ordered internal structure revealed only by X-ray crystallography, e.g. DNA, insulin, penicillin.
liquid crystals substances, e.g. cyanobiphenyl, cholesteryl benzoate, existing in a stable form intermediate between solid and liquid states (p.15) at normal temperatures, e.g. −10°C to +60°C; application of an electric field (p.165) disturbs the randomly oriented molecules and alters the optical properties of

body
e.g. metal cube

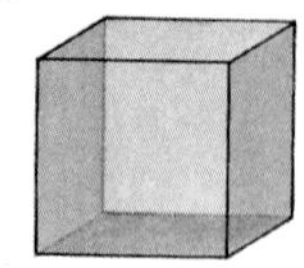

lamina
e.g. triangular lamina
area = ½h × b

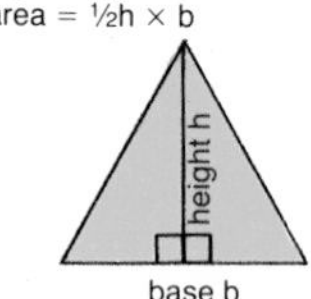

crystal lattice

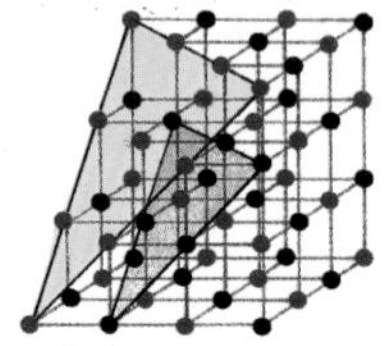

● Cl−
● Na+
□ Cl− plane
□ Na+ plane

crystal lattice of sodium chloride: lattice planes parallel to cube face have equal numbers of Na^+ and Cl^- ions. Lattice planes coloured have ions of one type only

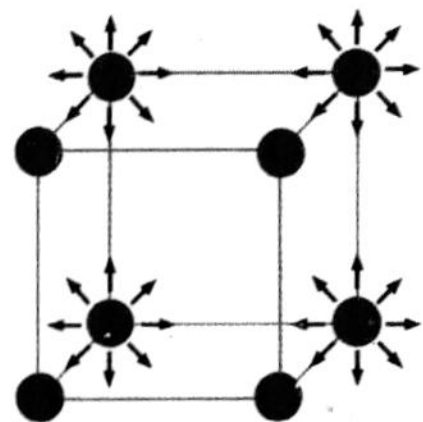

lattice vibrations: take place in all directions around atom (shown in one plane only)

mean rest position
(mean distance separating 2 vibrating atoms in crystal lattice)
mean rest position
line of centres
mean rest position

ordered internal structure of crystal lattice: atoms or molecules maintain same mean rest position at different temperatures

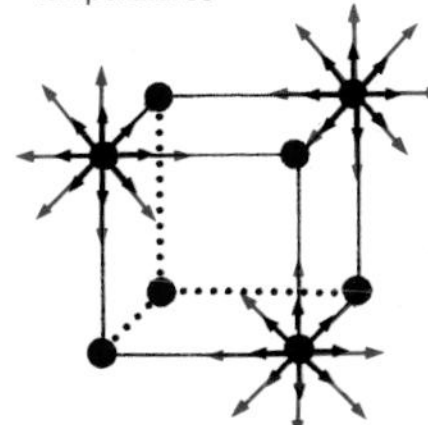

● atom or molecule in mean rest position
← vibrations at temperature T_1
→ vibrations at higher temperature T_2

the crystal (↑) giving it a clear colour and sharply defined appearance; used in small scale digital displays in watches and pocket calculators, and in a matrix arrangement as the colour display elements of a small-scale pocket TV tube, activated by a connecting matrix of silicon transistors integrated into a single silicon chip (p.227) circuit.

lattice vibrations describes simple harmonic vibration (p.50) of atoms or molecules in a crystal lattice due to thermal energy (p.28); occurs in random (p.14) directions about a mean rest position.

mean rest position the centre about which atoms or molecules in a crystal lattice vibrate due to thermal energy (p.28).

internal energy the sum of all the kinetic (p.27) and potential energies (p.27) of the atoms or molecules in a system; the internal energy of a system cannot be determined, only changes in its value can be measured.

internal kinetic energy the total kinetic energy (p.27) of all atoms or molecules in a given sample of substance having constant mass; includes vibrational (↓) and translational (↓) energy, both depending on temperature (p.116).

molecular vibrational energy the energy of an atom or molecule vibrating about a mean rest position (↑) in a crystal lattice; the energy of a liquid or gas molecule whose atoms vibrate with respect to each other; vibrational energy includes kinetic and potential energy.

translational energy the kinetic energy (p.27) of an atom or molecule in a liquid or gas, free to move from place to place because cohesive forces (p.14) are weak or negligible.

molecular potential energy the energy of molecules due to the intermolecular forces; this depends on the separation of the molecules, i.e. the volume of the substance.

ordered internal structure the structure possessed by a material or substance with crystalline structure (↑); individual atoms or molecules remain in the same rest positions (↑) in the crystal lattice even when thermal energy (p.28) varies due to temperature (p.116) variation. **order** (*n*).

disorder (*n*) the property of randomness (p.14) in a physical system. **disordered** (*adj*).

randomness (*n*) a lack of internal organization or order (p.13) in a physical system, e.g. atoms or molecules of a gas (↓) are in a constant state of motion due to their translational energy (p.13) but their movements have no preferred direction. **random** (*adj*).

random movement the movement of atoms or molecules of a substance without any preferred direction.

Brownian movement the continuous irregular motion of particles (p.7) of approximate diameter 10^{-3} mm or less when held in suspension in a fluid (↓); observable for smoke particles in air and originally for pollen grains in water; the effect is due to constant bombardment of suspended particles by molecules of the fluid in random motion (↑), offering evidence of the existence of molecules and confirming the kinetic theory of matter (p.143).

cohesion (*n*) the property of atoms or molecules of a substance to attract each other without chemical bonding. **cohesiveness** (*n*), **cohere** (*v*), **cohesive** (*adj*).

cohesive forces attractive forces between atoms or molecules of the same substance not chemically bonded to each other.

van der Waals' forces an alternative name for cohesive forces (↑).

adhesion (*n*) the property of atoms or molecules of one substance to attract atoms or molecules of another substance to which they are not chemically bonded. **adhesiveness** (*n*), **adhere** (*v*), **adhesive** (*adj*).

adhesive forces attractive forces between atoms or molecules of one substance and atoms or molecules of another substance not chemically bonded to it.

incompressible substance a material or substance which resists changes in volume due to changes in external pressure, e.g. matter in the solid or liquid state (↓). **incompressibility** (*n*).

state of matter this specifies whether a material or substance is a solid (↓), liquid (↓) or gas (↓).

phase of matter an alternative name for state of matter (↑).

change of state the process by which a material or substance changes from one state of matter (↑) to another without a change in temperature (p.116); accompanied by a change in volume and in the degree of randomness (↑) in the internal structure.

random movement of gas molecules in a cubical container; no preferred direction; all directions possible

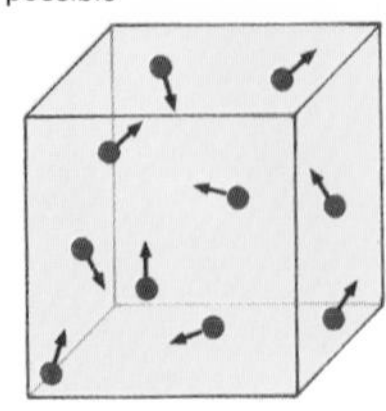

Brownian movement: irregular random paths followed by particles in suspension

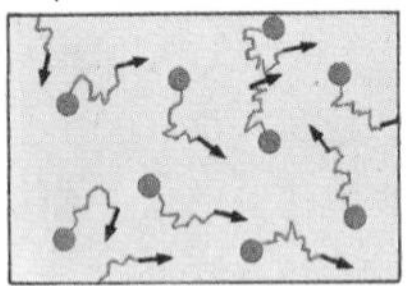

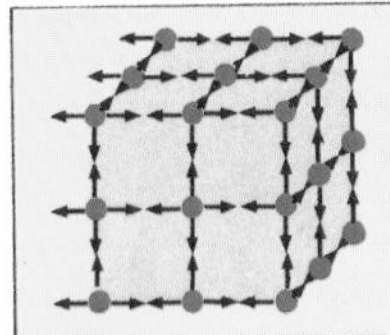

solid state

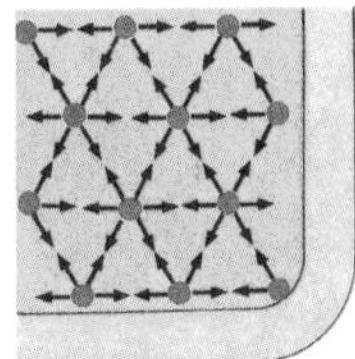

liquid state

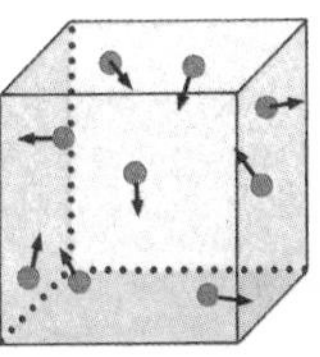

gaseous state

solid state a state of matter (↑) characterized by an ordered internal crystalline structure (p.12).

liquid state a state of matter (↑) characterized by greatly reduced internal order (p.13) in its crystalline structure; chemical bonds in the crystal lattice are weakened by increased vibrational thermal energy (p.28); the crystal lattice extends across only a few crystal unit cells (p.12) in any direction and groups of chemically bonded atoms or molecules can move past each other into new positions, giving the liquid its characteristic fluid flow (↓) property; cohesive forces (↑) are strong. **liquefy** (*v*).

gaseous state a state of matter (↑) characterized by the absence of any ordered internal structure (p.13) as vibrational thermal energy (p.28) exceeds the chemical bond strengths; individual atoms or molecules show independent random movements (↑) throughout their containing volume; cohesive forces (↑) are weak or negligible and a gas has the property of fluid flow (↓).

fluid (*n*) liquid (↑) or gaseous (↑) matter and sometimes powders. **fluid** (*adj*).

fluid flow the property of fluid (↑) matter to move in the direction from high to low pressure when a pressure gradient (↓) is applied.

pressure gradient pressure (p.39) difference applied between 2 points; pressure change per metre between 2 points; units are newton/metre (Nm^{-1}).

pressure gradient
pressure difference between AA′ and BB′ $= (p_1 - p_2)$
pressure gradient along section of pipe $= \frac{(p_1 - p_2)}{l}$ (Nm^{-1})

pipe carrying fluid

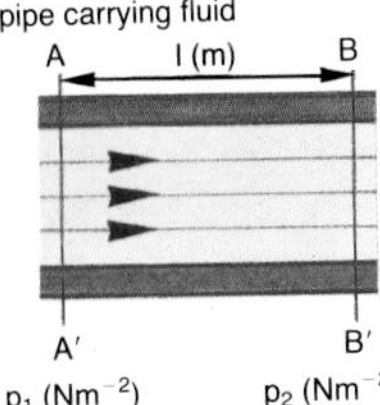

melting (*n*) the change of state (↑) from solid (↑) to liquid (↑) occurring at a constant temperature (p.116). **melt** (*v*), **melting** (*adj*), **molten** (*adj*).

melting point the characteristic temperature (p.116) at which a pure substance changes state from solid (↑) to liquid (↑); usually given at standard atmospheric pressure (p.40).

freezing point the characteristic temperature at which a pure substance changes state from liquid to solid at standard atmospheric pressure; identical with melting point (↑).

evaporation
evaporation from liquid surface: molecules or atoms with high kinetic energy can pass through the liquid surface; occurs at all temperatures

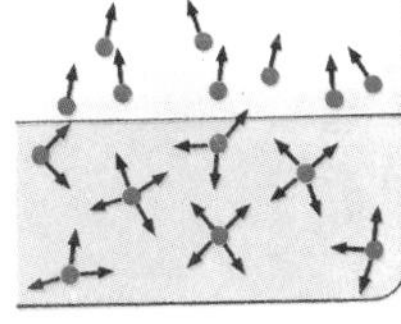

evaporation (*n*) the escape of molecules from the surface of a solid (p.15) or liquid (p.15); it is possible at all temperatures (p.116). **evaporate** (*v*), **evaporated** (*adj*).

vapour (*n*) a gas (p.15) which can be liquefied (↓) by increase of pressure (p.39) alone; a gas below its critical temperature (p.140). **vaporous** (*adj*).

vaporization (*n*) an alternative name for evaporation (p.15) from the liquid state (p.15). **vaporize** (*v*).

boiling (*n*) the change of state (p.14) from a liquid (p.15) to a vapour (↑) at the boiling point (↓); it occurs when the saturation vapour pressure (p.18) of the liquid equals the external pressure (p.39); it is characterized by the formation of vapour bubbles throughout the liquid. **boil** (*v*), **boiling** (*adj*).

boiling point the characteristic temperature (p.116) at which a pure substance changes state from a liquid (p.15) to a vapour (↑) by boiling (↑) at standard atmospheric pressure (p.40); the temperature at which the saturation vapour pressure (p.18) of a liquid equals the standard atmospheric pressure.

condensation (*n*) a change of state (p.14) from a vapour (↑) to a liquid (p.15) or solid (p.15). **condense** (*v*), **condensed** (*adj*).

liquefaction (*n*) a change of state (p.14) from vapour (↑) to liquid (p.15). **liquefy** (*v*), **liquefied** (*adj*), **liquefiable** (*adj*).

supercooled liquid material or substance existing in the liquid state (p.15) at a temperature below its melting point (p.15), e.g. sodium thiosulphate below 32.4°C.

latent heat a change in the internal energy (p.13) of a physical system without a change in the temperature (p.116), e.g. during a change of state (p.14); the heat energy absorbed by a material or substance during melting (p.15) or boiling (↑) increases disorder (p.13) in the internal structure and makes the change of state possible; similarly heat energy given out during condensation or solidification reduces disorder and makes it possible for the material or substance to assume a more ordered internal structure (p.13).

specific latent heat of fusion is the quantity of heat required to change 1 kilogram of a specific solid (p.15) substance at its melting point (p.15) to liquid (p.15) at the same temperature; units are joule/kilogram ($J\,kg^{-1}$); kilojoule/kilogram ($kJ\,kg^{-1}$) e.g. specific latent heat of fusion of ice at 0°C (273K) = $334\ kJ\,kg^{-1}$ ($334\ J\,g^{-1}$).

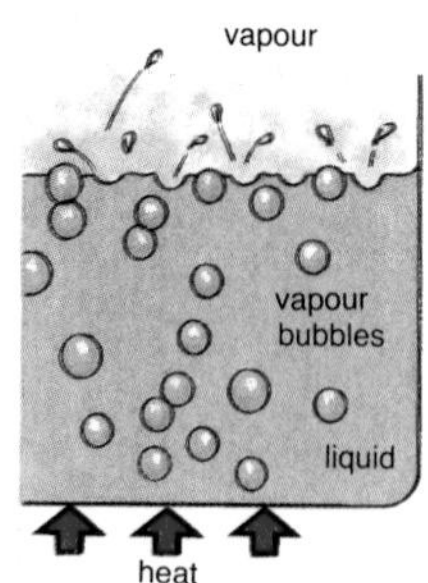

boiling
vapour bubbles rise to the liquid surface and burst

supercooled liquid

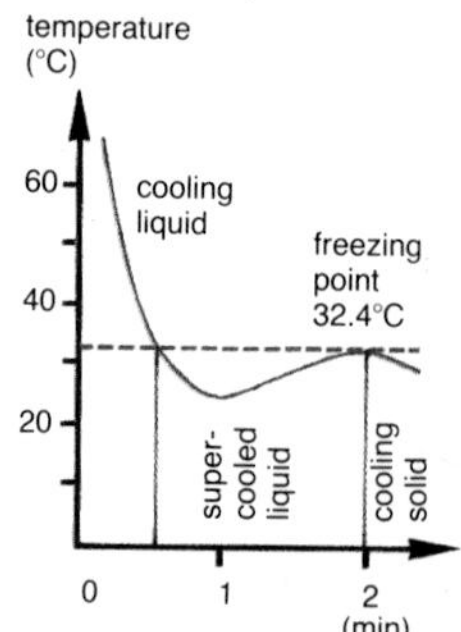

supercooling sodium thiosulphate below 32.4°C

specific latent heat of vaporization the quantity of heat required to change 1 kilogram of a specific liquid (p.15) at its boiling point (↑) to a vapour (↑) at the same temperature; units are joule/kilogram ($J\,kg^{-1}$); kilojoule/kilogram ($kJ\,kg^{-1}$), e.g. specific latent heat of vaporization of water at 100°C = $2260\,kJ\,kg^{-1}$ ($2260\,J\,g^{-1}$).

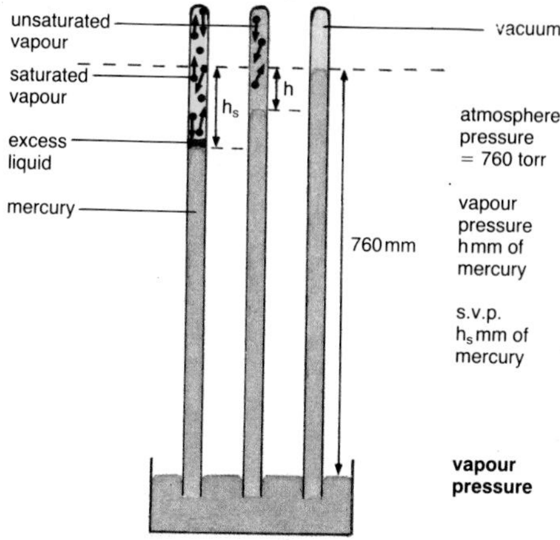

vapour pressure

vapour pressure the pressure (p.39) exerted by a vapour (↑) alone or the partial pressure (↓) exerted by a vapour in non-combining mixture with other gases, e.g. water vapour in air.

partial pressure individual pressure (p.39) exerted by a vapour (↑) or gas (p.15) component in a non-combining mixture with other gases or vapours, e.g. water vapour pressure (↑) in air; total pressure of a gas/vapour mixture is given by Dalton's Law (↓) of partial pressures.

Dalton's Law of partial pressures states that the total pressure of a non-combining gaseous mixture is equal to the sum of the individual partial pressures (↑) of its components, each of which exerts the same pressure as it would if it alone occupied the containing volume at that temperature (p.116).

Dalton's Law of partial pressures

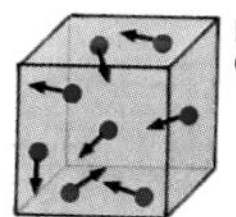

partial pressure due to gas 1 = p_1

partial pressure due to gas 2 = p_2

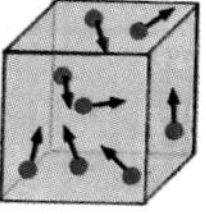

total pressure due to both gases 1 and 2 in same volume = $p_1 + p_2$

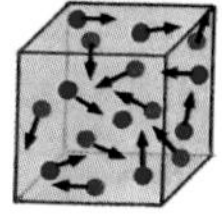

saturated vapour condition in an enclosed volume of space when it contains the maximum possible amount of a particular vapour (p.16) either alone or in a non-combining mixture with other gases or vapours; the vapour pressure (p.17) alone or the partial pressure (p.17) of vapour has its maximum value; the vapour is in surface contact with its liquid and a dynamic equilibrium condition (↓) exists in which the number of molecules leaving the liquid surface per second by evaporation (p.15) equals the number of vapour molecules re-entering the liquid surface per second by condensation (p.16); a saturated vapour does not conform with the Gas Laws (p.138) or show ideal gas behaviour (p.138), and saturation vapour pressure (↓) depends only on temperature (p.116).

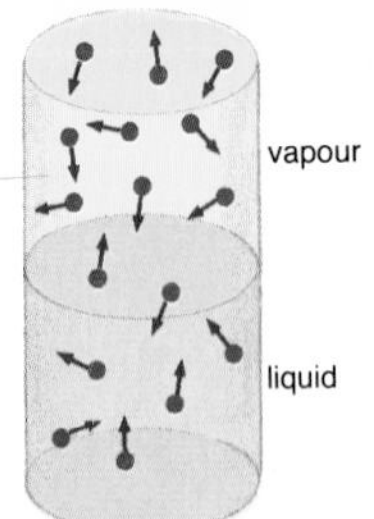

saturated vapour
rate of evaporation of molecules equals rate of condensation

dynamic equilibrium condition a situation in which the overall external conditions remain the same, although constant change occurs within a physical system, e.g. continuous exchange of molecules between a saturated vapour (↑) and its liquid in surface contact with it.

unsaturated vapour the condition in an open or enclosed space before the space becomes saturated with vapour (p.16) exerting saturation vapour pressure (↓); a vapour far from its saturation (↑) condition, conforms well with the Gas Laws (p.138) and they can be applied to it in making calculations, e.g. under most conditions the atmosphere (↓) contains some unsaturated water vapour, which can be treated as an ideal gas (p.138) for calculation purposes.

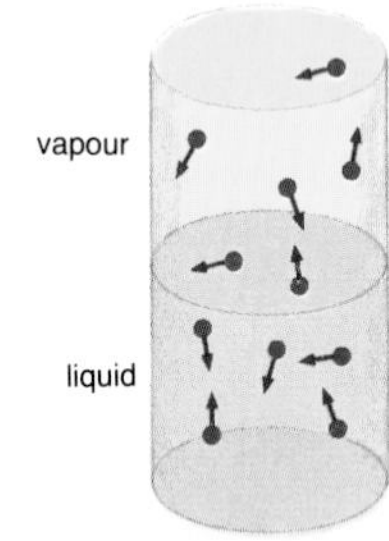

unsaturated vapour
rate of evaporation of molecules exceeds rate of condensation

saturation vapour pressure denoted by s.v.p., it is the maximum pressure exerted by a saturated vapour (↑) in contact with its liquid; its value depends only on temperature (p.116); a liquid boils (p.16) when its s.v.p. equals the external pressure applied, e.g. water boils at 100°C (373K) at normal atmospheric pressure.

atmosphere (*n*) any surrounding gas (p.15); the Earth's atmosphere is a gaseous mixture with composition by volume: 78% nitrogen, 21% oxygen, 1% carbon dioxide, water vapour and inert gas traces; the pressure (p.39) at sea level on the Earth's surface is approximately 760 torr (p.39), air density (p.7) at the Earth's surface is 1.293 kg/m^3 at s.t.p. (p.138); atmospheric pressure (p.40) is measured by barometer (p.39); standard pressure of 1 atmosphere = 1.01325 $\times 10^5$ newton/m^2 (N m^{-2}). **atmospheric** (*adj*).

atmosphere of the Earth
○ Appleton-F-layer
□ Kenelly-Heaviside-E-layer

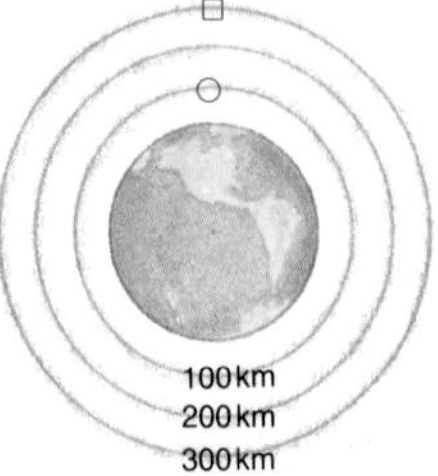

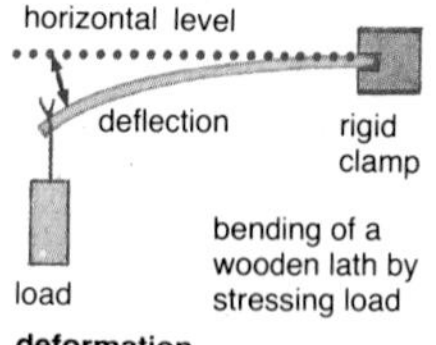

deformation

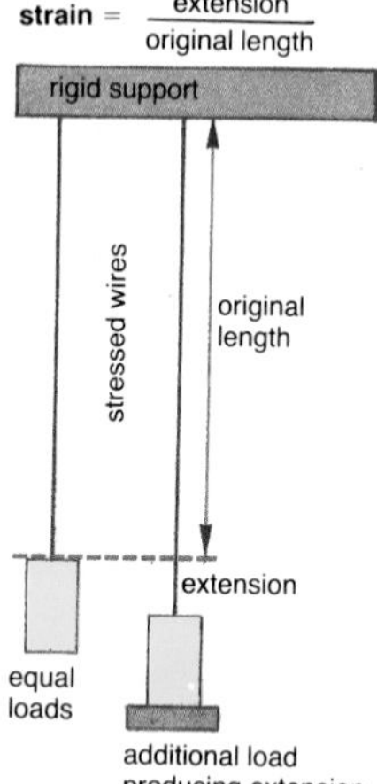

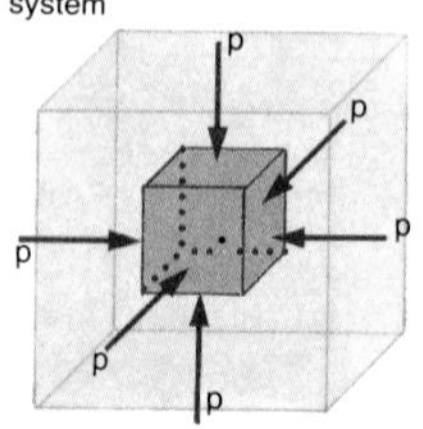

p external deforming forces of applied pressure (newton/m^2)

external pressure p applied in all directions (newton/m^2) reduces large volume V by small volume dV without change of shape
bulk modulus = pV/dV (Nm^{-2})

deformation (*n*) a change in shape or volume, or both, of a physical system, caused by external force (p.26). **deform** (*v*), **deforming** (*adj*), **deformed** (*adj*).

creep the slow permanent deformation (↑) of a crystal (p.12) or other specimen under constant stress (↓).

deforming force an external force (p.26) acting on a physical system to produce deformation (↑).

stress (*n*) the deforming force (↑) per metre^2 applied to a physical system; units are newton/metre^2 (Nm^{-2}), kilogram force/metre^2 (kgf m^{-2}). **stress** (*v*), **stressed** (*adj*).

strain (*n*) measure of deformation (↑) in a physical system produced by application of stress (↑); change in a physical system compared with the original physical state of that system, e.g. change in length of a stressed (↑) wire compared with its original length; value is a ratio and has no units. **strain** (*v*), **strained** (*adj*).

elasticity (*n*) a property of a physical system enabling it to return to its original physical state after the removal of stress (↑), e.g. a steel wire can return to its original length on removal of stress applied within the elastic limit (↓), a gas regains its original volume after compression according to Boyle's Law (p.137). **elastic** (*adj*).

elastic limit the limit beyond which a physical system loses its elasticity (↑) and does not recover its original physical state after deformation (↑).

modulus of elasticity the ratio of stress (↑) ÷ strain (↑) for a body obeying Hooke's Law (p.20), e.g. Young's modulus (p.20).

bulk modulus the modulus of elasticity (↑) for a compressible physical system undergoing a change in volume without a change in shape.

isothermal bulk modulus the bulk modulus (↑) for a gas (p.15) under isothermal conditions (p.140), value is p (Nm^{-2}), where p = gas pressure.

adiabatic bulk modulus the bulk modulus (↑) for a gas (p.15) under adiabatic conditions (p.142), its value is γp (Nm^{-2}), where p = gas pressure, and γ = ratio of $C_p \div C_v$; it is characteristic of the gas concerned; C_p and C_v are the principal specific heats (p.142) of the gas.

tensile stress the longitudinal stress (↑) producing extension (↓) in a rigid body or physical system.

extension (*n*) an increase in length of a wire or spring, when subjected to tensile stress (↑). **extend** (*v*), **extended** (*adj*), **extensible** (*adj*).

longitudinal strain the change in length δL compared with the original length L of a wire or spring subjected to tensile stress (p.19); δL (mm or m) ÷ L (mm or m) has no units.

load[1] (*n*) a mass (p.7) suspended from a wire, spring or other support, so that its weight (p.32) acts as a stress (p.19). **loading** (*n*), **load** (*v*), **loaded** (*adj*).

Hooke's Law states that the strain (p.19) produced in a material or physical system subjected to stress (p.19) is directly proportional to the stress applied, provided that a certain stress is not exceeded; stress ÷ strain = constant, where the constant is the modulus of elasticity (p.19) of the material; for a wire or spring subjected to tensile stress (p.19), extension is directly proportional to load (↑) up to a certain load.

limit of proportionality the point where the stress/strain graph ceases to be linear; for a steel wire the elastic limit (p.19) occurs for stress (p.19) only slightly greater than that at the limit of proportionality, but for rubber the elastic limit occurs at much greater stress than at the limit of proportionality.

yield point a point just beyond the elastic limit (p.19) of a material subjected to tensile stress (p.19), at which increases in load (↑) cause proportionately larger increases in extension (p.19) than are predicted by Hooke's Law (↑); it arises from internal changes in the structure of the material from which it does not totally recover on removal of stress.

plastic deformation the non-recoverable extension (p.19) of a material beyond the yield point (↑).

ductility (*n*) the property of certain metals, when in the form of a wire, to tolerate plastic deformation (↑), e.g. mild steel. **ductile** (*adj*).

brittleness (*n*) the property of certain materials to break due to growth of cracks, e.g. glass, carbon steel. **brittle** (*adj*).

ultimate tensile stress the maximum load (↑) a material can tolerate before breaking; its value for mild steel wire is $4.5 \times 10^8\,\mathrm{Nm^{-2}}$.

Young's modulus of elasticity the modulus of elasticity (p.19) for a material subjected to tensile stress (p.19) within its limit of proportionality (↑); denoted by E; longitudinal stress ÷ longitudinal strain = constant (E). E = tensile stress ($\mathrm{Nm^{-2}}$) ÷ extension δL/original length L; units of E are newton/m^2 ($\mathrm{Nm^{-2}}$); its value for mild steel is $2 \times 10^{11}\,\mathrm{Nm^{-2}}$.

Hooke's Law
deformation of a steel wire up to its elastic limit E and characteristic behaviour under increasing stress beyond E

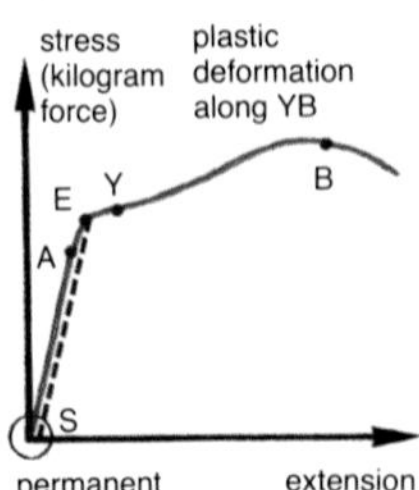

A – limit of proportionality
E – elastic limit
Y – yield point
B – ultimate tensile stress

Young's modulus
for a steel wire: load W produces strain δL/L

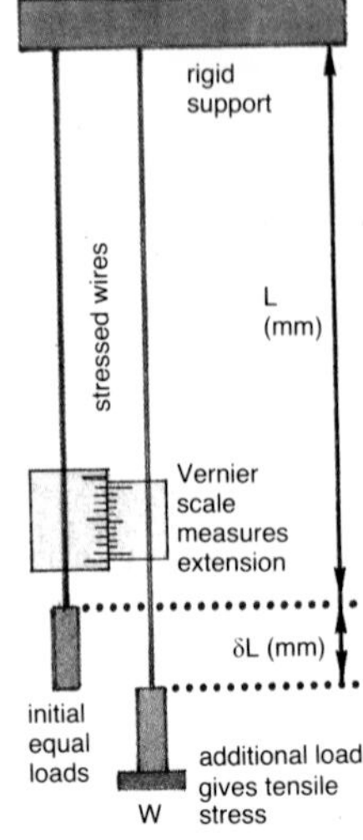

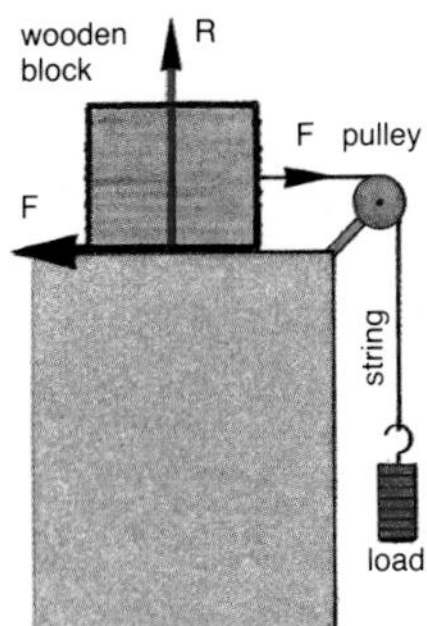

F total frictional force in plane of contact
R resultant reaction over supporting surface

static friction
value of force F increases to maximum before wooden block slips

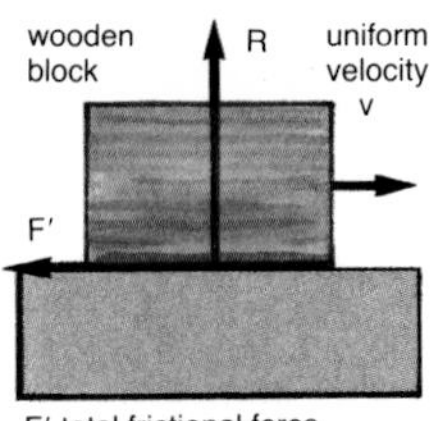

F′ total frictional force in plane of contact
R resultant reaction over supporting surface

kinetic friction
value of frictional force is F′ for uniform velocity v of block relative to supporting surface

friction (*n*) force (p.26) in the plane of contact between 2 surfaces, acting parallel to the surfaces in a direction opposing relative motion (↓) between them. **frictional** (*adj*).

frictionless (*adj*) having negligible friction (↑), e.g. highly polished smooth surfaces.

solid friction friction (↑) between 2 solid surfaces.

static friction the frictional force (↑) acting in the plane of contact between 2 solid surfaces before relative motion (↓) begins.

relative motion the movement of a body (p.12) or surface with respect to another body, or surface, which can either be stationary or itself moving.

limiting frictional force the maximum value of the static frictional force (↑) as relative motion (↑) begins.

normal reaction a force (p.26) exerted on a body by the surface supporting it; it acts perpendicular to the plane of contact between body and surface.

coefficient of static friction a characteristic of 2 specified solid surfaces in contact, before relative motion (↑) occurs between them; the ratio of the limiting frictional force (↑) F ÷ normal reaction (↑) R: F/R = a constant μ; μ has its value in the range 0.2–0.6 for wood on metal; the value is a ratio, having no units and it is independent of the areas of the surfaces in contact.

kinetic frictional force the value of the force of friction (↑) between 2 solid surfaces having uniform relative motion (↑) between them.

coefficient of kinetic friction a characteristic of 2 specified solid surfaces in contact, with uniform relative motion (↑) between them; ratio kinetic frictional force (↑) F′ ÷ normal reaction (↑) R; F′/R = constant μ′; μ′ has values less than the coefficient of static friction (↑) μ for the specified surfaces; the value is a ratio and it is independent of the area of the surfaces in contact; it is also independent of the relative motion (↑) if low, but increases if the relative motion is high.

coefficient of dynamic friction an alternative name for the coefficient of kinetic friction (↑).

lubrication (*n*) the use of oil, grease, graphite, or other suitable material to reduce kinetic frictional forces (↑), e.g. in machines with moving parts. **lubricant** (*n*), **lubricate** (*v*), **lubricating** (*adj*).

fluid friction friction (↑) arising within a moving fluid (p.15) during fluid flow; it increases with the relative velocity of the fluid.

laminar flow describes fluid flow (p.15) due to a pressure gradient (p.15) acting parallel to the direction of flow; the fluid can be considered to be made up of parallel layers in contact, moving in the direction of flow; the relative velocity between adjacent layers causes fluid friction (p.21) to arise in the surfaces of contact between layers, opposing their relative motion (p.21).

streamlined flow an alternative name for laminar flow (↑) in fluids; streamlines are drawn to illustrate diagrammatically the boundaries of layers moving at different velocities.

viscosity (*n*) a property of fluids by which they exert internal fluid friction (p.21) during flow. **viscous** (*adj*).

turbulence (*n*) non-laminar fluid flow (↑) is disturbed by pressure gradients perpendicular to the direction of flow. **turbulent** (*adj*).

drag (*n*) the fluid frictional (p.21) force exerted by a fluid (p.15) on the surface of a solid body (p.12) with which it has relative motion (p.21), e.g. aircraft flight relative to an airstream.

streamlined shape the shape for a solid body which best allows laminar flow (↑) to be maintained when it is in relative motion (p.21) with a fluid, e.g. aircraft wings and body.

aerofoil (*n*) a solid surface of streamlined shape (↑) such that streamlined flow (↑) is maintained in the airstream in relative motion (p.21) with it; the upper aerofoil surface is more strongly curved to give faster airflow across it and to generate lift (↓).

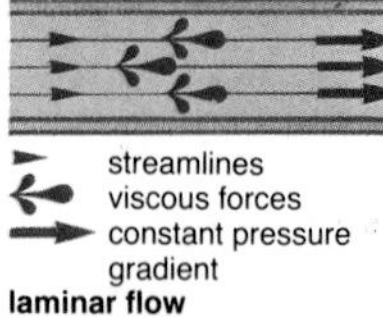

laminar flow

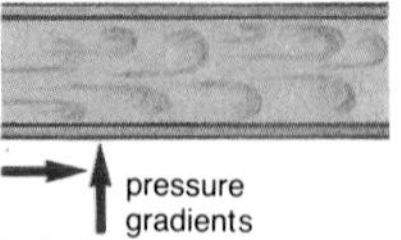

turbulence

aerofoil

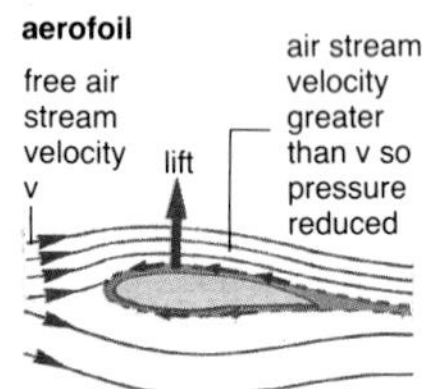

air stream velocity less than v so pressure increased

streamlines
drag forces
very narrow boundary layer

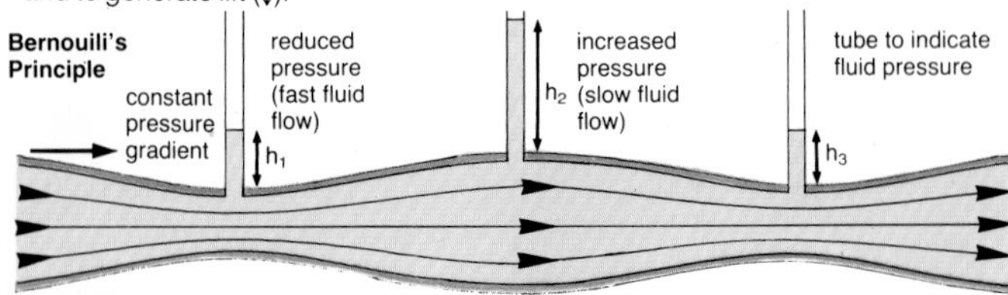

Bernouilli's Principle states that the total energy of a fluid (p.15), in streamlined flow (↑) at a constant rate ($kg\,s^{-1}$) remains constant; when the fluid flows faster, kinetic energy (p.27) increases and potential energy (p.27), due to fluid pressure (p.39), decreases; streamlines are drawn closer together to indicate faster flow.

lift (*n*) an upward force (p.26) generated perpendicular to an aerofoil (↑) surface when air is moving in streamlined flow (↑) relative to it; the difference

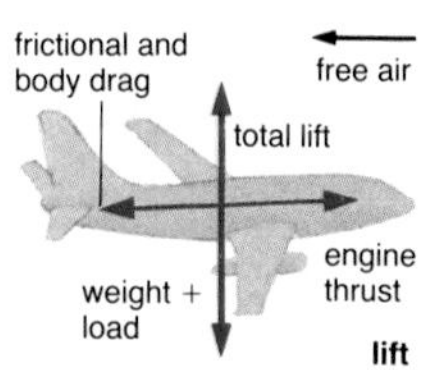

lift

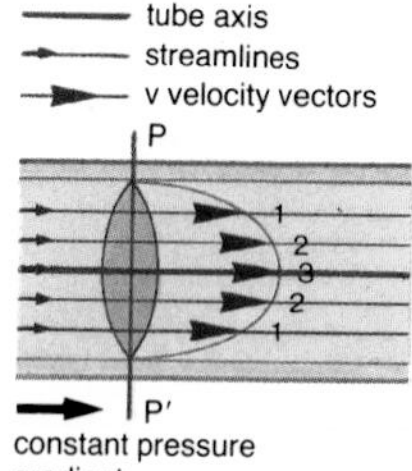

PP′ section across tube perpendicular to liquid flow

velocity gradient dv/dr (s^{-1}) at any specified point

velocity gradient

in airstream velocity across upper and lower aerofoil surfaces causes a pressure difference, according to Bernouilli's Principle (↑), and the resulting upward pressure generates lift. **lift** (*v*).

thrust (*n*) a forward horizontal force (p.26) provided by aircraft engines; it must exceed drag (↑) to maintain flight. **thrust** (*v*).

Newton's viscosity formula relates fluid frictional (p.21) force F (newton), acting between fluid layers in laminar flow (↑), to the velocity gradient (↓) dv/dr ($second^{-1}$) at the point considered, and to the area of contact A ($metre^2$) between the moving layers: $F \propto A \times dv/dr$ so $F = constant\ (\eta) \times A.dv/dr$; η is constant for a particular liquid at a specified temperature (p.116); η is the coefficient of viscosity (↓).

velocity gradient is the velocity (p.30) change per metre dv/dr for laminar fluid flow (↑) measured perpendicular to the direction of flow at the point concerned; the units are $ms^{-1}m^{-1}$; s^{-1} (↓).

coefficient of viscosity the constant (η) in Newton's viscosity formula (↑); characteristic of a specific fluid (p.15) at a specified temperature (p.116); its units are newton second/$metre^2$ (Nsm^{-2}) or kg/metre/second ($kgm^{-1}s^{-1}$); $1 Nsm^{-2} = 1$ dekapoise; for water at 10°C: $\eta = 1.3 \times 10^{-3}$ dekapoise = 0.013 poise (↓). η decreases with temperature increase.

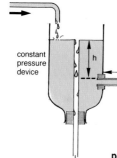

test liquid inlet

constant pressure device

h

l

tube of radius a (m)

pressure gradient h/l (Nm^{-2}) and liquid flow direction

coefficient of viscosity

measuring cylinder ($VmLs^{-1}$)

poise the practical unit for the coefficient of viscosity (↑); 1 newton second/$metre^2$ (Nsm^{-2}) = 1 dekapoise = 10 poise.

Poiseuille's viscosity formula relates the volume of liquid flowing per second (mLs^{-1}) through a tube of radius a (metre), due to a pressure gradient (p.15) p/l newton/$metre^2$/metre ($Nm^{-2}m^{-1}$), to the coefficient of viscosity η (↑); under laminar flow (↑) conditions; volume/second (mLs^{-1}) $= \pi pa^4/8\eta l$; it can be used to determine η.

terminal velocity the velocity with which a body moves through a fluid when the resultant force on the body is zero, e.g. falling raindrops.

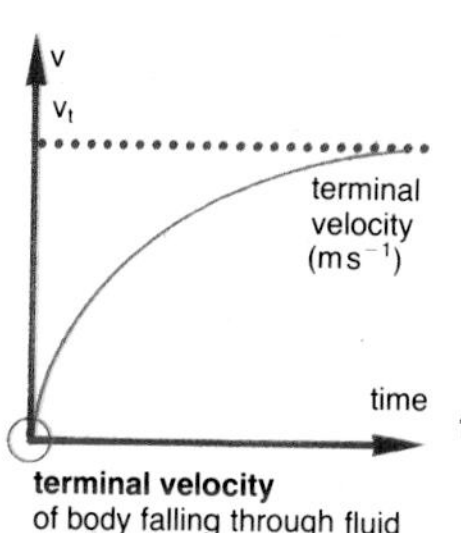

terminal velocity of body falling through fluid

free surface energy describes the potential energy (p.27) possessed by molecules in a liquid surface above that of other molecules within the liquid; surface molecules experience resultant inward forces due to cohesion (p.14) against which work (p.26) would have to be done to remove molecules from the liquid surface, e.g. by evaporation (p.15); the surface energy is a minimum for minimum surface area of a given volume of liquid, so small liquid droplets, e.g. mercury on glass, raindrops in air, aniline floating in water, have a spherical shape; free surface energy is defined as the work done (joule) per metre^2 to increase the surface area of the liquid under isothermal (p.140) conditions; it is equal in value to the surface tension (↓) and has the same units.

interatomic force the force between atoms; for distances greater than one atomic diameter the force between atoms is attractive (and falls off with distance) but for distances less than one atomic diameter the force is repulsive and increases steeply as separation decreases; thus liquids and solids are incompressible (p.14).

potential energy of atoms in a solid or liquid since work must be done to either increase or decrease the separation between atoms in a solid or liquid, their equilibrium separation represents the position of minimum separation.

surface tension cohesive forces (p.14), acting parallel to a liquid surface between surface molecules, cause the liquid surface to behave like an elastic skin in tension, having minimum area and minimum surface energy (↑); surface tension is defined as force per metre (newton metre^{-1}) acting perpendicularly to, and on either side of, any line in the liquid surface; it is denoted by T or γ; units are $\mathrm{N\,m^{-1}}$; its value depends on the medium in contact with the surface, e.g. for water in contact with air at 20°C, $T = 7.26 \times 10^{-2}\mathrm{N\,m^{-1}}$; for mercury in contact with air at 20°C, $T = 46.5 \times 10^{-2}\mathrm{N\,m^{-1}}$; for water in contact with olive oil at 20°C, $T = 2.1 \times 10^{-2}\mathrm{N\,m^{-1}}$; T is equal in value to the free surface energy (↑) and has the same units.

liquid film a very thin layer of liquid either with 2 surfaces, both in contact with air, e.g. soap film, or with one surface in contact with another liquid, e.g. oil film on water.

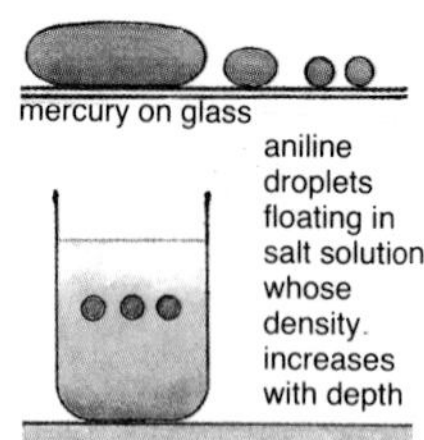

free surface energy

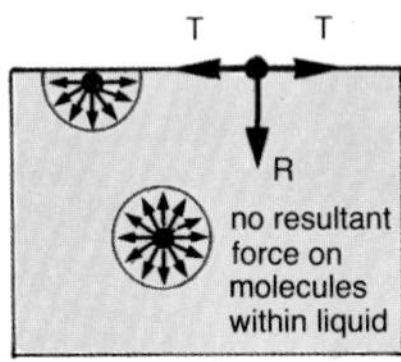

T cohesive forces parallel to surface
R resultant of forces acting downwards on surface molecules

surface tension

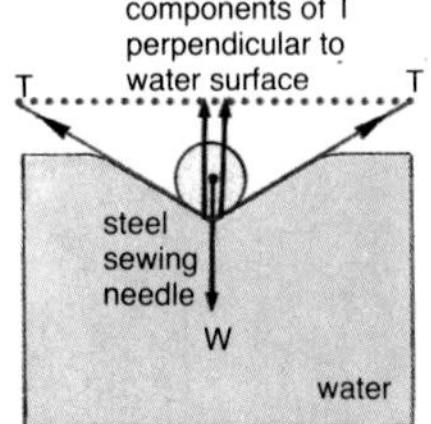

forces can support weight W of a floating needle

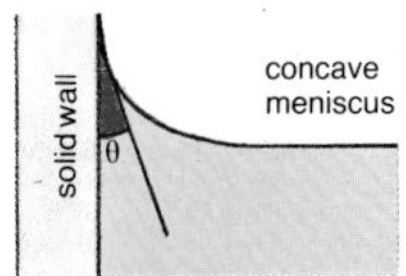

acute angle of contact

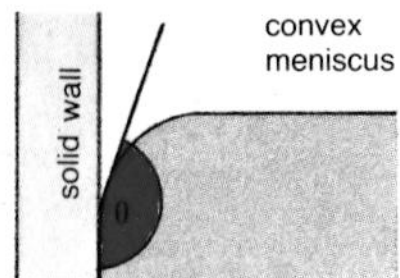

obtuse angle of contact
e.g. mercury on glass

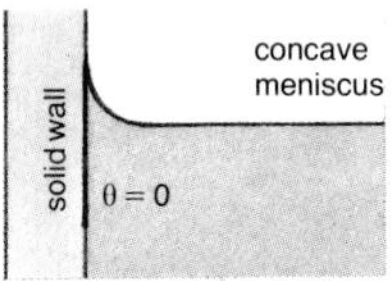

zero angle of contact
e.g. water on clean glass

capillarity

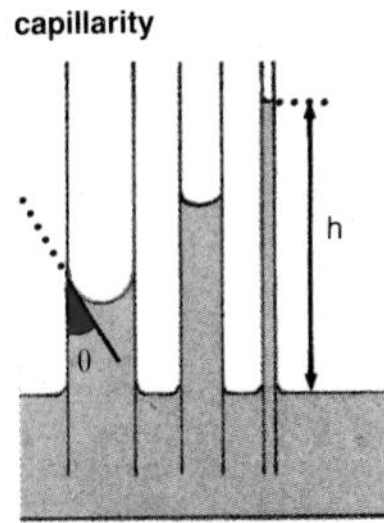

capillary rise height h measured from bottom of concave meniscus to free liquid surface

depth h measured from top of convex meniscus to free liquid surface

capillary depression

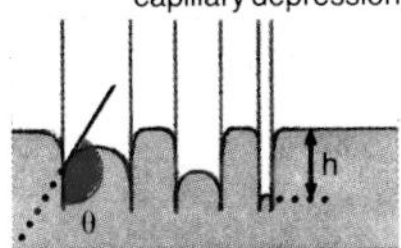

bubble (*n*) a spherical liquid film (↑) with air inside and outside, e.g. soap bubble; an excess pressure of the inside over the outside of the bubble exists given by: $p = 4T/r$ ($N m^{-2}$).

angle of contact may be measured for a liquid in contact with a supporting solid surface, usually in air, e.g. mercury or water on glass; its value depends on equilibrium between cohesive forces (p.14) for liquid molecules and adhesive forces (p.14) between liquid and glass; a tangent to the liquid surface where the liquid touches the glass makes an angle of contact θ with the glass, where θ is measured through the liquid; θ can be acute, obtuse or have value zero; when $\theta = 0$ the liquid spreads over the glass and wets it, e.g. water and ethanol on clean glass; for mercury on glass, $\theta = 137°$.

capillary tube a glass tube of fine bore or small radius.

capillarity (*n*) an effect demonstrated when a capillary tube (↑), or fibrous material such as blotting paper, loosely woven cloth or cotton wool fibres, is dipped into a liquid; if the adhesive forces (p.14) exerted by the tube or material are stronger than the cohesive forces (p.14) of the liquid molecules for each other, the liquid rises up the capillary; whereas if the cohesive forces are stronger than the adhesive forces, the liquid is depressed down the capillary, e.g. mercury in a glass tube; for capillary effects in a glass tube, the surface curvature of its liquid meniscus (↓) depends on the angle of contact (↑); capillary rise can be used to determine surface tension (↑) and to investigate its variation with temperature; surface tension decreases as temperature rises; the height (h) of capillary rise or depth (h) of capillary depression is inversely proportional to the tube radius (r). **capillary** (*adj*).

meniscus (*n*) the spherically curved interface between a liquid in a capillary tube (↑) and the air in contact with it; for liquids with an angle of contact (↑) zero, e.g. water and ethanol on clean glass, the radius of curvature of the meniscus equals the radius of curvature of the capillary bore; for non-zero angle of contact the radii are not equal; for liquids showing capillary (↑) rise effects, e.g water, the meniscus is concave upwards; for capillary depression, e.g. mercury, the meniscus is convex upwards; excess pressure (p.39) acts across the curved interface towards its centre of curvature; the value of excess pressure is $2T/r$ ($N m^{-2}$).

physical field a volume of 3-dimensional space throughout which forces (↓) can act and in which energy (↓) is available, e.g. gravitational field (p.42), electric field (p.165); a field can be represented by lines whose direction is given by a convention related to the particular force.

force (*n*) a force shows the presence of energy in a physical field (↑); it can alter the position of a body in space and can change the momentum of a body by causing it to accelerate (p.30); it can enable work (↓) to be done; units of force are newton (N), force is a vector quantity (p.30). **force** (*v*).

newton the absolute unit of force (↓) based on Newton's 2nd Law of Motion (p.36); the force which produces an acceleration of l metre/second^2 ($m s^{-2}$) when acting on a mass of l kilogram (kg); denoted by N.

physical quantity something that can be observed and measured, e.g. mass (p.7), electric current (p.152), energy (p.26).

work (*n*) work is done when energy (↓) is exchanged between a field (↑) and a body (p.12) or particle (p.7) experiencing forces (↑); energy is gained by the body when it moves in the direction of the field and this does work on it, e.g. an electron accelerated in an electric field acquires energy; energy is lost by the body when it moves against the direction of the field and does work against the field, e.g. an electron decelerated in an electric field loses energy; the unit is the joule (J); work is a scalar quantity (p.30); work done on or by a body is defined as the product of the force F acting on the body and the distance s moved by the body, in or against the direction of the field; for force F (newton) and distance s (metre), work done is Fs (newton metre, joule). **work** (*v*), **working** (*adj*).

joule the absolute unit of work (↑) and energy (↓); work done when the point of application of a force of 1 newton (N) moves through a distance of 1 metre (m) in the direction of the force; 1 joule (J) = 1 newton metre (Nm).

energy (*n*) the physical quantity present in a field (↑) enabling work (↑) to be done when the field forces act on a body or particle; energy and work are aspects of the same phenomenon and are measured in the same units, joule; whenever work is done in a field an identical amount of energy is expended, energy is a scalar quantity (p.30). **energize** (*v*), **energetic** (*adj*).

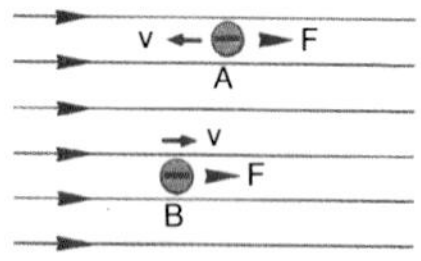

field
work done in an electric field. Electron at A, moving with velocity v against field forces, experiences retarding force F. At B, when moving with field forces, it experiences accelerating force F.

force

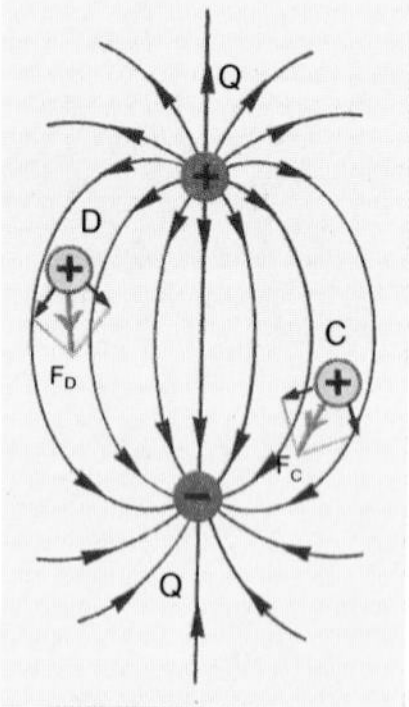

electric field forces

point charges at C and D in electric field of charges Q experience resultant forces F_C and F_D due to field

F moves from A to B through 1 metre

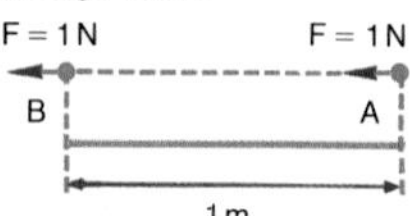

unit of **work** – **joule**

power (*n*) the rate at which work (↑) is done; the rate at which energy (↑) is expended, the unit of power is the watt (W) (↓); 1 watt (W) = 1 joule/second (Js^{-1}). **power** (*v*), **powerful** (*adj*), **powered** (*adj*).

watt the unit of power (↑); 1 watt (W) = 1 joule/second (Js^{-1}).

horsepower a former practical unit of power (↑) used in the British Imperial System, it is not now used for scientific purposes; 1 horsepower (H.P.) = 550 foot-pounds-weight/second; 1 H.P. = 746 watt (approximately).

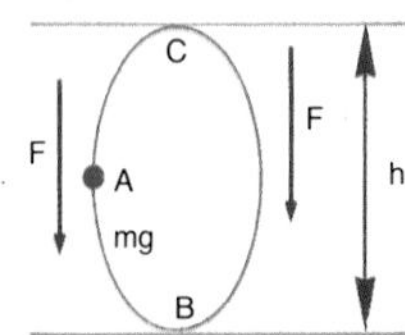

conservative field
e.g. Earth's gravitational field. Work done in taking mass m around closed path ACBA = zero. Work done in raising m through BC = −mgh and in m falling through CB = +mgh

conservative field a field (↑) in which a body experiencing field forces (↑) can follow any closed path, returning to its original position, without the performance of work (↑), e.g. gravitational field, electric field.

conservation of energy a practical law or principle stating that energy (↑) cannot be created or destroyed but can only change from one form to another; the total energy of a closed or given physical system is constant, though energy transformations are possible within it, e.g. in a hydroelectric power scheme potential energy (↓) of a water head changes to kinetic energy (↓) of falling water, then to rotational kinetic energy (↓) of a water turbine, then to electrical energy (p.28) from a turbo-generator, then to light (p.28), heat (p.28) and mechanical energy (p.28) during power (↑) consumption.

potential energy a form of energy (↑) due to the position or state of a body or physical system, e.g. if a body of mass m (kg) is raised by a height h (m) then its increase in potential energy in the Earth's gravitational field is mgh (joule) where g (ms^{-2}) is the acceleration due to gravity; potential energy is increased when a spring is stretched or compressed; work (↑) is done when potential energy, e.g. of a wound spring, is released as kinetic energy (↓).

kinetic energy a form of energy (↑) possessed by a body because of its motion; a body of mass m (kg) and velocity v (metre/second) (p.30) has kinetic energy $\frac{1}{2}mv^2$ (joule).

rotational kinetic energy = $\frac{1}{2}I\omega^2$ joule

moment of inertia = $I\,kg\,m^2$

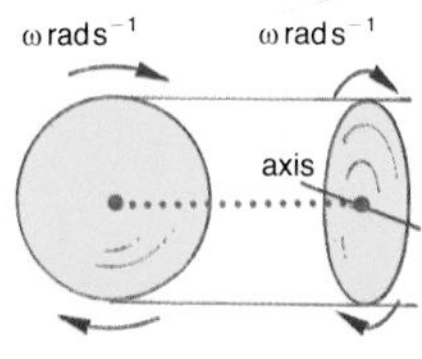

rotational kinetic energy

rotational kinetic energy the kinetic energy (↑) possessed by a body rotating about an axis, e.g. flywheel; a flywheel of moment of inertia I (kg metre²) (p.36) and angular velocity ω ($second^{-1}$) has rotational energy $\frac{1}{2}I\omega^2$ (joule).

rotation (*n*) circular motion (p.46) of a body (p.7) around a specified axis of rotation passing through the body, e.g. the Earth's rotation about its geographic axis.

vibrational energy the energy (p.26) possessed by a vibrating body or physical system interchangeable between kinetic (p.27) and potential (p.27) forms, e.g. vibrating pendulum, vibrating spring.

mechanical energy forms of energy (↑) which can be possessed by bodies or physical systems because of their position or motion, e.g. vibrational (↑), rotational (p.27), kinetic (p.27) and potential (p.27) energy.

sound (*n*) a form of vibrational energy (↑) transmitted through a medium as compression waves (p.56) of frequencies audible to the human ear; the source of sound is always a body (p.12) or physical system undergoing forced vibration (p.102).

electromagnetic energy energy (p.26) transmitted through a medium or free space as electromagnetic waves (p.55) over the continuous range of wavelengths of the electromagnetic spectrum.

light (*n*) electromagnetic energy (↑) of wavelength range approximately 4×10^{-7} m to 8×10^{-7} m stimulating light-sensitive cells of the retina of the human eye to produce the perception (p.78) of vision. **lighting** (*n*), **light** (*v*), **alight** (*adj*), **lighted** (*adj*).

infrared radiation (*n*) electromagnetic energy (↑) of wavelength range approximately 8×10^{-7} m to 10^{-3} m.

heat (*n*) that form of energy transferred between bodies due to their difference in temperature; heat may be transferred by infrared radiation (↑), conduction (p.123) or convection (p.124). **heating** (*n*), **heat** (*v*), **hot** (*adj*).

thermal energy heat (↑) possessed by a body, substance or physical system as molecular vibrational energy (p.13).

electrical energy a form of energy (p.26) manifested as stored electric charge (p.165) in a capacitor (p.171), or as electric charge flowing in a circuit (p.158) as electric current (p.152) and producing electrical heating, electromagnetic and electrochemical effects of current.

electricity (*n*) an alternative name for electrical energy (↑). **electric** (*adj*), **electrical** (*adj*).

chemical energy energy (p.26) stored in a chemical system, e.g. primary cell (p.148), or in a fuel, e.g. oil, coal, gas; can be released as heat (↑) by combustion of fuel and as electrical energy (↑) from a primary cell by allowing the cell components to react.

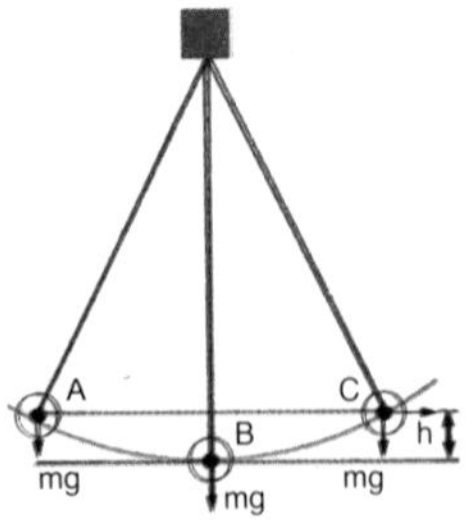

potential energy at A and C
kinetic energy at B

vibrational energy

oscillating spring
restoring force F = Ma
where acceleration a $= -ky$
F ∝ y and is always directed towards the normal rest position

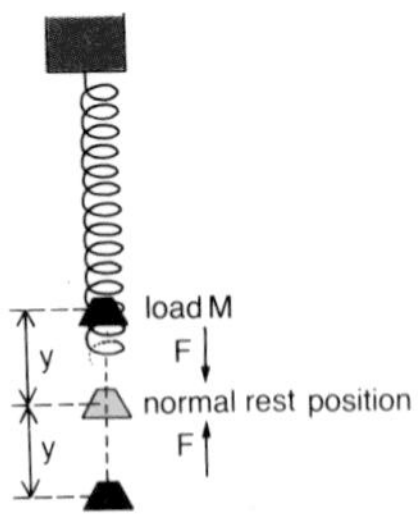

position at extension y

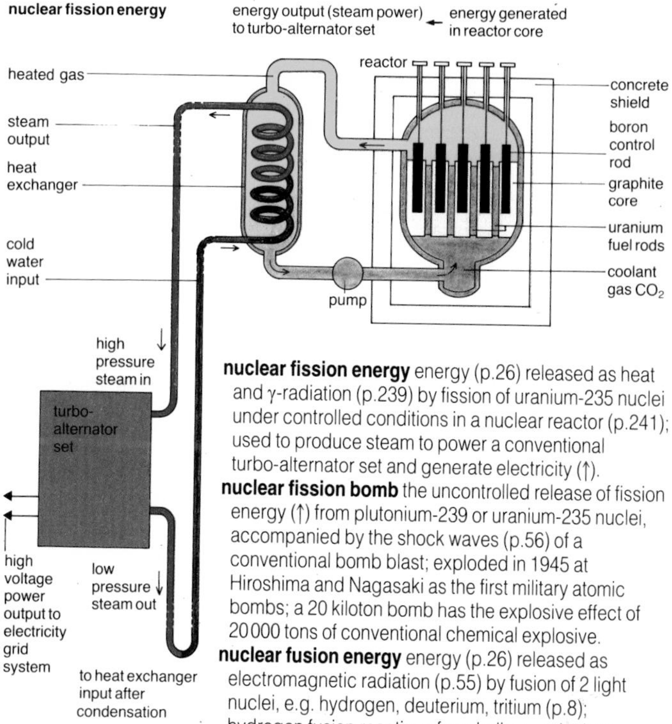

nuclear fission energy

nuclear fission energy energy (p.26) released as heat and γ-radiation (p.239) by fission of uranium-235 nuclei under controlled conditions in a nuclear reactor (p.241); used to produce steam to power a conventional turbo-alternator set and generate electricity (↑).

nuclear fission bomb the uncontrolled release of fission energy (↑) from plutonium-239 or uranium-235 nuclei, accompanied by the shock waves (p.56) of a conventional bomb blast; exploded in 1945 at Hiroshima and Nagasaki as the first military atomic bombs; a 20 kiloton bomb has the explosive effect of 20000 tons of conventional chemical explosive.

nuclear fusion energy energy (p.26) released as electromagnetic radiation (p.55) by fusion of 2 light nuclei, e.g. hydrogen, deuterium, tritium (p.8); hydrogen fusion reactions form helium and release solar energy (↓); the uncontrolled release of fusion energy, accompanied by the shock waves (p.56) of a conventional bomb blast, occurs on explosion of a hydrogen bomb; controlled release of fusion energy for everyday consumption is still at the experimental stage, e.g. J.E.T. (Joint European Torus) research project, U.K.; Tokamak, U.S.S.R.

solar energy nuclear fusion energy (↑) released in the interior of the sun, giving an estimated core temperature (p.116) of up to 40 million K; radiated to Earth mainly as heat (↑) and light (↑) with some ultraviolet radiation (p.74) and a small proportion of X-radiation (p.100); ultimately the source of all energy in the solar system (p.43) and the means of sustaining all forms of life on Earth.

vector quantity a physical quantity for which the direction, as well as the magnitude, must be specified, e.g. displacement (↓), velocity (↓), acceleration (↓), force (p.26).
vector (*n*) a straight line drawn on a diagram to represent the magnitude and direction of a vector quantity (↑). **vectorial** (*adj*).
scalar quantity a physical quantity for which magnitude only, not direction, need be specified, e.g. distance, speed (↓), work, energy (p.26).
fixed reference point a point acting as the origin of measurement for vector quantities (↑) involving distance, e.g. displacement (↓), velocity (↓), acceleration (↓).
displacement[1] (*n*) distance measured from a fixed reference point (↑) in a specified direction; vector quantity (↑); units are cm, m, km. **displace** (*v*), **displaced** (*adj*).
speed (*n*) the rate of change of distance with respect to time for a moving object; scalar quantity (↑); its units are metre/second ($m s^{-1}$), kilometre/hour ($km h^{-1}$); denoted by distance s ÷ time t or ds/dt using calculus notation. **speed** (*v*), **speedy** (*adj*).
velocity (*n*) the rate of change of displacement (↑) with respect to time for a moving object; speed (↑) in a specified direction, e.g. 60 km/hour in a North-West direction; a vector quantity (↑); measured in units of $m s^{-1}$, $km h^{-1}$; denoted by s/t or ds/dt using calculus notation.
acceleration (*n*) the rate of change of velocity (↑) with respect to time for a moving object; vector quantity (↑); its units are metre/second^2 ($m s^{-2}$); kilometre/hour/second ($km h^{-1} s^{-1}$); denoted by velocity (*v*) ÷ time (t), or dv/dt or d^2s/dt^2 using calculus notation. **accelerate** (*v*), **accelerating** (*adj*).
deceleration (*n*) the rate of decrease of velocity (↑) with respect to time for a moving object; negative acceleration (↑) during slowing down; its units are metre/second^2 ($m s^{-2}$); kilometre/hour/second ($km h^{-1} s^{-1}$). **decelerate** (*v*), **decelerating** (*adj*).
retardation (*n*) an alternative name for deceleration (↑). **retard** (*v*).
uniform velocity a velocity (↑) which does not vary with time; equal distances are travelled in equal time intervals in a specified direction; the displacement/time graph is linear with a constant gradient.

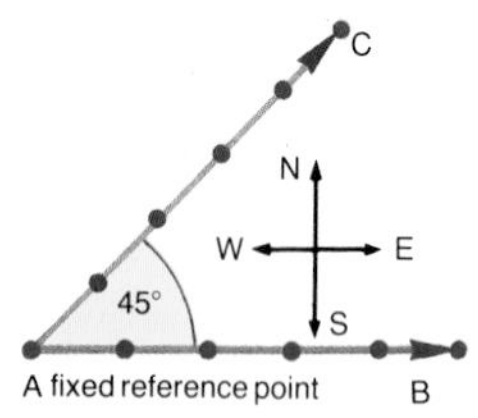

displacement
C is 5 km NE of A
AC is 5 cm
B is 5 km E of A
AB is 5 cm

velocity vector representing velocity 60 $km h^{-1}$ NW

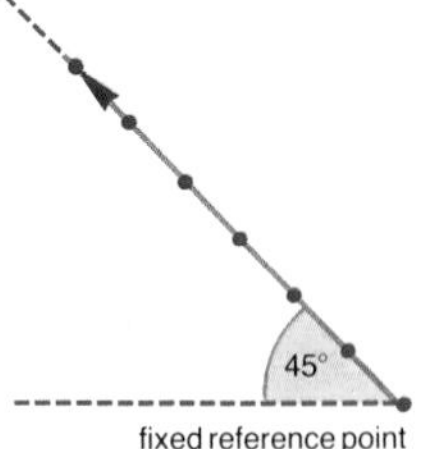

uniform velocity

displacement time graph for uniform velocity

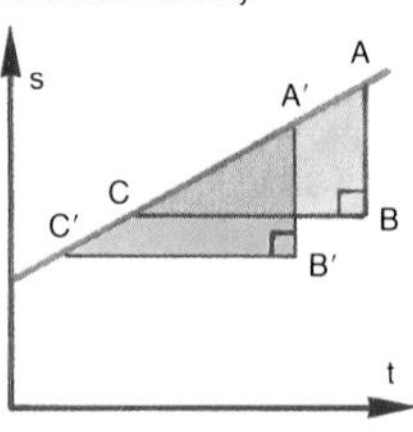

gradient of graph $= \frac{AB}{BC} = \frac{A'B'}{B'C'}$

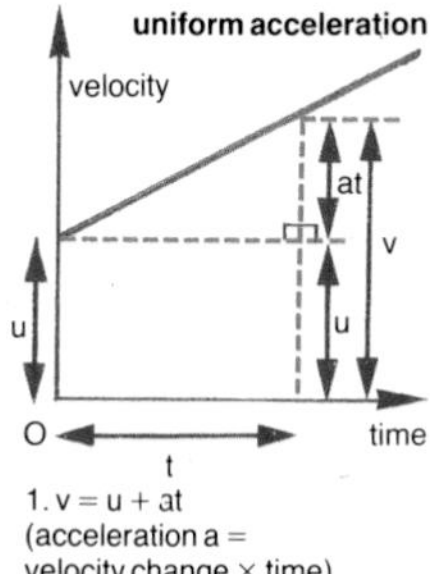

1. $v = u + at$
(acceleration a = velocity change × time)

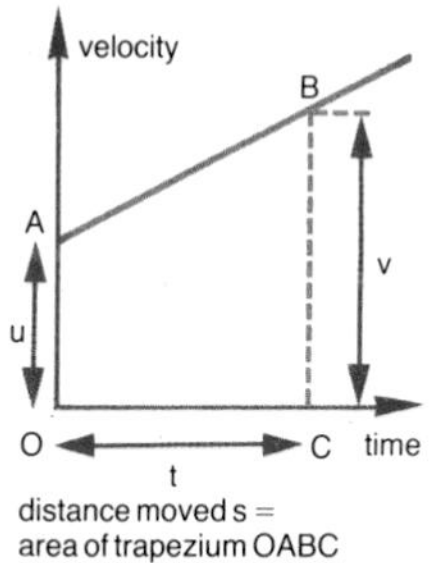

distance moved s = area of trapezium OABC
2. $s = \frac{(v + u)\,t}{2}$

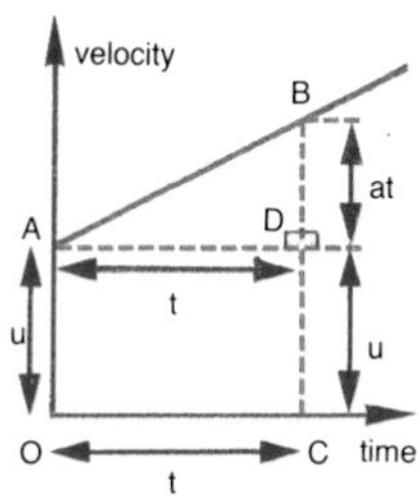

distance moved s = area OADC + area ABD
3. $s = ut + \frac{1}{2}at^2$

from 1. $\frac{(v - u)}{a} = t$

substituting t in 2.
$s = \frac{(v + u)(v - u)}{2a}$
$v^2 = u^2 + 2as$

uniform acceleration an acceleration (↑) which does not vary with time; a moving object undergoes equal changes in velocity (↑) in equal time intervals; the velocity/time graph is linear with a constant gradient.

constant velocity velocity (↑) which cannot be varied by an experimenter, e.g. velocity of electromagnetic waves in free space (p.55).

constant acceleration acceleration (↑) which cannot be varied by an experimenter, e.g. acceleration g due to Earth's gravity (p.32).

initial velocity the velocity (↑) at the instant of zero time, t = 0, when acceleration (↑) begins or a measurement is commenced; denoted by u ($m\,s^{-1}$).

final velocity the velocity (↑) at the instant of time when acceleration (↑) ceases or measurement is completed; denoted by v ($m\,s^{-1}$).

average velocity for uniform acceleration (↑), this is the arithmetic mean of initial and final velocities (↑); its value is $(v + u)/2$.

relative velocity the velocity (↑) of a body or particle (p.7) with respect to another body or particle, which may itself be moving or stationary; for bodies A and B in relative motion (p.21), with velocities v_A and v_B: the relative velocity is $(v_A + v_B)$ for bodies travelling in opposite directions and $(v_A - v_B)$ for bodies travelling in the same direction.

equations of motion these are equations describing uniformly accelerated motion (↑) of a body or particle (p.7) travelling in a straight line; for initial velocity u ($m\,s^{-1}$) (↑), final velocity v ($m\,s^{-1}$) (↑), displacement s(m) (↑), acceleration a ($m\,s^{-2}$) (↑) and time interval t (s) between measurements of u and v.

equations of motion	$v = u + at$ $s = (v + u)t/2$ $s = ut + \frac{1}{2}at^2$ $v^2 = u^2 + 2as$

weight[1] (*n*) the force (p.26) on a body due to the gravitational attraction of the Earth; weight W = mass m × acceleration g due to the Earth's gravity (p.32); mass (p.7) is constant for a particular body, but g, and hence W, varies slightly over the Earth's surface, depending on distance from the Earth's centre and on the latitude; measured in newtons (N) (p.26). **weigh** (*v*).

weight[2] (*n*) mass (p.7) used to apply force (p.26), due to the Earth's gravitational attraction acting on it, e.g. load (p.20) on a stressed (p.19) wire or spring, counterpoise weights on the pan of a beam balance. **weight** (*v*), **weighted** (*adj*).

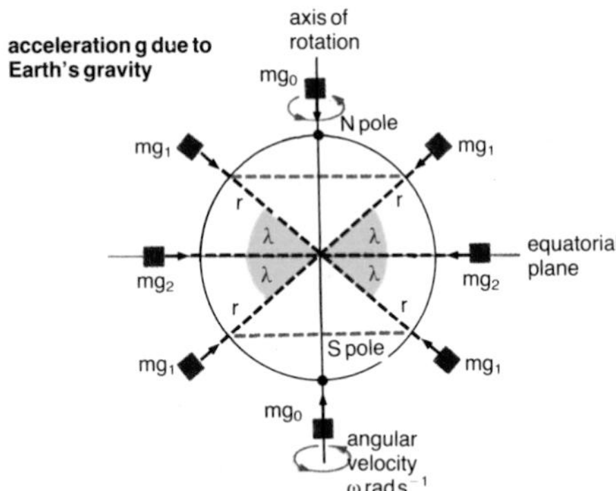

λ = latitude
r = Earth's radius
g_0 = maximum value of free fall acceleration

mg_0 for $\lambda = 90°, \cos\lambda = 0$
$mg_1 = m(g_0 - r\omega^2)\cos^2\lambda$
$mg_2 = m(g_0 - r\omega^2)$ for $\lambda = 0, \cos\lambda = 1$

acceleration g due to Earth's gravity the acceleration (p.30) of a body falling freely near the Earth's surface in the Earth's gravitational field (p.26); g = 9.81 metre/second^2 ($m s^{-2}$) or newton/kilogram ($N kg^{-1}$) at sea level and latitude 45°; g varies with the distance from the Earth's centre and is greater at the poles, where the Earth is slightly flattened, than at the equator; g also varies with latitude due to the effect of centripetal force (p.47) caused by the Earth's rotation on its axis; it shows maximum reduction in value at the equator and no change due to centripetal effects at the Earth's poles; the variation in g affects the weight (↑) of a body at different points on the Earth's surface.

trajectory (*n*) the path of a body or particle fired by force, e.g. a bullet from a gun; a particle projected in a uniform field follows a parabolic path.

projectile (*n*) a body or particle fired by force, e.g. a bullet from a gun. **project** (*v*).

moment of a force product of the force acting to produce a turning effect and the perpendicular distance of its line of action (↓) from the point or axis of rotation; force (newton) × distance (m) = moment (Nm).

torque (*n*) alternative name for moment of a force (↑).

line of action direction along which a force (p.26) acts.

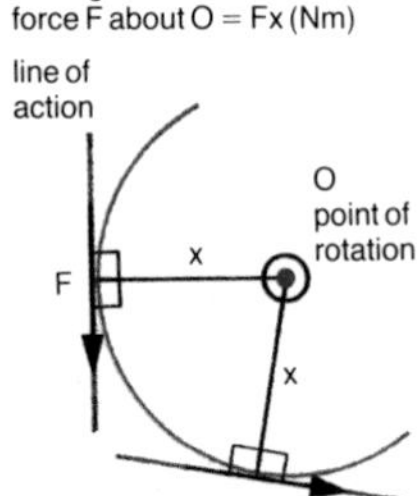

moment of a force

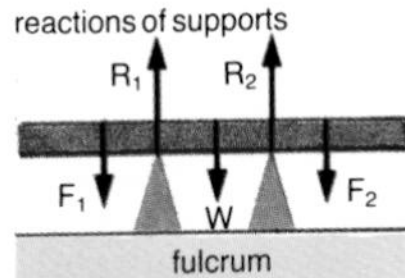

static equilibrium of pivoted bar
W = weight of bar acting through centre of gravity
$R_1 + R_2 = F_1 + W + F_2$
static equilibrium

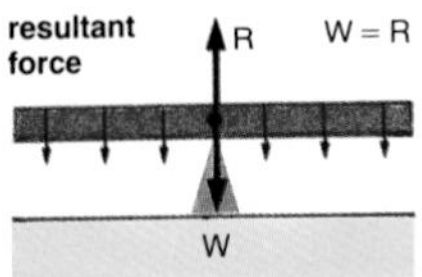

bar pivoted at centre of gravity
W = resultant force representing weight of bar
R = reaction of pivot

equilibrium (*n*) situation of balance between opposing physical effects, e.g. static equilibrium (↓) of forces; dynamic equilibrium condition (p.18) for a saturated vapour; temperature equilibrium between a body and its surroundings.

static equilibrium situation in which 2 or more forces (p.26) act on a body (p.12) in such a way that they have no resultant (↓) effect, e.g. balanced beam; body supported on a surface.

fulcrum (*n*) point at which a body in static equilibrium (↑) is pivoted (↓) on a knife-edge support (↓) or supported by a suspension.

pivot (*v*) to support a body in static equilibrium (↑) using a pointed or knife-edge support at the appropriate point. **pivot** (*n*), **pivoted** (*adj*).

resultant force the single force (p.26) having the same effect as 2 or more forces acting together at a point.

resultant vector the single vector (p.30) having the same effect as 2 or more vectors acting together at a point.

parallelogram of forces if 2 forces (p.26) are represented in magnitude and direction by the adjacent sides of a parallelogram, their resultant force (↑) is represented both in magnitude and direction by the diagonal of the parallelogram drawn from the point where the adjacent sides meet.

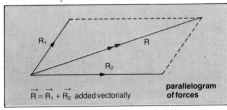

R1
R
R2

$\vec{R} = \vec{R_1} + \vec{R_2}$ added vectorially

parallelogram of forces

parallelogram of vectors
showing components R_1 and R_2 and resultant R

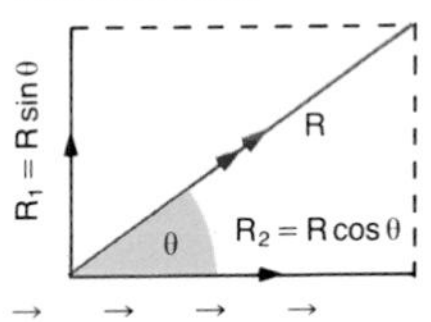

$\vec{R} = \vec{R\cos\theta} + \vec{R\sin\theta}$
added vectorially

parallelogram of vectors if 2 vectors (p.30) are represented in magnitude and direction by the adjacent sides of a parallelogram, their resultant vector (↑) is represented both in magnitude and direction by the diagonal of the parallelogram drawn from the point where the adjacent sides meet, e.g. parallelogram of forces (↑), velocities, accelerations.

vector addition using the parallelogram of vectors (↑) to find the resultant vector (↑) of 2 or more vectors.

component vector one of 2 or more vectors which together form a single resultant vector (↑).

vector resolution the method of using a right-angled parallelogram of vectors (p.33) to obtain 2 perpendicular component vectors (p.33) of a single vector.

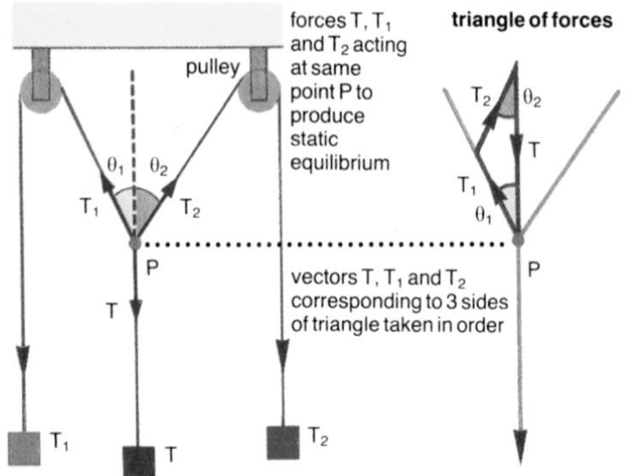

triangle of forces if 3 forces (p.26) acting in the same plane at a point are in equilibrium (p.33) they can be represented in magnitude and direction by the 3 sides of a triangle taken in order.

stable equilibrium a static equilibrium (p.33) in which a body, after being slightly disturbed, returns to its original position due to the turning moment (p.32) of its own weight (p.32).

unstable equilibrium a static equilibrium (p.33) in which a body, on being slightly disturbed, takes a different position of static equilibrium, due to the turning moment (p.32) of its own weight (p.33).

neutral equilibrium a static equilibrium (p.33) in which a body can be moved in such a way that the turning moment (p.32) of its weight (p.33) remains zero; the body's position can be altered while its equilibrium situation remains the same.

centre of gravity the point in a body (p.12) through which the resultant force (p.33) of the Earth's gravitational attraction on all parts of the body acts; the point around which the body's mass (p.7) is evenly distributed.

principle of moments when a body is in static equilibrium (p.33) the sum of the clockwise moments (p.32) about any point in the body equals the sum of the anti-clockwise moments about that same point.

unstable equilibrium weight W acts outside supporting apex of cone

stable equilibrium weight W acts within area of base of cone

neutral equilibrium weight W acts within area of supporting surface

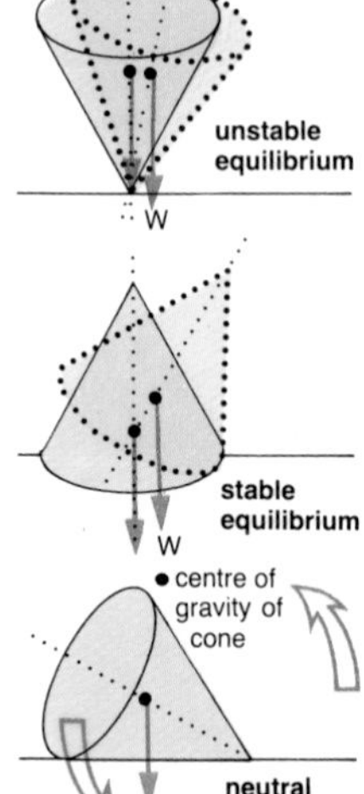

moment of a couple

line of action
line of action
point of rotation
F
F
x
x
O
turning moment of couple 2F about O = 2Fx (Nm)

couple (*n*) 2 equal and opposite parallel forces (p.26) acting at diametrically opposite points on a body (p.12) so as to produce rotation of the body about a point or about an axis through its centre.

moment of a couple product of one of the parallel forces acting to produce a turning moment (p.32) and the perpendicular distance between the 2 forces of the couple: force (newton) × distance (m) = turning moment (N m).

machine (*n*) a device by means of which energy (p.26) input at one point can give work (p.26) output at another point, e.g. lever (↓), pulley system; according to the law of conservation of energy (p.27), work output cannot exceed energy input; in practice it is less than the energy input because the efficiency of machines (p.36) is less than 100%. **machinery** (*n*), **machine** (*v*).

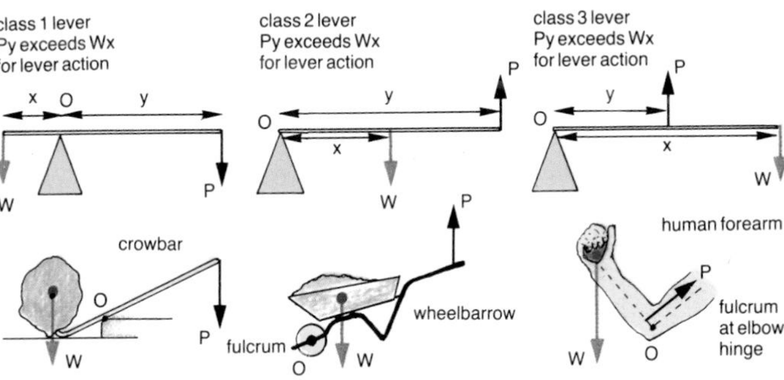

lever

lever (*n*) a simple machine (↑) using opposing turning moments (p.32) about a fulcrum (p.33) for its effect; there are 3 classes of lever according to the position of the fulcrum with respect to the load (↓) and effort (↓). **leverage** (*n*), **lever** (*v*).

effort describes the force (p.26) used to apply energy input to a machine (↑); its units are newton, kilogram-force.

load^2 (*n*) force (p.26) external to a machine (↑) which the work output must overcome; its units are newton (N); kilogram-force (kg f). **loading** (*n*), **load** (*v*), **loaded** (*adj*).

mechanical advantage a ratio of load (↑) ÷ effort (↑) for a machine (↑); ratio has no units.

velocity ratio a ratio of the distance moved by the effort (↑) per second ÷ distance moved by the load (↑) per second for a machine (↑); ratio has no units.

efficiency of machine the ratio of work output ÷ energy input for a machine (p.35); the ratio has no units and is expressed as a percentage; its value is always less than 100% in practical machines because moving parts are never totally without friction (p.21); it is also defined as [mechanical advantage (p.35) ÷ velocity ratio] × 100%.

inclined plane a simple machine (p.35) in which the load (p.35) is moved through a small distance by a much smaller effort (p.35) moving through a larger distance; a practical application is the screw.

inertia (*n*) the property of a body (p.12) to resist changes in its state of rest or of uniform motion (↓) in a straight line as described in Newton's Law 1 of Motion (↓); measured by the mass (p.7) of the body; inertial mass of a body is identical in value to its mass in experiencing gravitational forces (p.42), e.g. a vehicle moving from rest is driven in low gear so that high engine torque (p.32) at low road speed gives the required force to overcome inertia. **inertial** (*adj*).

impressed force an external force (p.26) applied to a body (p.12) or physical system.

action (*n*) an alternative name for impressed force (↑).

reaction (*n*) all forces are interactions; if object A applies a force on object B then object B applies an equal reaction force on A; the response of a body (p.12) or physical system having inertia (↑) to the action (↑) of an impressed force.

uniform motion the motion of a body travelling at uniform velocity (p.30) without acceleration (p.30).

linear momentum (*n*) a physical quantity defined by the product of mass m (p.7) × velocity v (p.30) for a body (p.12); units are $kg\,m\,s^{-1}$ or newton-second; it is a vector quantity (p.30).

Newton's 3 Laws of Motion 1. A body will remain in its state of rest or of uniform motion (↑) in a straight line unless an impressed force (↑) acts on it. e.g. a spacecraft travelling between the Earth and the Moon will pass through a region surrounding a gravitational neutral point (p.44), where it will have unaccelerated uniform motion (↑) because no resultant force (p.33) acts on it; its inertia (↑) causes it to continue at a uniform velocity (p.30),

2. The rate of change of a body's momentum (↑) is directly proportional to the impressed force (↑) acting on the body and takes place in the direction in which the impressed force acts; this law is used to define the newton (↓) as the absolute unit of force. e.g. a

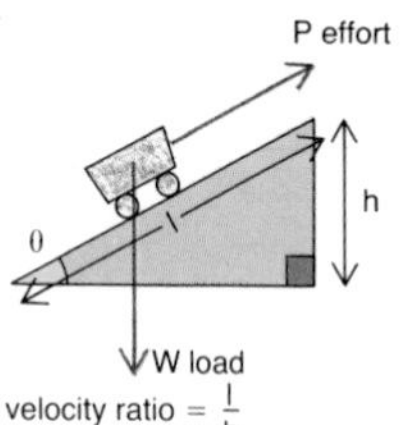

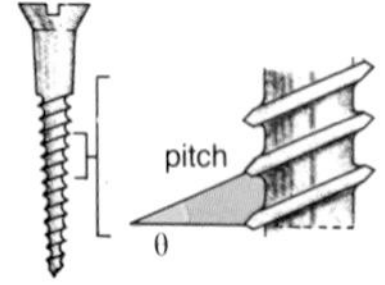

inclined plane

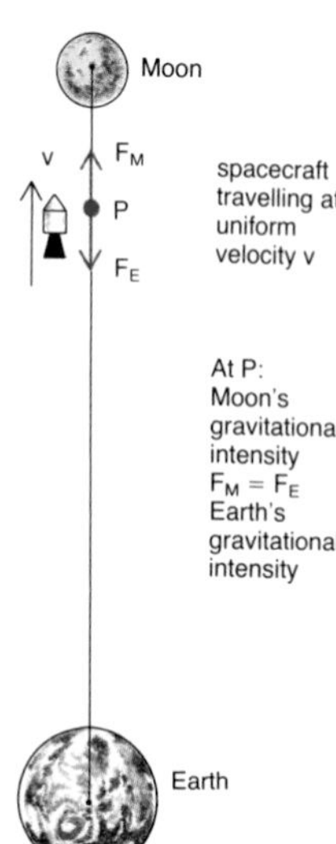

Newton's Laws of Motion 1
gravitational neutral point

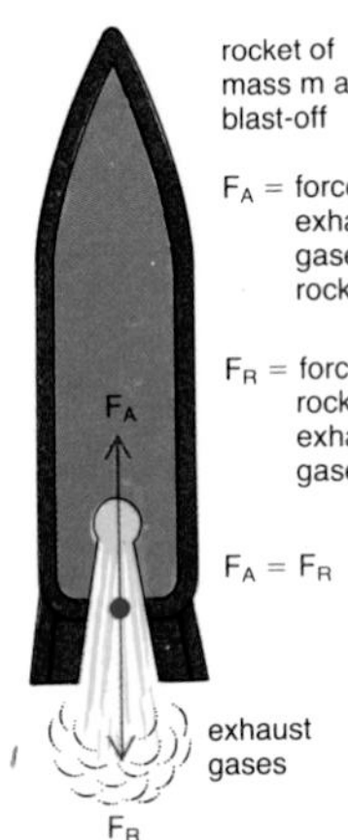

rocket moves up with a specific momentum and exhaust gases move down with the same momentum; total momentum is conserved

Newton's Laws of Motion 3
action and reaction

spacecraft landing on return to Earth must have its velocity reduced to a low value so that change of momentum on impact of touchdown or splashdown does not cause it to disintegrate,

3. Action (↑) and reaction (↑) are equal and opposite, e.g. a rocket leaving the Earth at blast-off is propelled upwards by the equal and opposite reaction force created when the exhaust gases from fuel combustion are discharged downwards through the booster rocket tail, conserving linear momentum (↑) along the line of flight. A gun recoils on firing as the bullet leaves the muzzle: bullet mass × bullet velocity = gun mass × recoil velocity, conserving linear momentum on firing.

newton absolute unit of force (p.26) defined from Newton's Law 2 of Motion: impressed force (↑) $F \propto$ rate of change of momentum (↑) or $F \propto$ rate of change of (mass m × velocity v) so $F = km(dv/dt)$ using calculus notation; therefore $F = kma$ where a = acceleration (p.30) constant $k = 1$ when $F = 1$, $m = 1$ kg and $a = 1\,\text{m s}^{-1}$ so $F = ma$ and $F = 1$ newton (N); force giving acceleration $1\,\text{m s}^{-2}$ to mass 1 kg.

impact (*n*) the physical effect of an impulsive force acting for a very short time, e.g. bullet on a target, elastic collision (↓) between gas molecules and with their containing vessel.

collision (*n*) the simultaneous meeting of 2 or more bodies (p.12), at least 1 of which must be moving; the bodies exert equal and opposite impulsive (↓) forces on each other and linear momentum is conserved (↓); kinetic energy is also conserved in elastic collisions (p.38). **collide** (*v*), **colliding** (*adj*).

conservation of linear momentum when 2 or more bodies collide, their total momentum (↑), measured in a specified direction along any straight line passing through the point of collision, is unaltered in value by the collision, provided that no forces external to the system of colliding bodies act at the same time. **conserve** (*v*), **conservative** (*adj*).

impulse (*n*) for a constant force (p.26) F (newton) acting for a time t (seconds), the impulse of the force = Ft (Ns); for a variable force F (N) acting for t (s), the impulse $= \int_0^t F.dt$ using calculus notation; since $F = m(dv/dt)$ from Newton's Law 2 of Motion (↑): the impulse $= \int_0^t m.(dv/dt)dt = \int_{v_1}^{v_2} mdv = [mv]_{v_1}^{v_2} = m(v_2 - v_1) =$ momentum change; if F is large and t small, other forces can be neglected and the total effect of F is measured by its impulse, e.g. hammer blow.

Newtonian mechanics a system of mechanics developed on the assumptions of Newton's 3 Laws of Motion (p.36); it describes motion and the forces affecting it for bodies whose velocities are low compared with the velocity of light $3 \times 10^8\,\mathrm{m\,s^{-1}}$, e.g. moving vehicles, aircraft, spacecraft, planets in the solar system (p.43); for bodies or particles (p.7) whose velocity exceeds about 10% of the velocity of light the system of relativistic mechanics (↓) is used.

relativistic mechanics a system of mechanics based on Einstein's theory of relativity; describes the variation of mass with velocity observed in bodies or particles (p.7) travelling at velocities exceeding about 10% of the velocity of light $3 \times 10^8\,\mathrm{m\,s^{-1}}$, e.g. β-particles (p.239) emitted from atomic nuclei; high energy particles in particle accelerators (p.194).

elastic collision a collision (p.37) in which linear momentum is conserved (p.37), and also kinetic energy (p.37), e.g. collisions between gas molecules and their container giving rise to gas pressure (p.39).

inelastic collision a collision (p.37) in which linear momentum is conserved (p.37) but not kinetic energy (p.27); some energy is lost as heat (p.28) or as work done (p.26) in causing physical damage, e.g. collisions between moving vehicles.

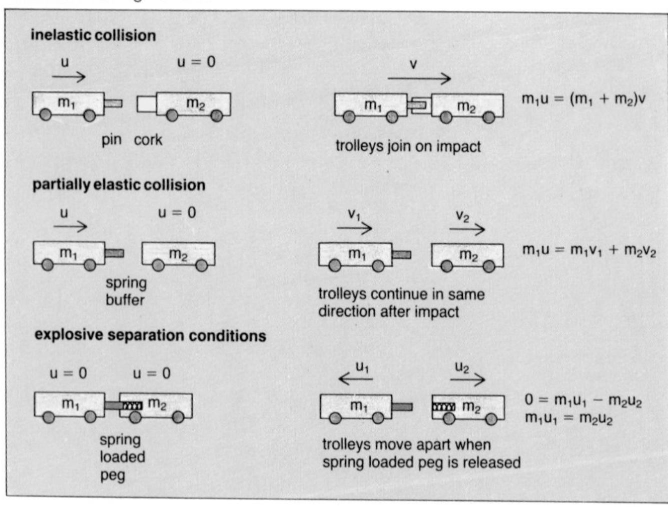

pressure (*n*) a force (p.26) exerted perpendicularly/metre^2 on a surface by a solid, liquid or gas; units of pressure are pascal (↓) or newton/metre^2. **press** (*n*), **pressurization** (*n*), **press** (*v*), **pressurize** (*v*), **pressurized** (*adj*).

pascal the SI unit of pressure (↑); 1 pascal (Pa) = 1 newton/m^2 (Nm^{-2}).

fluid pressure the pressure (↑) exerted by a liquid or a gas; a difference in pressure between 2 points in a fluid causes fluid flow (p.15); fluid pressure acts in all directions at any point in the fluid; it increases in direct proportion to the depth of the point below the fluid surface, e.g. water pressure is greater at the sea bed than it is at the sea surface, air pressure is greater at sea level on Earth than it is in the upper atmosphere (p.18); the pressure is the same for all points at the same horizontal level below the fluid surface; its value is $\rho g h$ (Nm^{-2}) where ρ ($kg m^{-3}$) = fluid density (p.7); g = acceleration due to Earth's gravity (p.32) = 9.81 (ms^{-2}; Nkg^{-1}); h (m) = depth of point P below fluid surface; total pressure at P = $\rho g h$ + atmospheric pressure (p.40) acting on fluid surface.

manometer (*n*) a device for the measurement of gas pressure; the liquid column in a U-tube measures the pressure (↑) difference between the gas in its container and atmospheric pressure (p.40) outside the manometer; excess fluid pressure (↑) is measured in mm of liquid from the difference in levels h between the U-tube limbs; the liquid can be mercury, water or light oil. **manometric** (*adj*).

barometer (*n*) an instrument for measuring accurately the pressure (↑) of the Earth's atmosphere (p.18), e.g. Torricellian (↓), Fortin and aneroid barometers (p.40). **barometric** (*adj*).

Torricellian barometer a simple barometer (↑) made by filling a tube of about 1 metre length with mercury and inverting it into a dish of mercury without allowing air to enter; the mercury falls from the top of the tube to a steady vertical height of about 760 mm, which the fluid pressure (↑) of the Earth's atmosphere (p.18) can support at the Earth's surface.

Torricellian vacuum the space above the mercury column in a Torricellian barometer (↑); a partial vacuum containing only mercury vapour (p.16).

torr a unit of low pressure (p.39) used in vacuum technology; 1 torr = 1 mm of mercury = 133 Nm^{-2}.

pressure
exerted by weight of
a solid of weight 50 newton

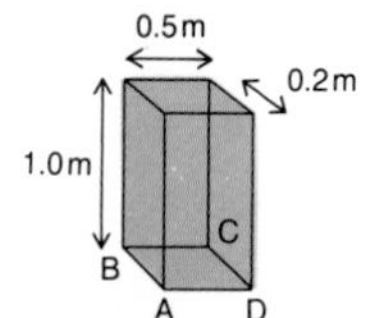

pressure over area ABCD
= $50/0.1 Nm^{-2} = 500 Nm^{-2}$

fluid pressure
variation with depth
pressure at P = $\rho g h$ (Nm^{-2})
ρ = liquid density ($kg m^{-3}$)
g = acceleration due to Earth's gravity (ms^{-2})

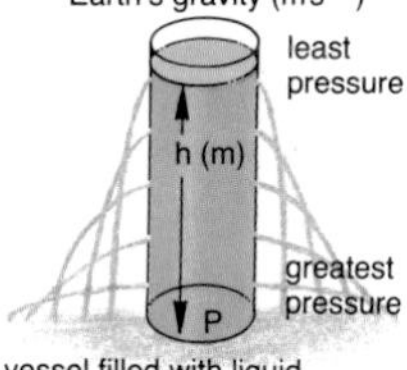

vessel filled with liquid
has holes at different levels

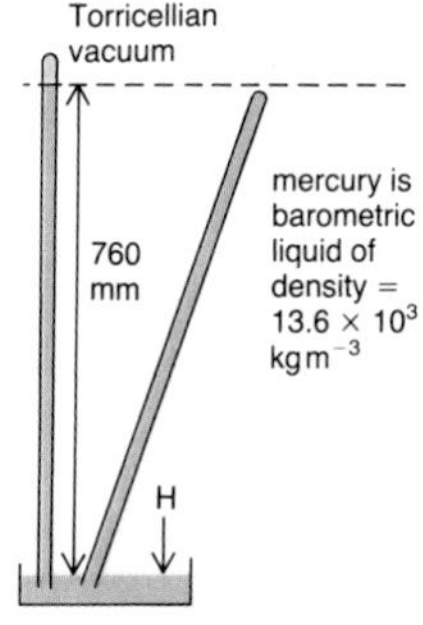

H = atmospheric pressure
= $hg\rho$ (Nm^{-2})
where h = 760 mm = 0.76 m
(vertical height above free liquid surface)
and $\rho = 13.6 \times 10^3$ ($kg m^{-3}$)
H = 760 mm of mercury
= $1.013 \times 10^5 Nm^{-2}$
Torricellian barometer

rotary vacuum pump first stage pump used in a vacuum system to reduce the pressure of a system from atmospheric pressure (p.40) 760 torr (p.39) to approximately 0.01 torr; rotor blades rotate, sweeping air towards the outlet valve and drawing in more air from the vacuum system; this air is then isolated by the following rotor blade and swept out of the pump; continuous pumping produces evacuation of the vacuum system to a pressure limit 0.01 torr; pump oil lubricates (p.21), cools and seals the pump from air.

atmospheric pressure the fluid pressure (p.39) exerted by the Earth's atmosphere (p.18) and measured by a barometer (p.39); it is measured in units of newton/metre^2 (Nm^{-2}) and mm of mercury (mm Hg); at sea level on the Earth's surface, over a long period of time, the average value is 760 mm Hg or $1.013 \times 10^5 Nm^{-2}$ or 760 torr (p.39); atmospheric pressure gradually decreases with the height above surface by an amount of approximately 10 mm Hg per 120 metres of ascent in the lower atmosphere.

standard atmospheric pressure a value of $1.01325 \times 10^5 Nm^{-2}$ for the atmospheric pressure (p.40) agreed as a standard with which other pressures can be compared; this is the pressure at the bottom of a mercury column 760 mm high, measured at sea level at a specified temperature of 0°C (273 K) and at a specified latitude of 45°; $1.01325 \times 10^5 Nm^{-2} = 1$ atmosphere (atm).

standard atmosphere an alternative name for standard atmospheric pressure (↑).

aneroid barometer a barometer (p.39) without mercury or other liquid; a partially evacuated and sealed metal box which responds to changes in atmospheric pressure (p.40) by slight expansions and contractions in its size; movements of the box are transmitted to a pointer moving over a calibrated scale.

rotary vacuum pump

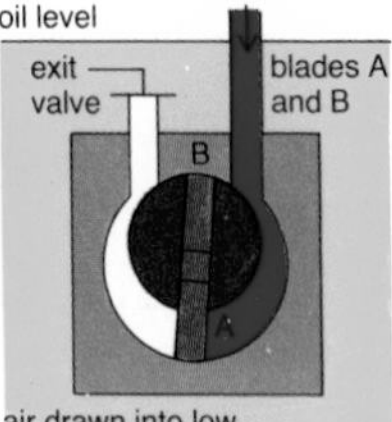

air drawn into low pressure side of pump

continuous pumping action

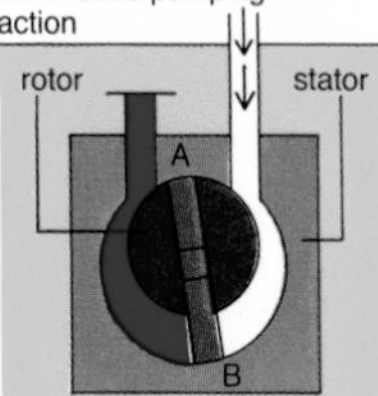

rotor blade B isolates air sample and pushes it towards low pressure side of pump

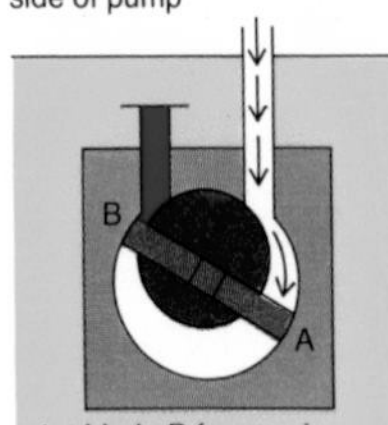

rotor blade B forces air sample through valve and oil out of pump

aneroid barometer

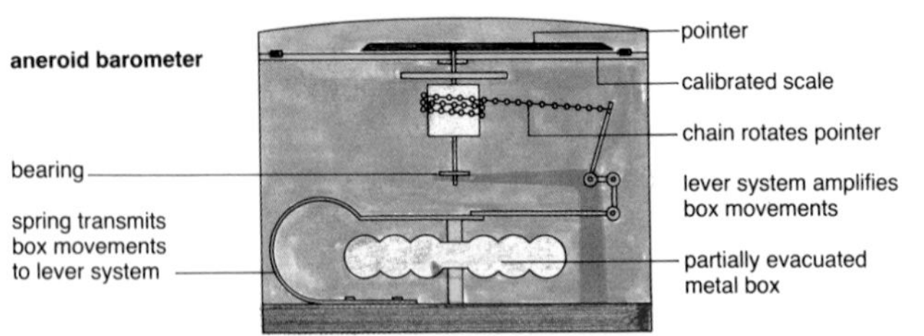

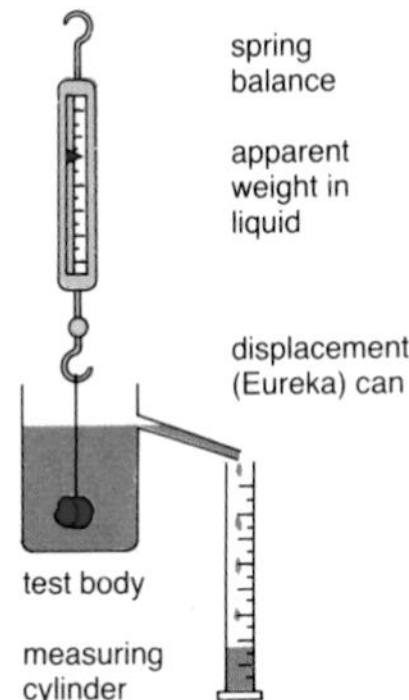

hydrometer principle

constant weight W of hydrometer is supported by liquid upthrust

$W = h_1A\rho_1 = hA\rho = h_2A\rho_2$
$W/A = h_1\rho_1 = h\rho = h_2\rho_2$
so $\rho_1 = h\rho/h_1$ and $\rho_2 = h\rho/h_2$
so ρ_1 and ρ_2 can be found since ρ is known

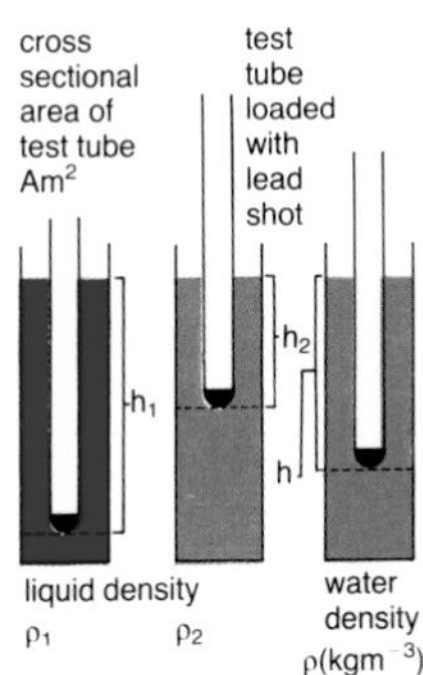

upthrust (*n*) an upward force (p.26) exerted on a body (p.12) immersed completely or partially in a fluid; due to fluid pressure (p.39); the weight (p.32) of the body is apparently decreased by an amount equal to the upthrust; apparent weight in fluid = true weight − upthrust; upthrust of the air is usually regarded as negligible but a correction can be made to beam balance readings if required.

Archimedes' Principle states that when a body is immersed completely or partially in a fluid, it experiences an upthrust (↑) from the fluid equal to the weight of fluid displaced (↓) by the body.

displacement² (*n*) the moving out of place of an equal volume of fluid by a body completely or partially immersed in it. **displace** (*v*).

flotation (*n*) the floating of a body on the surface of a liquid or when surrounded by a fluid; fluid upthrust (↑) supports the body's weight and Archimedes' Principle (↑) states that the weight of fluid displaced (↑) equals the body's weight. **float** (*v*).

hydrometer (*n*) an instrument used to measure the specific gravity (p.7) of liquids; the hydrometer floats vertically; upthrust (↑) supports its weight; the length h (mm) of stem below the liquid surface is inversely proportional to the relative density ρ ($kg\,m^{-3}$); $h \propto 1/\rho$; the stem is calibrated directly in relative density values.

gravitation (*n*) a property of matter by which every particle (p.7) having mass (p.7) exerts an attractive force (p.26) on every other particle having mass in the universe; the value of the attractive force can be calculated from Newton's Inverse Square Law of Gravitation (↓). **gravity** (*n*), **gravitate** (*v*), **gravitational** (*adj*).

Inverse Square Law of Gravitation a practical law expressing the variation in gravitational (↑) attractive force F (newton), between 2 point particle masses m_1 and m_2 (kg), with distance r (m) between them; F is directly proportional to the product m_1m_2 of the masses and inversely proportional to the square of the distance between them; $F \propto m_1m_2/r^2$ so $F = Gm_1m_2/r^2$, where G is the universal gravitational constant (p.42) for all matter; the value of $G = 6.67 \times 10^{-11}\,N\,m^2\,kg^{-2}$; the law can be shown to extend to massive bodies, as well as being applicable to point particle masses, the mass of the body being considered to be concentrated at its centre of gravity.

universal gravitational constant denoted by G, is the constant of proportionality in the equation expressing Newton's Inverse Square Law of Gravitation (p.41); its value is $G = 6.67 \times 10^{-11}\ Nm^2kg^{-2}$.

gravitational field the field (p.26) surrounding any massive body or particle in which any other massive body or particle experiences gravitational (p.41) forces; the gravitational field is conservative (p.27).

gravitational field intensity
variation of gravitational potential V and field intensity g with distance x from centre 0 of body of mass M

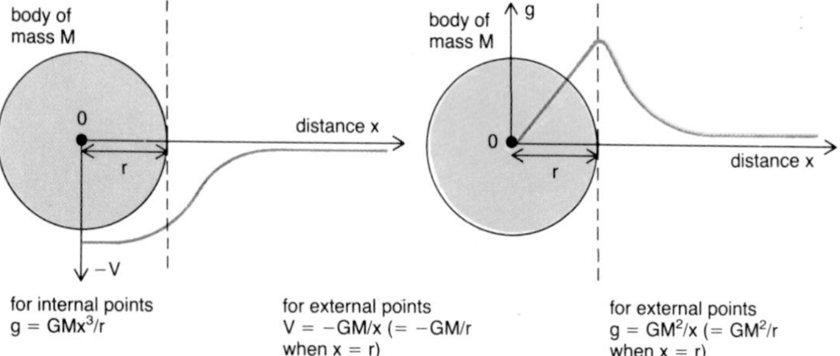

gravitational field intensity a physical quantity whose magnitude in a given direction at a specified point in a gravitational field (↑) gives the strength of the field at that point; it is a vector quantity (p.30); it is the force F (newtons) exerted on a mass of 1 kilogram (kg) placed at the specific point; its units are Nkg^{-1} and it is also given by the variation $-dV/dr$ of gravitational potential V (↓) with distance r at the point.

gravitational field strength an alternative name for gravitational field intensity (↑).

gravitational force the force on a mass in a gravitational field (↑).

gravitational field of Earth for a body of mass m (kg), at a distance r (m) from the centre of the Earth's mass M (kg), the attractive force F (N) due to Earth's gravity (p.41) is given by the Inverse Square Law of Gravitation (p.41) such that $F = GMm/r^2$; but the attractive force F (N) is also the weight (p.31) of the body of mass m, so $F = mg = GMm/r^2$, where g = acceleration due to Earth's gravity (p.32) (ms^{-2}; Nkg^{-1}); thus, for any value of m the value of $g = GM/r^2$ and so g varies according to an inverse square law for points external to the Earth; for

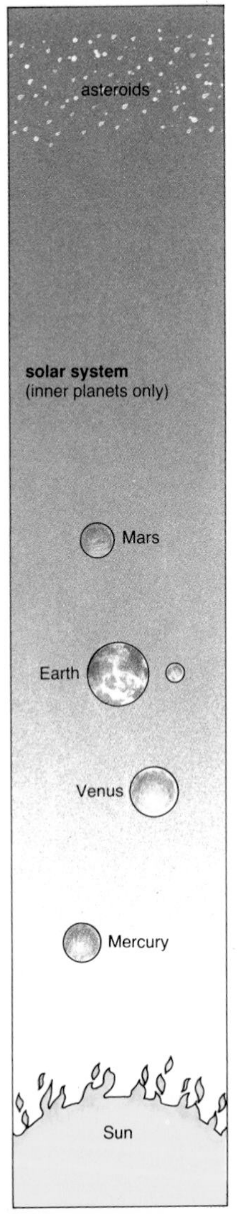

internal points g is directly proportional to r; for m = 1 kg: $F = g\ (N\,kg^{-1}) = GM/r^2\ (N\,kg^{-1})$, giving field intensity value.

gravity (*n*) the individual property of gravitation (p.41) possessed by a particular body, e.g. the Earth's gravity, the Sun's gravity, the Moon's gravity.

gravitational potential denoted by V, a scalar quantity defined as the work done in bringing a 1 kg mass from infinity, where potential is defined to be zero, to the specific point; since gravitational force is always attractive the potential at a point is a negative quantity, e.g. the potential at a distance r from body of mass $M = -GM/r$, where G = universal gravitational constant.

solar system the sun (↓) and 9 major planets (p.44) revolving in orbit around it at different distances with different time periods of revolution; a group of about 1500 minor planets or asteroids, of diameter less than 500 km, revolve in orbits between the orbits of Mars and Jupiter.

sun (*n*) the central star about which the planets (p.44) of the solar system (↑) revolve; one of about 10^{11} stars in our galaxy (↓); about 40×10^{12} km from next nearest star; its estimated surface temperature (p.116) is about 6000°C and core temperature up to 40×10^6 °C; the source of solar energy is thermonuclear fusion (p.29) of hydrogen gas to helium gas; the sun is the ultimate source of all energy on Earth. **solar** (*adj*).

galaxy (*n*) a group of many millions of stars evolving from a common origin early in the history of the universe; the universe is made up of countless galaxies, separated by large amounts of space, and constantly moving further apart as the evolution of the universe proceeds; our galaxy contains about 10^{11} stars of which the sun (↑) is one. **galactic** (*adj*).

big bang theory of universe the theory that the universe, originally in a highly dense state, exploded outwards 10^{10} years ago; evidence for the expanding universe is seen in the red shift (p.111) of distant galaxies and in the microwave background radiation observed.

supernova (*n*) a star that explodes as a result of running out of nuclear fusion fuel (p.29); they are a probable source of the heavy elements in the universe.

pulsar (*n*) a highly dense, rapidly spinning star which emits highly directional pulses of radiation; believed to be remnant of supernova (↑) explosion.

neutron star (*n*) a highly condensed star with density approximately 5×10^{10} kg m^{-3}; most of the electrons and protons have interacted to form neutrons; pulsars (p.43) are neutron stars.
quasar (*n*) quasi-stellar object; quasars have large red shifts (p.111) suggesting that they are at enormous distances; this being so, their brightness suggests that their luminosity is comparable to that of a galaxy.
black hole (*n*) an astronomical object with such a high gravitational field (p.42) that neither matter nor light may escape from its influence.
planet (*n*) a body revolving in orbit around the sun, or around a planet in the solar system (p.43).
moon (*n*) the single planet (↑) of the Earth, at 3.84×10^5 km distance, with a time period of revolution 27.3 days around the Earth; satellite (↓) of Jupiter or of other member planet of the solar system (p.43). **lunar** (*adj*).
satellite (*n*) a natural moon (↑) of a planet (↑) in the solar system (p.43); artificial moon of the Earth put into orbit from Earth by rocket systems.
synchronous satellite a satellite (↑) whose period of revolution in orbit around the Earth is 24 hours, the same time as for one rotation of the Earth on its axis, so that the satellite appears stationary at the same point in the sky, e.g. Telstar.
parking orbit refers to the orbit of a synchronous satellite.
moon's gravity the individual gravity (p.41) of the moon; the moon's mass is 0.0123 of the Earth's mass and the moon's diameter is 0.273 of the Earth's diameter, so the moon's gravitational force (p.42) at its surface is approximately ⅙ of the Earth's gravitational force; acceleration due to the moon's gravity is approximately 1.64 ms^{-2}.
gravitational neutral point a region in space where gravitational fields are superimposed so that field strength vectors are equal and opposite; a body passing through this region experiences no resultant gravitational force and is weightless.
weightlessness (*n*) a condition experienced by astronauts in spacecraft in orbit around the Earth; centripetal force (p.47) is provided by the Earth's gravitational attractive force (p.42) and centripetal acceleration (p.47) equals the acceleration g due to Earth's gravity (p.32), so that persons and objects in the orbiting spacecraft are falling freely (↓) and experience

satellite
satellites in Earth orbit

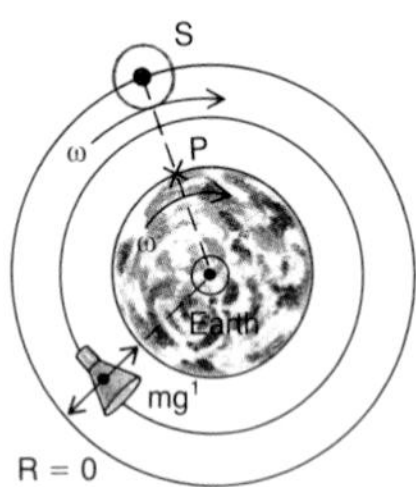

For astronaut falling freely, reaction R from spacecraft is zero. Synchronous satellite S has same angular velocity ω as Earth on its axis and always appears above P.

none of the sensations of weight (p.31) and no reaction (p.36) from the floor of the spacecraft; also a condition experienced by astronauts in a spacecraft passing through gravitational neutral points without acceleration. **weightless** (*adj*).

free fall fall of a body (p.12) towards the Earth's surface at an acceleration g due to Earth's gravity (p.32) experienced as the condition of weightlessness (↑) by astronauts in a spacecraft in Earth orbit, when persons, bodies and the spacecraft itself all have centripetal acceleration g.

escape velocity the initial velocity (p.31) required by an object so that its kinetic energy (p.27) is sufficient to do the work of overcoming gravitational attractive forces and allowing the object to escape from a gravitational field (p.42); for the Earth its value is approximately $11\,km\,s^{-1}$; for the moon its value is $2.4\,km\,s^{-1}$; in general, escape velocity = $\sqrt{(2GM)/R}$ where G = universal constant of gravitation (p.42), M = mass of object escaped from and R = distance from centre of object escaped from.

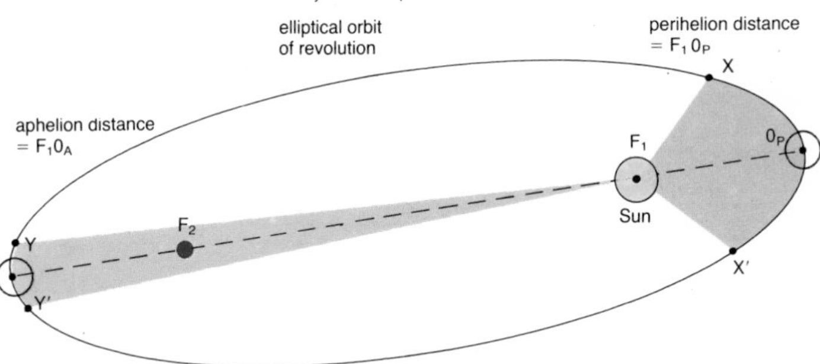

1. foci of ellipse are F_1 (Sun) and F_2

2. vector line F_1X sweeps area $F_1XX'F_1$ in same time as vector line F_1Y sweeps area $F_1YY'F_1$

3. mean distance r of planet from sun at F_1 is $(0_PF_1 + 0_AF_1)/2$; orbital period is T; $T^2 \propto r^3$

Kepler's 3 Laws

Kepler's 3 Laws describe the revolution of the planets around the sun in the solar system (p.43):

1. A planet's orbit is an ellipse with the sun at one focus;
2. The vector (p.30) line joining the sun to an orbiting planet moves across equal areas in equal times;
3. The square of the time period (T) of revolution of a planet around the sun is directly proportional to the cube of its mean distance (r) from the sun; $T^2 \propto r^3$.

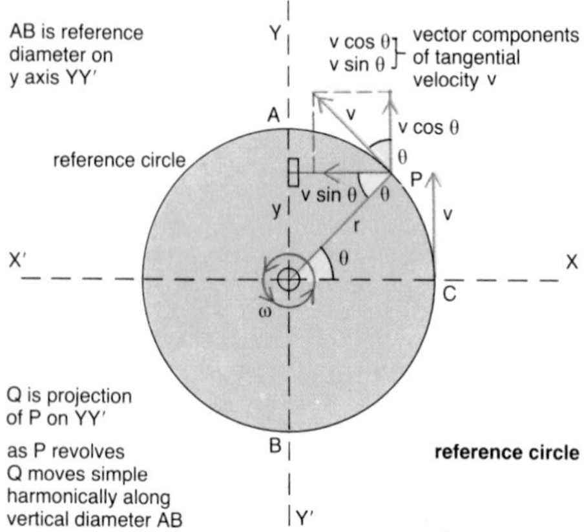

reference circle

reference circle a circle around which we consider a particle (p.7) to revolve at uniform angular velocity (↓) and constant speed and from which we can study the particle's motion.

circular motion movement of a body (p.12) or particle (p.7) in a circular path so that it revolves (p.8) or rotates around the circle's centre.

radius vector a line joining the centre of a reference circle (↑) to the particle revolving round the circumference.

rotating vector a radius vector (↑) rotating at uniform angular velocity (↓) around a reference circle (↑); used for diagrammatic representation of a physical quantity varying simple harmonically (p.50), e.g. a.c. quantities (p.218).

rotating phasor an alternative name for a rotating vector (↑); it is used for a.c. quantities (p.218) with phase difference (p.59).

radian (*n*) the unit of measurement for angles in a 2-dimensional plane; the angle subtended at the circle centre by an arc equal in length to the circle radius; 1 radian = 57.3°; it is a dimensionless (p.12) unit, being the ratio of 2 linear measurements.

angular velocity the angle per second through which a radius vector (↑) rotates around a reference circle (↑); it is denoted by ω, and measured in units of radian/s ($\text{rad}\,\text{s}^{-1}$); $\omega = \theta/t$, where θ = angle through which the radius vector rotates in time t; $\omega = d\theta/dt$ using calculus notation.

component of v at C acting towards O = zero
component of v at P acting towards O = $v \sin \theta$;

when θ is very small ($\delta\theta$):
rate of velocity change towards O = $\frac{v \sin \delta\theta}{\delta t} = v\frac{\delta\theta}{dt}$
radial acceleration a as $\delta t \rightarrow$ zero
$a = v\left(\frac{d\theta}{dt}\right) = v\omega$

radial acceleration

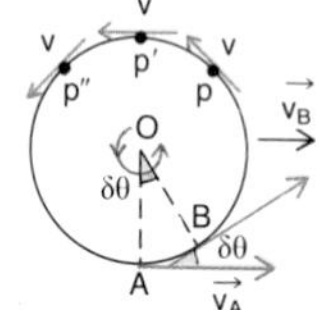

combining vectors:
$\vec{v}_B + (-\vec{v}_A) = \vec{\delta v}$ towards O
radial acceleration a towards O:
$a = \frac{\delta v}{\delta t} = v.\frac{\delta\theta}{\delta t}$
as $\delta t \rightarrow$ zero $\frac{\delta\theta}{\delta t} = \frac{d\theta}{dt} = \omega$
since $v = r\omega$
$a = v\omega = r\omega^2 = v^2/r$

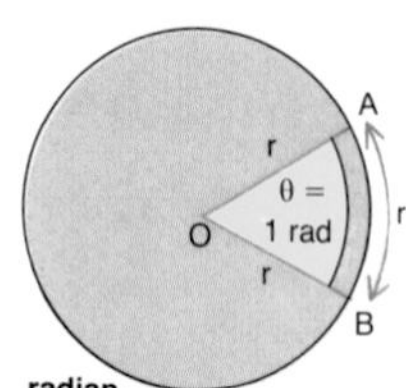

radian

θ radians (rad) = arc AB subtended at centre O ÷ radius r
∴ arc length = $r\theta$
2π radians = 360°
π rad = 180°
1 rad = 57.3°
radian: angle subtended at circle centre O by arc AB of length equal to circle radius r: arc AB/radius r = 1 (rad)

tangential velocity v changes direction continuously for particle revolving at uniform angular velocity ω

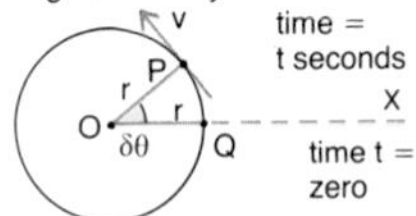

arc PQ = δs
speed $v = \delta s/\delta t$
since $\delta s = r\delta\theta$, $v = r\delta\theta/\delta t$
as $\delta t \rightarrow$ zero, $v = r\dfrac{d\theta}{dt} = r\omega$
relation between speed v and angular velocity ω for revolving particle P in reference circle centre O

centripetal force
$= mv^2/L =$ tension T in string

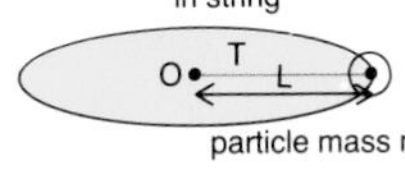

particle mass m on an inelastic string of length L, revolving in a horizontal circle, centre O, radius L

bank
$R\sin\theta = mv^2/r$
$R\cos\theta = mg$
$\tan\theta = \dfrac{v^2}{gr}$

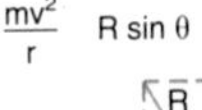

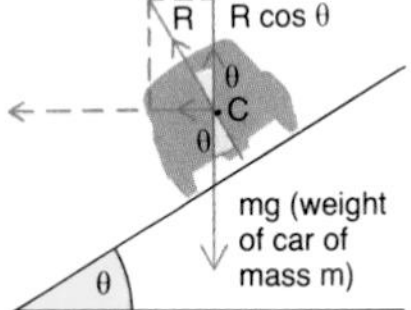

car on track banked at angle θ: combined reaction at wheel-ground surface is R; frictional forces are not shown; C is centre of gravity of car.

tangential velocity the instantaneous value of linear velocity (p.30) of a body or particle in circular motion (↑); the value of its speed (p.30) is constant but the direction of its velocity vector (p.30) is continuously varying; this direction is tangential to the circular path at any instant; it is denoted by v and measured in units of $m\,s^{-1}$; $v = r\omega$ where r = circle radius; ω = angular velocity (↑).

radial acceleration the acceleration (p.30) towards the centre of the circular path in which a body or particle revolves when in circular motion (↑); it arises from the change in direction of the radial component of its linear velocity; its value is $\omega^2 r$; $v\omega$; v^2/r measured in units of $m\,s^{-2}$; denoted by $d\omega/dt$ or $d^2\theta/dt^2$, using calculus notation.

centripetal force a body or particle of mass m in circular motion (↑) requires a radial force to maintain centripetal acceleration (↓); its value is $mr\omega^2$ or $mv\omega$ or mv^2/r newton (N); centripetal force = mass × centripetal acceleration, e.g. a body on a string moving in a horizontal or vertical circle, a car on a banked (↓) track, an aircraft making a circular turn; centripetal force must be maintained to keep the body in its circular path; if the centripetal force becomes zero, the body leaves the circular path in a tangential direction, with velocity v at the instant of release.

centripetal acceleration radial acceleration (↑) experienced by a body or particle revolving in a circular path as a result of centripetal force (↑).

bank (*v*) to incline at an acute angle; for a car on a circular banked track, the component of the normal reaction (p.21) towards the circle centre provides centripetal force (↑) and helps to prevent sideslipping of the car; for an aircraft banking to make a circular turn, the component of lift (p.22) towards the circle centre provides centripetal force.

centrifugal effects centripetal force (↑) maintains a body in circular motion; if the centripetal force becomes zero, the body will leave its circular path, taking a direction tangential to the circle at its point of release; the body's subsequent path will be a straight line, according to Newton's 1st Law (p.36), until some other force acts upon it, e.g. Earth's gravity; the body's movement away from the circle is due to its own inertia (p.36), not to the action of any other force, and the direction of movement is always tangential, not radial.

time period (*n*) the time interval between 2 specified consecutive events in a succession of similar periodic events, e.g. vibration of a simple pendulum (p.50).
periodic time alternative name for time period (↑).

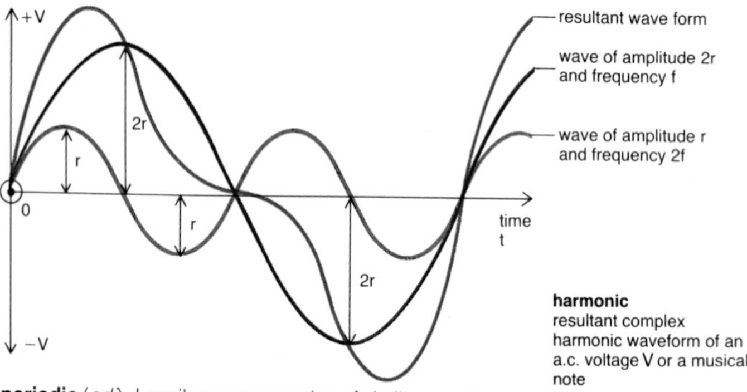

harmonic
resultant complex harmonic waveform of an a.c. voltage V or a musical note

periodic (*adj*) describes a succession of similar events with equal time intervals between them.
harmonic (*adj*) an alternative name for periodic (↑) when describing a continuously varying physical quantity whose pattern of variation is repeated at periodic time intervals, e.g. displacement y from the normal rest position (p.50) of a particle in simple harmonic motion (↓); variation of the value of an a.c. quantity with time.
harmonic motion a periodic (↑) motion of a particle or mechanical system; it can be complex but is always periodic (↑), e.g. waveforms (p.53) demonstrated by musical instruments.

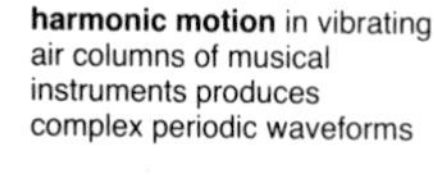

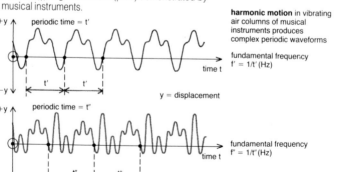

harmonic motion in vibrating air columns of musical instruments produces complex periodic waveforms

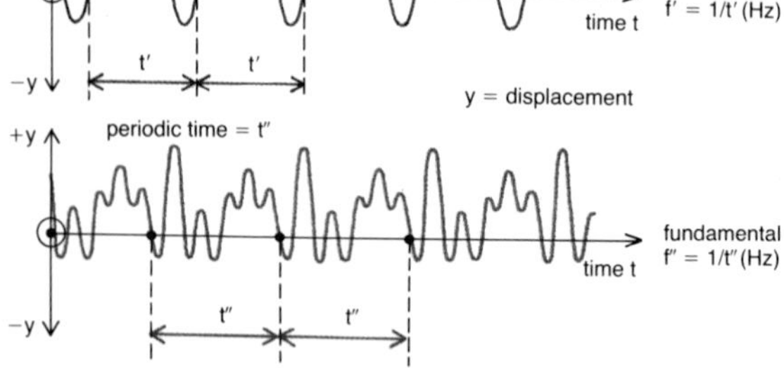

fundamental frequency
$f' = 1/t'$ (Hz)

fundamental frequency
$f'' = 1/t''$ (Hz)

harmonic variation in the value of a.c. voltage V

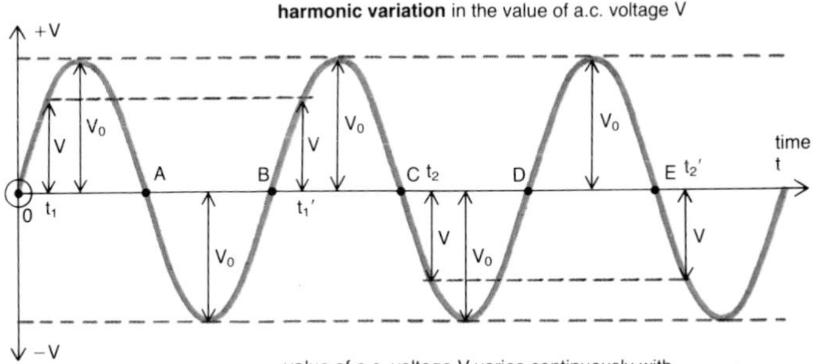

OAB, ABC, BCD are cycles of variation

value of a.c. voltage V varies continuously with time t. Variation is sinusoidal with form $V = V_0 \sin \omega t$. +V is instantaneous value at times t_1 and t_1'. −V is value at t_2 and t_2'.

harmonic variation variation in the value of a physical quantity which can be complex but is always periodic (↑), e.g. complex waveforms (p.53) in a.c. circuits (p.217) when several components of an a.c. quantity are present.

sine curve this shows the relationship between 2 variables x and y in the form $y = \sin x$; x can represent a variable involving time t, e.g. displacement $y = r \sin \omega t$ for simple harmonic motion (↑); x can also represent the distance from the origin measured along a wave profile curve (p.53). **sinusoidal** (*adj*).

sinusoidal variation the variation of a dependent variable y with respect to an independent variable x according to the relationship $y = \sin x$; graph is a sine curve (↑).

vibration
vibrating mechanical system:
cantilever (straight rod) clamped at one end

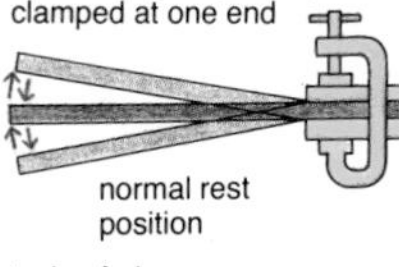

tuning fork

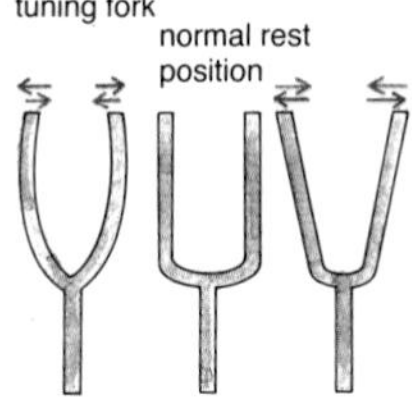

vibration (*n*) the harmonic motion (↑) of a particle or mechanical system. **vibrate** (*v*), **vibrating** (*adj*), **vibrational** (*adj*).

oscillation (*n*) an alternative name for vibration (↑); harmonic variation (↑) of the value of a physical quantity. **oscillate** (*v*), **oscillating** (*adj*).

restoring force when a mechanical system, having inertia (p.36) and elasticity (p.19) is displaced from its normal rest position (p.50) the system itself will offer an opposing force tending to restore its original situation; because of its acquired momentum it will overshoot the normal rest position; simple harmonic motion (p.50) then follows, e.g. oscillating spring (p.51).

amplitude (*n*) the maximum displacement in simple harmonic motion (p.50) or wave motion (p.52).

normal rest position the fixed reference point or position with respect to which a vibrating (p.49) particle or mechanical system undergoes simple harmonic (↓) displacement (p.30); for a particle it is located at the centre of the reference circle (p.46) and mid-point of the reference diameter.

simple harmonic motion a periodic (p.48) or harmonic (p.48) motion of a body, e.g. vibrating pendulum (↓), or of a mechanical system, e.g. oscillating spring (↓), in which the movement is along a straight line with respect to a fixed reference point in that line; the essential condition describing the motion is that linear acceleration (p.30) is directly proportional to the displacement (p.30) from the fixed reference point and is directed towards the fixed reference point; as a consequence of this, the period of a simple harmonic oscillator is independent of its amplitude (p.49); linear acceleration $a = -ky$, where y = displacement (p.30); k is the proportionality constant characteristic of that particular mechanical system; its value is ω^2 when related to the revolution of a particle around the reference circle (p.46) at angular velocity ω (p.46); the negative (−) sign $-ky$ indicates that the direction of displacement is away from the fixed reference point, while the direction of acceleration is towards it; the restoring force F (p.49) = ma, where m = mass (kg) of vibrating body or system, and F (newton) varies with the instantaneous value of $a = -\omega^2 y$; $y = r \sin \omega t$ where r = amplitude of vibration; the instantaneous value of linear velocity $v = \omega\sqrt{(r^2 - y^2)}$; time period $T = 2\pi/\omega$; frequency $f = 1/T$.

simple pendulum a spherical bob of small mass m (kg) is suspended by a light inextensible string of length l (m) from the point of suspension to the centre of gravity (p.34) of the bob; on displacement (p.52), through a small angle θ from its normal rest position (↑), and then released, the pendulum vibrates (p.49) about its normal rest position with simple harmonic motion (↑) of time period $T = 2\pi\sqrt{l/g}$ where g = acceleration due to Earth's gravity (p.32).

force constant the force per unit displacement for a material such as a wire or spring; units are newton/metre; Young's modulus (p.20) for a wire may be given as $E = kL/A$ where k = force constant ($N m^{-1}$), L = original length of wire (m) and A = area of cross-section of wire (m^2).

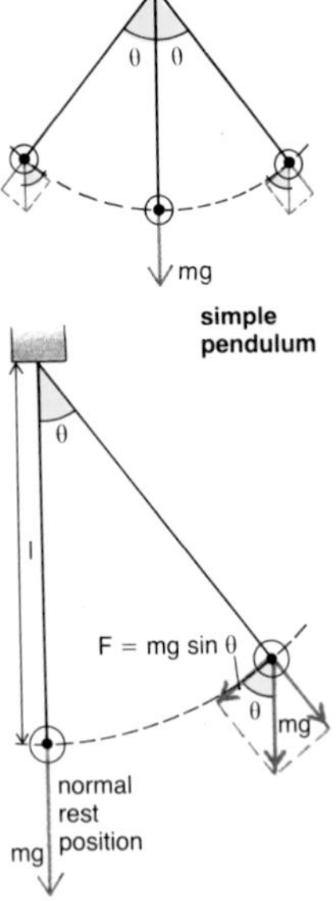

restoring force $F = ma$ where acceleration $a = -g \sin\theta$
$F = -mg\theta$ for small θ and is always directed towards the normal rest position

oscillating spring
restoring force F = Ma
where acceleration
a = −ky

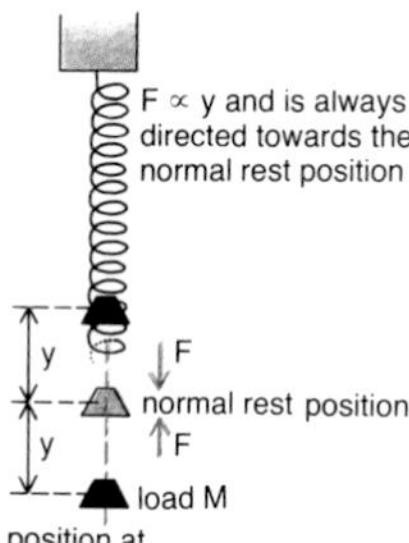

position at
extension y

angular momentum
conservation of angular
momentum Iω during
spin

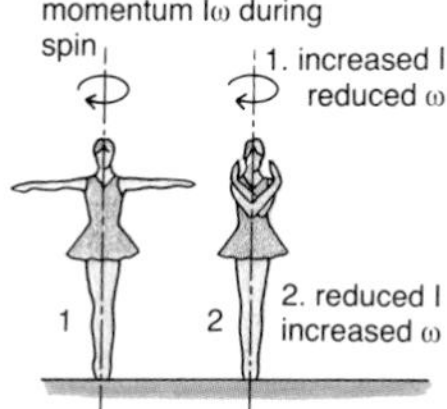

precession
Earth's axis of rotation
precesses around vertical
once every 25 800 years

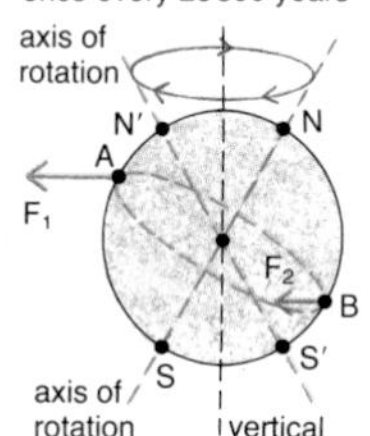

different attractive forces F_1 and F_2 exerted on Earth by Sun and Moon.

A and B are diametrically opposite points on Earth's equator where radius slightly exceeds polar radius

oscillating spring a suspended helical spring of mass m is loaded (p.20) with mass M (kg), given a small displacement (p.30) downwards and then released; it vibrates (p.49) simple harmonically (↑), with respect to the normal rest position (↑), with time period $T = 2\pi\sqrt{(M + m/3)/k}$ where k = force constant (↑) of the spring (newton/metre); for a light spring (m = 0), $T = 2\pi\sqrt{M/k}$.

rotational inertia the property of a rotating body (p.12) to resist changes in its angular velocity (p.46) around its axis of rotation; measured by the moment of inertia I (↓) of the body around that specified axis; it depends on mass (p.7), size and shape.

angular momentum the usual name for the total moment of momentum of all particles in a rotating body around the specified axis of rotation; angular momentum is conserved (↓) as the value Iω about the specified axis, in the absence of external forces, where I = moment of inertia (↓) and ω = angular velocity (p.46).

moment of inertia a physical quantity giving a measure of the rotational inertia (↑) of a body about a specified axis of rotation; it depends on the mass (p.7), size and shape of the body and is denoted by I with units of $kg\,m^2$.

conservation of angular momentum angular momentum Iω (↑) of a rotating system is conserved about a specified axis of rotation in the absence of external forces; if I is decreased ω will be correspondingly increased and vice versa, e.g a spinning skater or dancer varies the rotational angular velocity (p.46) by increasing or decreasing I, by holding the arms outwards or flat against the body respectively.

precession (*n*) descibes the situation when the axis of rotation itself revolves around the vertical in a conical path as a result of an applied couple (p.35) acting on a rotating body at a line perpendicular to its axis of rotation, e.g. spinning top, Earth's axis. **precess** (*v*).

gyroscope (*n*) a rotating wheel mounted so that the movement of its supports does not alter the direction of its axis of rotation; an applied couple (p.35) tending to alter the direction causes the wheel to precess (↑) about this direction; this effect can be used to stabilize the direction of travel of ships and aircraft and forms the basis of the non-magnetic gyroscopic compass, the wheel being driven electrically; it now replaces the magnetic compass needle (p.182) for most practical direction-finding purposes.

transmission (*n*) the passage of energy through a suitable medium. **transmit** (*v*), **transmitting** (*adj*).

propagation (*n*) an alternative name for energy transmission (↑). **propagate** (*v*), **propagating** (*adj*).

wave (*n*) a periodic or harmonic (p.48) disturbance by means of which energy is propagated (↑) through a transmitting medium; in material media the energy of the disturbance is transmitted by elastic deformation (p.19) of the transmitting material in a transverse or longitudinal direction, e.g. transverse water waves (p.54), longitudinal compression waves (p.56); electromagnetic energy (p.28) is propagated as a harmonic, 3-dimensional transverse wave in free space or in a suitable transmitting medium; for all waves the wave velocity v (ms^{-1}) = wavelength λ (m) × frequency f (s^{-1}) (p.54); for electromagnetic waves in free space the wave velocity $C_o = \lambda f$ where $C_o = 3 \times 10^8\ ms^{-1}$ for all electromagnetic radiations (p.55).

displacement[3] (*n*) the movement of a particle (p.7) or layer of transmitting (↑) medium from its normal rest position (p.50) as energy passes through that point or region, or when force is applied; usually the varying displacement (p.30) of the particle or layer is a harmonic motion (p.48). **displace** (*v*), **displaced** (*adj*).

wave motion the transmission (↑) of energy through a medium in a way that can be described by the wave equations.

wavetrain (*n*) a continuous sequence of energy waves (↑) passing from the energy source through the transmitting medium by straight line transmission (↑).

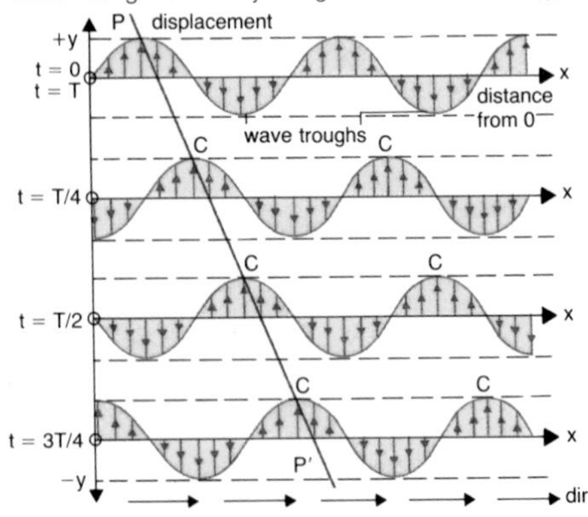

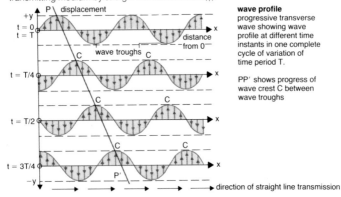

wave profile progressive transverse wave showing wave profile at different time instants in one complete cycle of variation of time period T.

PP′ shows progress of wave crest C between wave troughs

wave profile the graph showing the variation of displacement (↑) of a particle or layer of transmitting medium with distance from the energy source, at a specific time instant during the transmission of energy waves (↑).

waveform (*n*) a wave equation describing a particular wave motion (↑), e.g. simple harmonic waveform (↓); wave profile (↑) is characteristic of that wave; graph showing variation of displacement (↑) of a particle or layer of transmitting medium with time, at a specific point in the path of the wave.

waveform/ simple harmonic wave simple harmonic transverse wavetrain of waveform $y = r \sin \omega t$ showing variation of displacement y with t at any point in the transmitting medium, or wave profile showing variation of y with distance x from 0 at time t

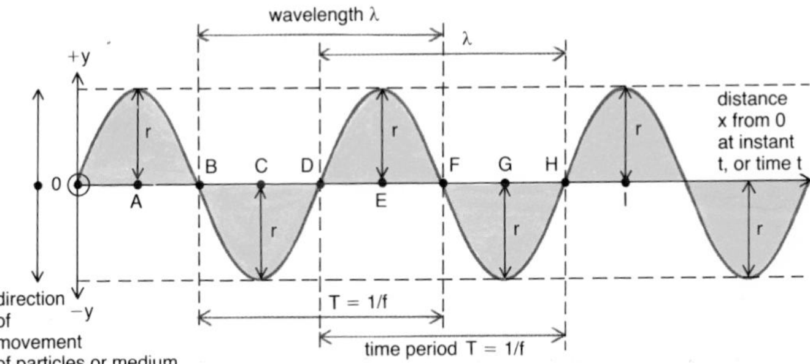

clarinet

fundamental frequency = 1/t′ (Hz)

+y, −y, 0, time t (s), t′, t′, time period

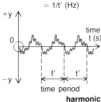

harmonic wave

saxophone

fundamental frequency = 1/t″ (Hz)

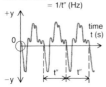

+y, −y, 0, time t (s), t″, t″

simple harmonic wave a wave (↑) containing one component only of disturbance contributing to displacement (↑); its waveform (↑) is a sine curve (p.49).

harmonic wave a wave (↑) whose waveform (↑) shows harmonic variation (p.49), though the wave profile (↑) may be complex, e.g. waveforms of notes produced by musical instruments containing overtones of the fundamental (p.104) frequency.

phase (*n*) this identifies one or more specific points on a harmonic (↑) wavetrain (↑), by reference to the angle $\omega t = \theta$ in the relationship $y = r \sin \theta = r \sin \omega t$ between displacement (p.30) and time t in simple harmonic variation (p.49), rather than by reference to displacement y or distance from the energy source measured along the wave profile (↑); particles or layers of transmitting (↑) medium at corresponding points on their cycles of harmonic variation are in phase; they differ in the value of angle $\theta = \omega t$ by 2π radians = 360°.

phase angle the value ωt in the relation $y = r \sin \omega t$ between displacement y (p.30) and time t in simple harmonic variation (p.49).

simple harmonic transverse wave of waveform $y = r \sin 2\pi f(t - x/v)$ of wavelength λ (m), frequency f (Hz), time period T (s) and velocity v ($m s^{-1}$)

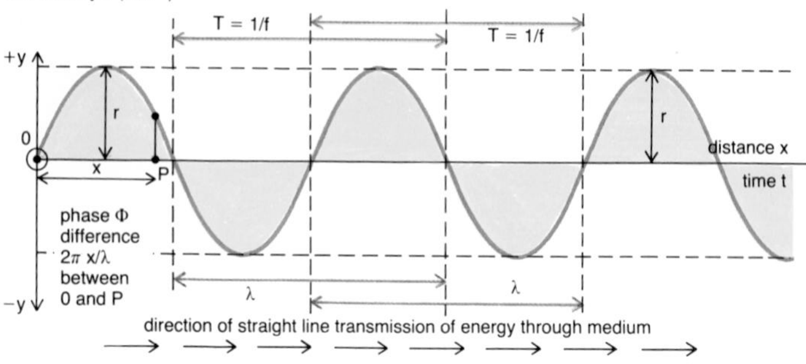

wavelength/frequency

wavelength (*n*) the distance between any 2 adjacent points vibrating in phase (p.53) measured along the wave profile (p.53) in the direction of energy transmission (p.52); denoted by λ; common units used are km; m; μm; nm.

frequency (*n*) the number of complete wavelengths (↑) passing any given point in the transmitting medium in 1 second; denoted by f; the units used are hertz (Hz) (↓); s^{-1}; cycles/s; the number of complete vibrations or oscillations (p.49) completed in 1 second; $f = 1/T$ where T = periodic time (s) (p.48). **frequent** (*adj*).

hertz (Hz) the unit of frequency (↑); 1Hz = 1 cycle/second.

wave velocity the velocity (p.30) of transmission of energy waves (p.52) through the transmitting medium; denoted by v; units are $km s^{-1}$; $m s^{-1}$; its value depends on the characteristic transmission properties of the medium and on the wave frequency f (↑); for all waves the wave velocity v ($m s^{-1}$) = wavelength λ (m) × frequency f (s^{-1}).

wave properties in general these include straight line transmission (p.52), reflection, refraction (p.69), diffraction (p.61), interference (p.58) and attenuation (p.57) according to an Inverse Square Law; transverse waves (p.54) can be also be polarized (p.61) though longitudinal waves (p.56) cannot.

transverse wave a wave (p.52) in which the disturbance of particles or layers of transmitting medium is in a direction perpendicular to the direction of energy transmission (p.52), e.g. water waves, waves in a plucked or bowed string, electromagnetic waves (↓).

ripple tank a laboratory scale apparatus for the observation and investigation of simple harmonic waves (p.53) on water; an electric motor (p.202) drives a rod perpendicular to the water surface to provide a source of circular wavefronts (p.57), or a horizontal bar to provide a source of plane wavefronts; wave properties (p.54) of reflection, refraction (p.69), diffraction (p.61) and interference (p.58) can be demonstrated for plane and circular wavefronts.

ripple tank

lamp illuminates water surface
electric motor on bar
spherical knob dipper for circular wavefronts
water depth about 1 cm
bar dipper for plane wavefronts
wavefronts viewed on screen below illuminated water surface

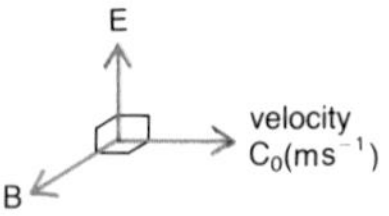

electromagnetic waves

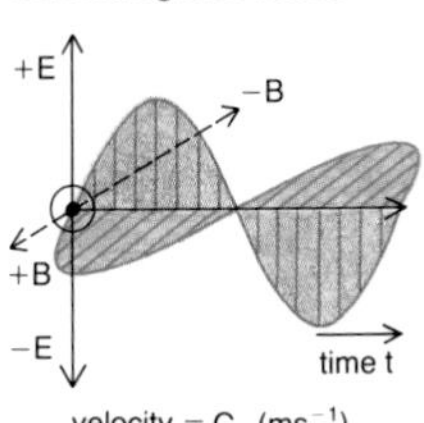

velocity = C_0 (ms^{-1})

direction of straight line transmission of electromagnetic energy through space

transverse electromagnetic wave showing vectors E and B perpendicular to each other and to transmission direction

electromagnetic waves electromagnetic energy (p.28) transmitted through a medium or free space as 2 perpendicular transverse wave (↑) components of the same frequency (p.54), the electric vector E representing the electric field strength (p.169) and the magnetic vector B representing flux density; the wave components are in phase (p.53) and have the same wave velocity (p.54) $V = 1/\sqrt{\mu_0\varepsilon_0} = 3 \times 10^8 \mathrm{m\,s^{-1}}$ in free space, where μ_0 = magnetic permeability (p.185) of free space $= 4\pi \times 10^{-7} \mathrm{H\,m^{-1}}$ and ε_0 = electric permittivity (p.168) of free space $= 8.85 \times 10^{-12} \mathrm{F\,m^{-1}}$; electromagnetic wave velocity equals the measured velocity of light C_0 (p.70) and the measured velocities of radio and radar waves (p.101) in free space, so these are assumed to be electromagnetic in nature; with other electromagnetic radiations (↓) of wavelength (p.54) range 10^{-12} m to 2 km they form a continuous electromagnetic spectrum (p.100).

electromagnetic radiation electromagnetic waves (↑) emitted in continuous wavetrains (p.52) from a source of electromagnetic energy (p.28).

longitudinal wave a wave (p.52) in which the disturbance of particles or layers of transmitting medium is in a direction parallel to the direction of energy transmission (p.52); the wave energy is transmitted by physical contact between particles or layers of the transmitting medium and so cannot be transmitted through free space or a vacuum; a wave originates due to pressure (p.39) changes transmitted through the medium, arising from varying longitudinal movement of the medium, e.g. compression wave (↓), sound waves (↓), shock wave (↓); effectiveness of transmission of the wave energy is increased with the density (p.7) of the medium, e.g. velocity of sound (↓) increases with density of medium.

compression wave a longitudinal wave (↑) in which the wave energy is transmitted through the medium as alternate compressions (↓) and rarefactions (↓) propagated (p.52) outwards from a source which is usually a vibrating mechanical system, e.g. tuning fork, vibrating string or wire; particles or layers of the medium vibrate (p.49) longitudinally with simple harmonic motion (p.50) around their normal rest positions (p.50) as the wave passes; a compression is an elastic deformation (p.19) of the medium, from which it recovers as a rarefaction follows; the velocity of the compression or sound (↓) wave in a solid medium $v\ (\mathrm{m\,s^{-1}}) = \sqrt{E/\rho}$, where E = Young's Modulus of elasticity ($\mathrm{N\,m^{-2}}$) (p.20) and ρ = density of medium ($\mathrm{kg\,m^{-3}}$).

sound wave a longitudinal compression wave (↑) in the range of audiofrequency (p.105), approximately 20 Hz to 20 kHz; the wave energy is detected by the human ear as sound (p.28); it provides a means of communication as speech; sound wave combinations can give the effects of both music and noise (p.106).

shock wave a progressive (↓) longitudinal wave (↑) with a single region of excessively high pressure above normal; it results from an explosion or from the impact of a supersonic aircraft on stationary air; a shock wave velocity exceeds the velocity of sound (p.102) in the transmitting medium; it can be observed as the Mach cone shock front (p.112) on a supersonic aircraft.

compression (*n*) region in a compression wave (↑) where pressure is temporarily slightly higher than normal atmospheric pressure.

rarefaction (*n*) region in a compression wave (↑) where pressure is temporarily slightly below normal.

compression wave

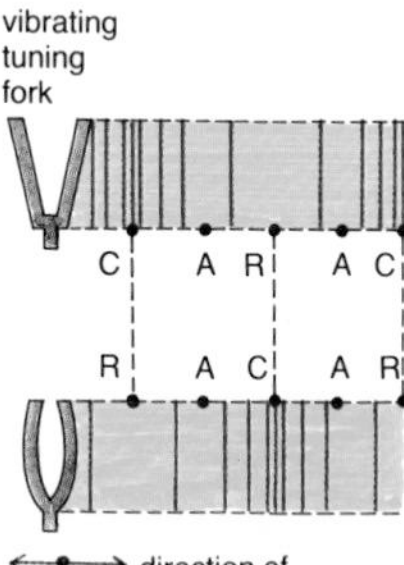

direction of movement of particles or medium

direction of straight line transmission through medium

C – compression
R – rarefaction
A – atmospheric pressure

longitudinal compression wave showing pressure variation along path of the wave at extremes of a cycle of vibration of tuning fork prongs

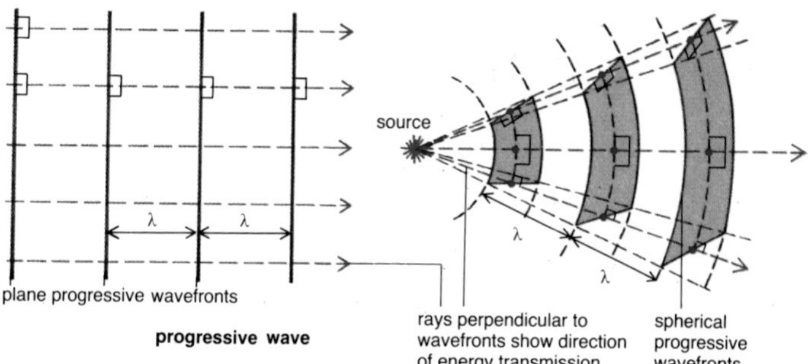

progressive wave

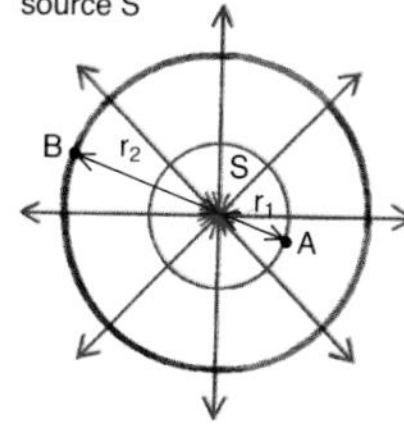

attenuation of wave energy intensity I with distance r from energy source S

radial energy emission from S in all directions (watt)
$I \propto 1/r^2$
(Inverse Square Law)

I_1 at A = $S/4\pi r_1^2$ (Wm^{-2})
I_2 at B = $S/4\pi r_2^2$ (Wm^{-2})
so $I_1/I_2 = r_2^2/r_1^2$

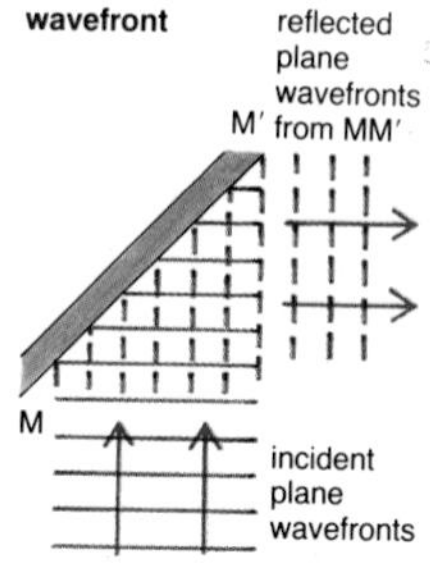

wavefront

plane surface reflection of plane water wavefronts

progressive wave a longitudinal (↑) or transverse (p.54) wave propagated (p.52) continuously outwards from the source of wave energy, through the transmitting medium, without any interruption to the wavetrain (p.52); the phase (p.53) of moving particles or layers of medium changes from point to point along the path of the wave; the phase of the electric and magnetic vectors in a transverse electromagnetic wave (p.55) changes similarly along the wave path.

attenuation (*n*) the gradual reduction of energy intensity I (↓) with distance r (m) from a source of wave energy; for a point source it follows an Inverse Square Law: $I \propto 1/r^2$ in free space or air; it follows an exponential law of absorption (p.128) for the transmission (p.52) of X-rays or γ-rays through solid media. **attenuate** (*v*), **attenuated** (*adj*).

energy intensity the energy per second passing normally through a plane area of 1 m^2 perpendicular to the direction of energy transmission (p.52) at a specified point in the wave (p.52) path, or at a specified distance from the energy source; it is expressed in units of joule/s/m^2 ($Js^{-1}m^{-2}$); watt/m^2 (Wm^{-2}); for attenuation (↑) in air or free space the energy intensity $I \propto 1/r^2$, where r = distance from the energy source (metre); for an energy wave (p.52) the intensity I at a point is ∝ $(\text{amplitude})^2$ (p.49).

wavefront (*n*) a surface, located in the path of a wave (p.52), over which all moving particles (p.7) are vibrating in phase (p.59); it can be plane, spherical or cylindrical; wavefronts are represented diagrammatically as being a distance 1 wavelength (λ) (p.54) apart along the wave profile (p.53).

superposition of waves when 2 or more waves (p.52) of the same nature, e.g. light (p.28) are transmitted simutaneously by the same medium, at the point and instant of crossing their individual disturbances of the medium will be superimposed on each other, giving rise to constructive (↓) or destructive interference (↓), e.g. Young's experiment (p.91).

principle of superposition when 2 or more wavetrains are superimposed (↑), the resultant amplitude (p.49) of the wave (p.52) at the point and instant of crossing is the algebraic sum of the individual amplitudes, e.g. for 2 individual displacements (p.52) y_1 and y_2, the resultant is $y_1 + y_2$; since the energy intensity I (p.57) at a point in the path of a wave is directly proportional to the $(\text{amplitude})^2$, then I is also modified, e.g. Young's experiment (p.91).

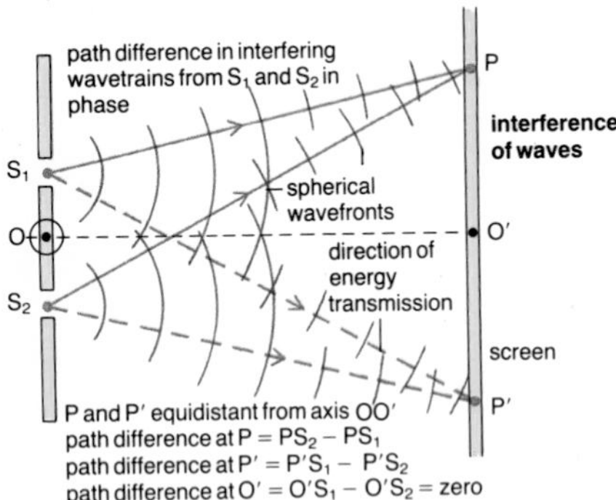

interference of waves

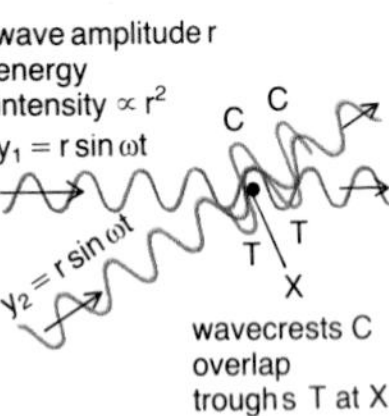

constructive interference between 2 wavetrains in phase at X

interference of waves the modification of amplitude (p.49) and energy intensity (p.57) occurring due to superposition of waves (↑). **interfere** (*v*), **interfering** (*adj*).

constructive interference amplitude modification giving a resultant amplitude greater than either individual amplitude; it occurs with maximum effect when the interfering (↑) wavetrains are in phase (↓), e.g. bright fringes in Young's experiment (p.91), bright rings in Newton's rings experiment (p.94).

destructive interference amplitude modification giving a resultant amplitude less than either individual amplitude; occurs with maximum effect when interfering (↑) wavetrains are out of phase (↓), e.g. dark fringes in Young's experiment (p.91), dark rings in Newton's rings experiment. (p.94).

wave amplitude r
energy intensity $\propto r^2$
$y_1 = r \sin \omega t$
$y_2 = r \sin \omega t$
C X C
T T
wavecrests C overlap troughs T at X
resultant amplitude = 0
energy intensity = 0

destructive interference between 2 wavetrains in antiphase at X

crest (*n*) a point of maximum positive displacement (p.52) on the wave profile (p.53) of a wavetrain (p.52); its value is wave amplitude + r (p.49).

trough (*n*) a point of maximum negative displacement (p.52) on the wave profile (p.53) of a wavetrain (p.52); its value is the wave amplitude − r (p.49).

in phase describes 2 or more points in the path of a harmonic wavetrain, having the same displacement (p.52) in the same direction at a given instant; the points have the same phase; all points on the same wavefront (p.57) are in phase.

in antiphase describes 2 points in the path of a harmonic wavetrain, having the same displacement (p.52) but in opposite directions at a given instant; the points have a phase difference (↓) $\pi = 180°$.

out of phase describes 2 points in the path of a harmonic wavetrain which are not in phase (↑); there is a phase difference (↓) between them; it is also used to describe points in antiphase (↑).

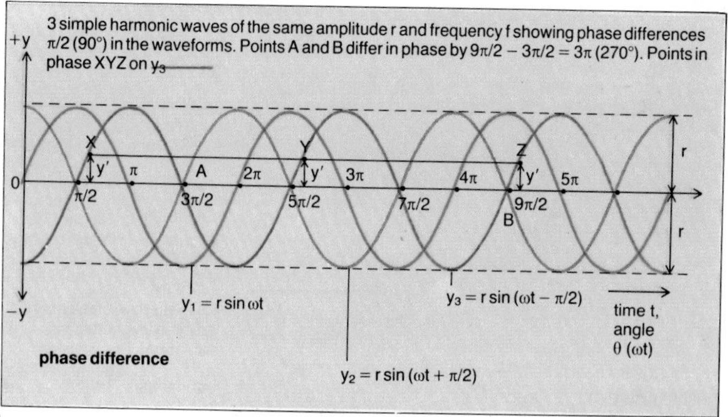

3 simple harmonic waves of the same amplitude r and frequency f showing phase differences $\pi/2$ (90°) in the waveforms. Points A and B differ in phase by $9\pi/2 - 3\pi/2 = 3\pi$ (270°). Points in phase XYZ on y_3

phase difference

phase difference a difference Φ in phase (p.53) between 2 points on the same harmonic wavetrain having different values of phase angle ωt (p.54); the difference in phase between corresponding points on 2 parallel wavetrains coming from different sources, or from the same source at different instants.

phase change a change in phase (p.53) introduced into a harmonic wavetrain, e.g. on reflection of light at an optically denser medium, as in Lloyd's mirror experiment (p.92) at the glass mirror surface.

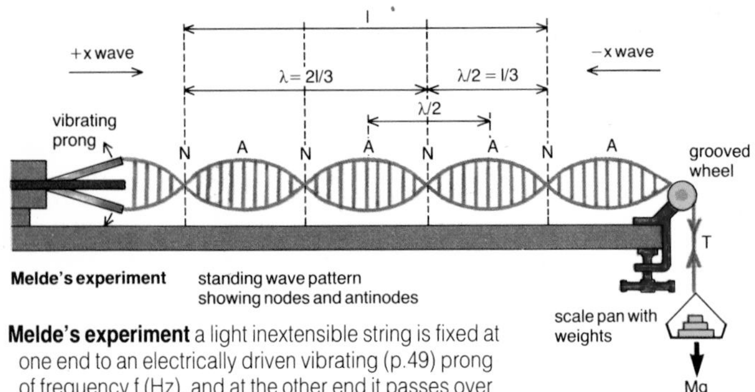

Melde's experiment standing wave pattern showing nodes and antinodes

Melde's experiment a light inextensible string is fixed at one end to an electrically driven vibrating (p.49) prong of frequency f (Hz), and at the other end it passes over a grooved wheel to a scale-pan carrying weights to hold the string in tension (↓); the vibrating prong causes transverse waves (p.54) to be transmitted (p.52) along the string towards the grooved wheel, where they are reflected and return through the string in the opposite direction to the advancing waves; with suitable adjustment of the tension T (newton) for a fixed string length l (metre), a series of n stationary loops can be established in the string; this standing wave pattern (↓) shows stationary waves (↓) with characteristic node (↓) and antinode (↓) points; the distance between any 2 consecutive nodes, or antinodes, is $\lambda/2 = l/n$, so $\lambda = 2l/n$; transverse wave velocity $v = \sqrt{T/m}$ for a string of mass m ($\text{kg}\,\text{m}^{-1}$); this apparatus can be used to demonstrate that the velocity of the waves (frequency f × wavelength λ) $\propto \sqrt{T}$ where $T = Mg$ for mass M (kg) in the scale pan.

tension (*n*) a force (p.26) acting in an inextensible string to oppose the stretching force due to a load (p.20). **tense** (*adj*).

node (*n*) a point of zero displacement (p.52) in a medium transmitting stationary waves (↓). **nodal** (*adj*).

antinode (*n*) a point of maximum displacement (p.52) in a medium transmitting stationary waves (↓).

standing wave pattern formed when a medium is transmitting stationary waves (↓); the wave profile (p.53) does not move through the medium, e.g. Melde's experiment (↑); the node (↑) and antinode (↑) positions for longitudinal (p.56) stationary waves (↓) in air columns can be made visible by the movement of dust particles, e.g. Kundt's tube.

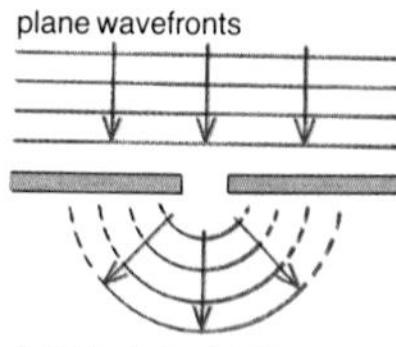

diffraction
of water waves at small circular opening

stationary waves describe a standing wave pattern (↑) which can be formed under appropriate conditions when a medium is transmitting 2 waves (p.52) of the same nature, with the same wave velocity v (p.54) and frequency f (p.54), in opposite directions through the medium; any 2 successive nodes (↑) are $\lambda/2$ apart; any 2 successive antinodes (↑) are $\lambda/2$ apart, where λ = wavelength (p.54) of the stationary wave; all points between any 2 successive nodes vibrate in phase (p.59); movement of the string carrying transverse stationary waves in Melde's experiment (↑) can be observed with a stroboscope (p.108).

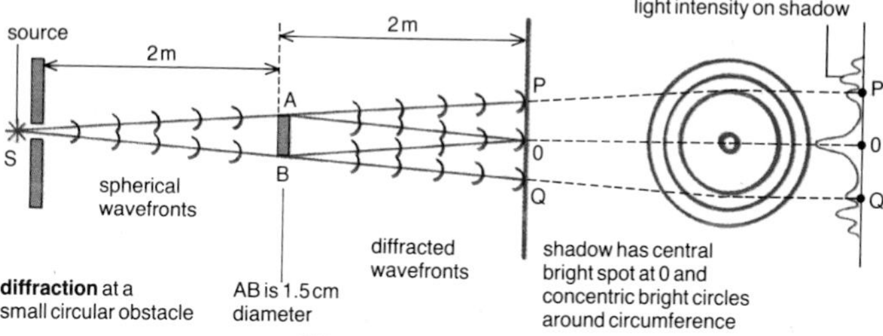

diffraction at a small circular obstacle

diffraction (*n*) when a progressive (p.57) wavefront (p.57) has its path obstructed by an aperture or obstacle whose dimensions (p.12) are close in value to those of the wavelength (p.54) waves spread round the aperture or obstacle, giving the effect that the wave path is deviated from its usual straight line transmission (p.52); it can be observed with light, (p.28) sound (p.28), water waves and electromagnetic waves (p.55) under appropriate conditions. **diffract** (*v*), **diffracted** (*adj*), **diffracting** (*adj*).

plane wavefronts

nearly plane wavefronts

diffraction
of water waves at wide opening

polarization (*n*) when a transverse wave (p.54) has its vibrations (p.49) restricted to one plane only it is polarized; it can be observed with light (p.28) or any transverse wave, but not for longitudinal waves (p.56). **polarize** (*v*), **polarized** (*adj*), **polarizing** (*adj*).

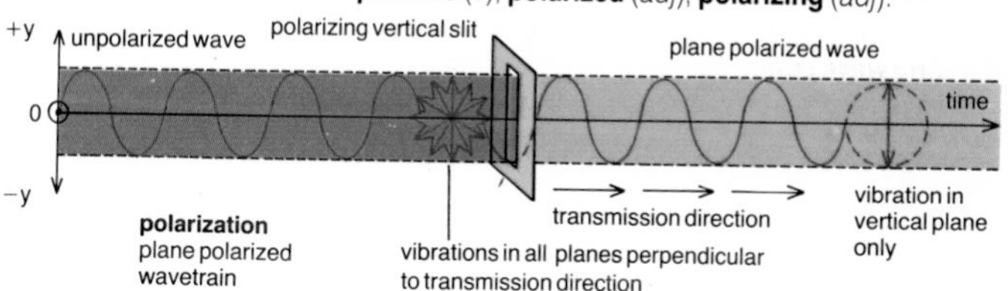

polarization
plane polarized wavetrain

luminous energy the energy (p.26) emitted from a light source (↓) as electromagnetic radiation (p.55) in the wavelength (p.54) range for visible light (p.28); it forms a region of the electromagnetic spectrum (p.100) over the wavelength range approximately 4×10^{-7} m (violet) to 8×10^{-7} m (red) and with average frequency (p.54) 10^{15} hertz (Hz); photon (p.234) energies are of 2 to 4 eV (p.228).

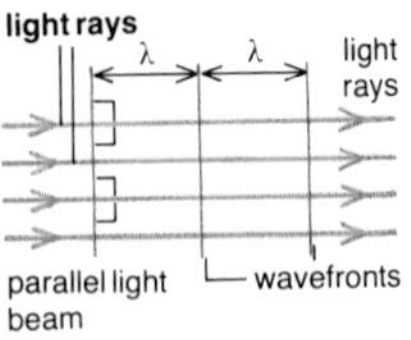

light source a source of luminous energy, e.g. the Sun, filament lamp, fluorescent tube.

luminous intensity a measure of the strength of a light source (↑), e.g. electric lamp; units are candela (cd) (p.244).

light rays straight lines drawn outwards from a light source perpendicular to the wavefronts (p.57) of light waves, to indicate the direction of transmission (p.52) of luminous energy (↑).

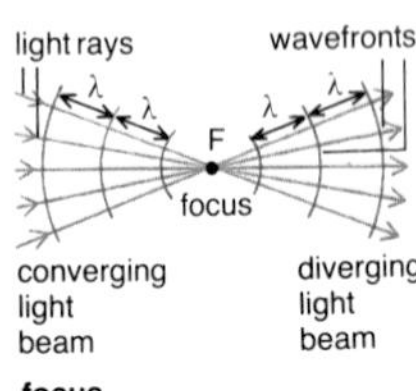

focus

focus (*v*) to converge rays of a light beam to a single point, e.g. the principal focus of a concave mirror (p.64) or of a converging lens (p.81); to adjust the positions of components of an optical system to obtain a sharply defined image of an object, e.g. as in a microscope (p.84) or telescope (p.84). **focus** (*n*), **focal** (*adj*), **focused** (*adj*), **focusing** (*adj*).

focal point a point through which all rays of a converging light beam pass; a focus (↑).

light reflection refers to the reflection by a surface of some of the luminous energy (↑) incident (↓) upon it; sometimes in equal proportions for all colours of the visible spectrum giving the surface a white or colourless appearance; otherwise in unequal proportions giving the surface the colour (p.76) reflected in the greatest proportion, e.g. a red surface reflects red light and absorbs other colours; it follows the laws of reflection (↓).

regular reflection

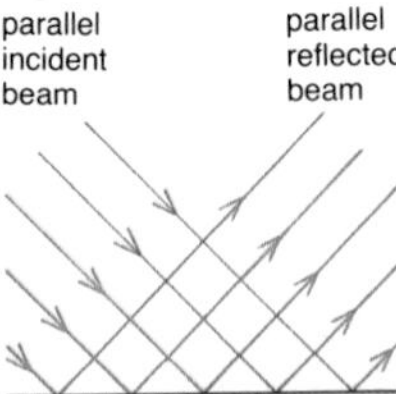

optical mirror a smooth, highly polished surface reflecting regularly (↑) 80–90% of the luminous energy (↑) incident (↓) upon it, e.g. the aluminium surface of a reflecting telescope (p.87), a curved or plane glass mirror with silvered backing layer.

incidence (*n*) the receiving of luminous energy (↑) by a surface, in the direction from the light source towards the surface. **incident** (*adj*).

diffuse reflection reflection by an uneven surface whose irregularities are large compared with the wavelength (p.54) of the waves (p.52) incident upon it.

diffuse reflection

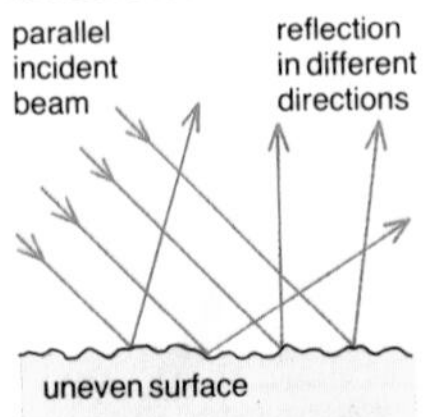

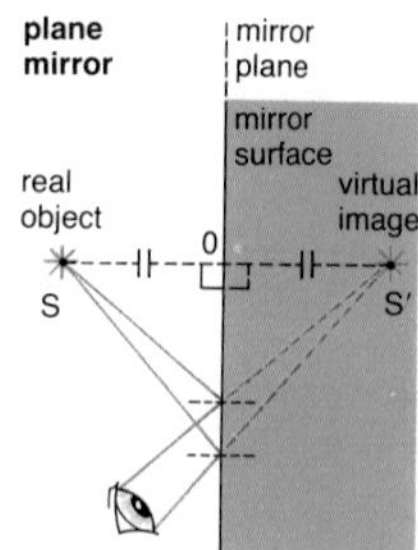

plane mirror image S′ of point object S showing S and S′ equidistant from mirror plane

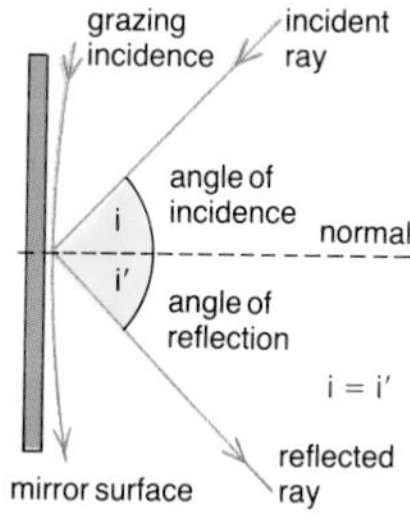

laws of reflection

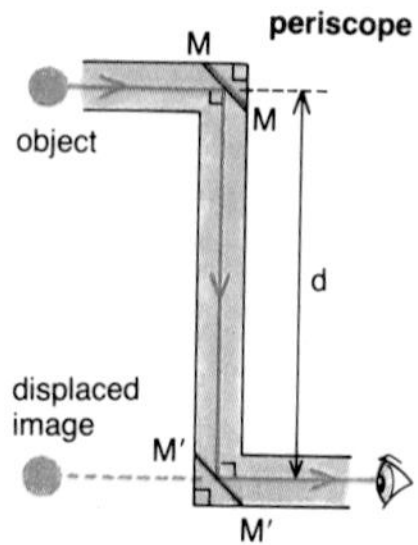

MM and M′M′ are plane mirrors (or reflecting prisms) placed at 45° to vertical axis of periscope tube. Image is displaced lateral distance d.

optics (*n*) the study of the nature of light (p.28) and its properties. **optical** (*adj*).

plane (*n*) 2-dimensional surface in space regarded as being of infinite extent. **planar** (*adj*).

plane mirror an optical mirror (↑) whose reflecting surface lies in a plane (↑).

normal (*n*) a straight line drawn perpendicular to a surface at a specified point, e.g. point of incidence (↑) of a light ray (↑) on an optical mirror (↑). **normal** (*adj*).

normal incidence incidence (↑) of energy waves (p.52) on a surface in a direction along the normal (↑); both the angle of incidence (↓) and the angle of reflection (↓) are zero.

angle of incidence the angle between the direction of incident energy on a reflecting (↑) or refracting (p.69) surface and the normal (↑) to that surface at the point of incidence, e.g. a light ray (↑) on a plane mirror (↑).

angle of reflection the angle between the direction of reflected energy from a surface and the normal (↑) to that surface at the point of incidence (↑), e.g. a light ray (↑) reflected from a plane mirror (↑).

laws of reflection refer to the reflection of energy waves (p.52) from a suitable reflecting surface, e.g. light waves (↑) from an optical mirror (↑): 1. the incident ray, reflected ray and normal (↑) are in the same plane: 2. the angle of reflection = angle of incidence; true for regular and diffuse reflection (↑) at plane and curved optical mirrors (↑).

real image the object is seen indirectly by means of light rays (↑) passing physically through its image, e.g. images from a concave mirror (p.64) or a converging lens (p.81); the image can be seen on a viewing screen.

virtual image the object is seen indirectly by means of light rays (↑) which do not pass physically through its image, but only appear to do so because of the directions the divergent light rays take, e.g. images from plane mirrors (↑); convex mirrors (p.64), or converging (p.81) and diverging lenses (p.81); the image cannot be seen on a screen.

lateral inversion the apparent left to right interchange of object points as seen in a plane mirror (↑) image.

periscope (*n*) an arrangement of 2 parallel plane mirrors (↑) inclined at 45° to the axis at opposite ends of their containing tube; the image can be seen displaced to a lower horizontal level than the object; an alternative arrangement uses internally reflecting prisms.

optical lever a sensitive means of detecting small movements of a plane mirror (p.63) by observing the deflection of the reflected ray; used in the mirror galvanometer (p.196).

parallax (*n*) the apparent relative movement of objects in different planes in space in front of the eye, when the eye position is changed; it is a possible source of error in scale readings.

parallax
church and tree in line when viewed from E. From E_1 church appears to right of tree. From E_2 church appears to left of tree

no-parallax absence of parallax (↑) between 2 objects in the same position or plane in space; it is used to locate mirror and lens images.

concave mirror an optical mirror (p.62) which is part of a sphere with a reflecting inner surface; for a narrow parallel light beam, incident close and parallel to the principal axis (↓), it gives a reflected light beam converging to a real principal focus (↓).

concave mirror F is real principal focus, 0 is mirror pole

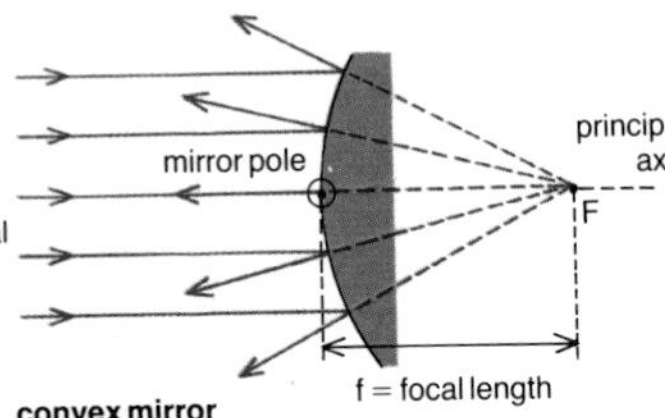

convex mirror
F is virtual principal focus, 0 is mirror pole

convex mirror an optical mirror (p.62) which is part of a sphere with a reflecting outer surface; for a narrow parallel light beam, incident close and parallel to the principal axis (↓), it gives a reflected light beam diverging away from a virtual principal focus (↓).

parabolic mirror an optical mirror (p.62) which is part of a parabolic surface with a reflecting inner surface; for a wide parallel light beam, incident parallel to the principal axis (↓), it gives a reflected light beam converging to a real principal focus (↓), e.g. reflecting telescope (p.87); for reversed light ray paths (↓), a point source of light at the principal focus (↓) gives a reflected beam parallel to the principal axis (↓), e.g. car headlamps, searchlight reflector.

parabolic mirror
wide parallel beam converged to real principal focus F.

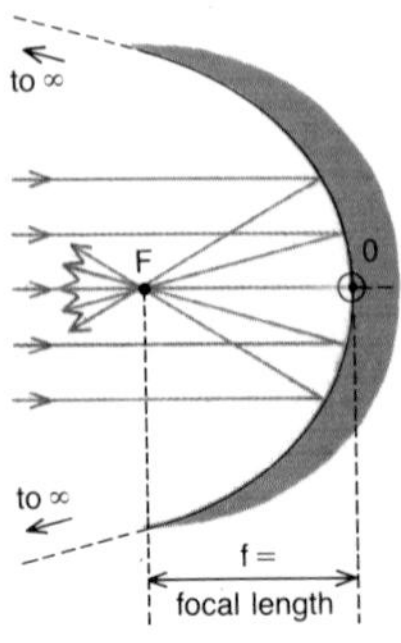

reversibility of light ray paths a light ray travelling through an optical system, or through several transmitting media, will travel the same path in reverse if turned through 180°; this is the path taking minimum time to travel, e.g. as from a parabolic mirror (↑) and reflector.

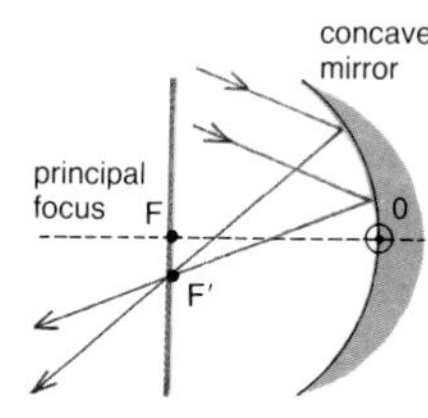

parallel beam incident obliquely is focused to F′ in principal focal plane

principal focal plane

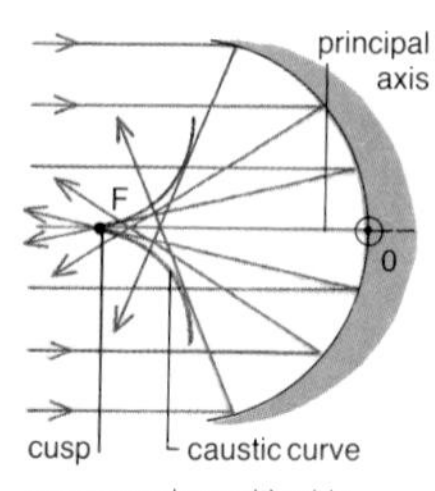

concave mirror with wide parallel incident light beam

OF = focal length f

caustic curve

mirror pole a central point of a concave (↑), convex (↑) or parabolic (↑) mirror through which the principal axis (↓) passes.

mirror principal axis an infinitely long straight line, passing through a mirror pole (↑) perpendicular to the mirror surface, or through the optical centre of a lens (p.81) perpendicular to its surface.

real principal focus a point on the principal axis (↑) of a concave (↑) or parabolic (↑) mirror, through which a narrow parallel light beam, incident close and parallel to the principal axis, is converged on reflection; a converging lens (p.81) has 2 real principal foci, located on the principal axis equidistant from the optical centre, through which a narrow parallel light beam, incident close and parallel to the principal axis, is converged on refraction.

virtual principal focus a point on the principal axis (↑) of a convex (↑) mirror away from which a narrow parallel light beam, incident close and parallel to the principal axis (↑), is diverged on reflection; a diverging lens (p.81) has 2 virtual principal foci, located on the principal axis (↑) equidistant from the optical centre, away from which a narrow parallel light beam, incident close and parallel to the principal axis, is diverged on refraction.

principal focal plane a plane perpendicular to a principal axis (↑) passing through a principal focus (↑).

mirror focal length the distance f (m) between a mirror pole (↑) and its principal focus (↑) measured along the principal axis.

caustic curve when a wide parallel light beam is reflected from a concave mirror (↑), tangents to the reflected rays lie on a caustic curve; for incidence parallel to the principal axis (↑) the cusp of the caustic curve is at the real principal focus (↑); when a wide parallel light beam, incident parallel to the principal axis is refracted by a converging lens (p.81), tangents to the refracted rays lie on a caustic curve with its cusp at the real principal focus (↑) of the lens; the caustic curve causes spherical aberration (p.83); since all rays of a wide parallel incident beam cannot be focused (p.62) to the same point and the image appears blurred.

erect image a virtual image (p.63) in which the object appears upright.

inverted image a real image (p.63) in which the object appears upside down.

radius of curvature the radius r of a spherical surface; for a concave or convex mirror (p.64), r is the distance from the mirror pole (p.65) to the centre of curvature measured along the principal axis (p.65); r = 2f, where f = focal length (p.65); used in the no-parallax (p.65) method for determination of f for a concave mirror.

linear size refers to the height of a mirror image, lens image or object measured perpendicular to the principal axis (p.65).

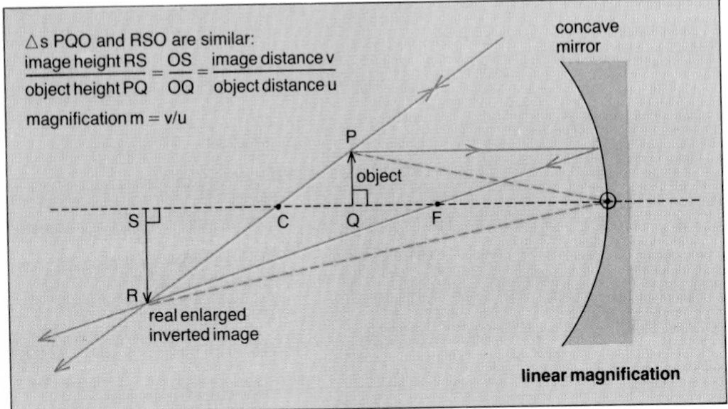

linear magnification

linear magnification the ratio of the linear size (↑) of the image/linear size of the object = image height/object height = image distance v/object distance u (↓) for a mirror image or lens image (p.82); denoted by m = v/u; m>1 for enlarged images; m<1 for diminished images; the ratio has no units.

object distance the distance measured along the principal axis (p.65) from the object to a fixed reference point in the optical system, e.g. a mirror pole (p.65), or optical centre of lens (p.81); denoted by u.

image distance the distance measured along the principal axis (p.65) from the image to a fixed reference point in the optical system forming it, e.g. a mirror pole (p.65) or optical centre of lens (p.81); it is denoted by v.

magnify (*v*) to cause an object to appear larger than its normal linear size (↑) or angular size (p.84), e.g. enlarged image. **magnified** (*adj*), **magnifying** (*adj*).

infinity (*n*) a number greater than any number to which we can assign a value; a physical quantity whose value is greater than any value we can assign to it, e.g. infinite distance; distance of the point of meeting of rays of a parallel light beam; location of a mirror image or lens image (p.82) formed by a parallel light beam; denoted by ∞; in practical situations rays of light entering an optical system from a distance of a few metres can be regarded as parallel, and a point a few metres distant approximates to infinity; for an electric field (p.165) or a magnetic field (p.180), points beyond the region of influence of the field forces are regarded as being at infinity. **infinite** (*adj*).

conjugate foci describes corresponding object and image positions for a mirror image or lens image (p.82), interchangeable because of the reversibility of light ray paths (p.64); values of u and v are interchangeable in the mirror formula (↓) or lens formula (p.82).

self-conjugate focus

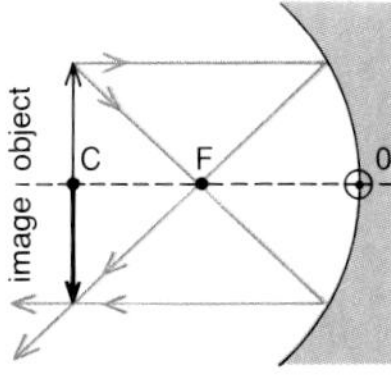

object and image coincide with no parallax at centre of curvature C. Magnification = 1 since v = u

self-conjugate focus a conjugate focus (↑) for u = v in the mirror formula (↓), for which linear magnification (↑) m = v/u = 1; object and image are both located at the centre of curvature of a concave mirror; used in determining the value of r = 2f, where r = radius of curvature (↑) and f = focal length for the mirror.

mirror formula the relation between object distance u, image distance v (↑) and focal length f for concave or convex (p.64) mirror; using the Real is Positive sign convention (↓): $1/v + 1/u = 1/f$; it is used in obtaining the value of f graphically from measurements of u, v and linear magnification m (↑).

sign convention arbitrary rules concerning the assigning of a positive or negative sign to distances u, v and f in the mirror formula (↑) and in the lens formula (p.82); it enables a distinction to be made between real and virtual images (p.63) and real and virtual principal foci (p.65) in the calculation of distances u, v and f; 2 principal sign conventions in common use are the Real is Positive sign convention, on which real object and image distances and real focal lengths have a + value, and virtual distances have a − value, and the Cartesian sign convention, on which distances are all referred to the origin of graphical axes at the centre of the optical system.

virtual object an image through which light rays do not physically pass, but which can act as an object for an optical mirror (p.62) or lens (p.81), e.g. telephoto lens (p.89).

shaving mirror a concave mirror (p.64) used with the face as the object, located between the mirror pole (p.65) and the real principal focus (p.65), giving a virtual, enlarged and upright image (p.63) of the face located behind the mirror surface.

driving mirror convex mirror (p.64) receiving light rays (p.62) from a wide field of view (↓) in front of its surface; images are virtual, diminished and upright, and are located behind the mirror surface between the mirror pole (p.65) and the virtual principal focus (p.65); distances between virtual images seen in the mirror are relatively small compared with the larger distances separating real objects in the field of view.

field of view the space in front of an optical system from which it can receive light rays (p.62), e.g. field of view is wide for a driving mirror (↑) but narrow for a Galilean telescope (p.84).

diagrammatic image location a method for locating an image of an object in an optical mirror (p.62) or lens by a diagrammatic construction drawn to scale; it uses light rays (p.62) of known path; for a concave or convex mirror (p.64): 1. Ray entering the mirror close and parallel to the principal axis (p.65) is reflected to emerge through the principal focus (p.65): 2. Reverse ray enters through the principal focus and emerges parallel to the principal axis: 3. Ray passing through the centre of curvature is reflected along its reverse path; rays of known path are also used for lens image construction (p.82).

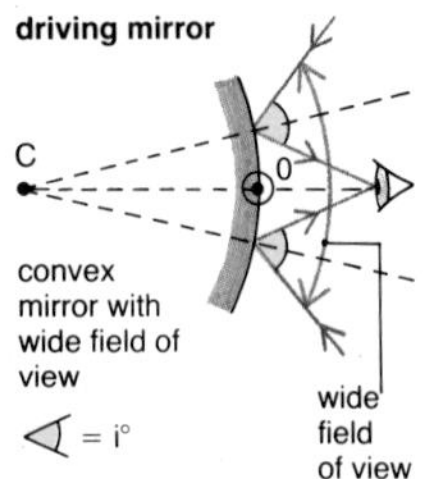

field of view

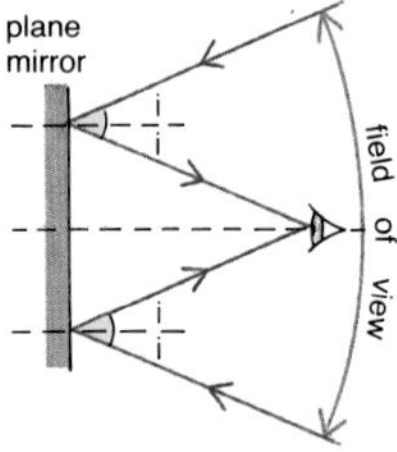

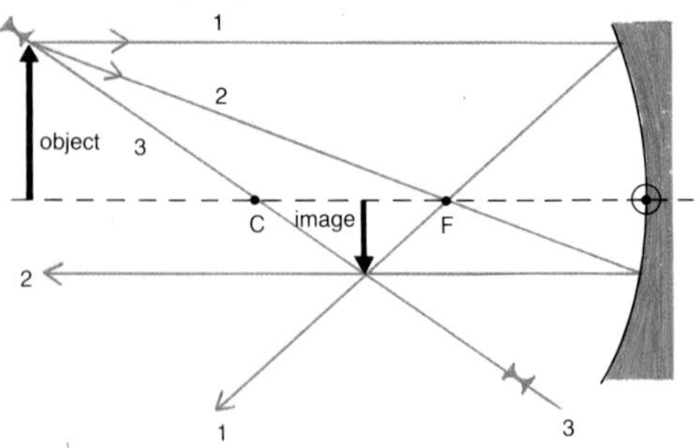

1. parallel ray emerges through F
2. ray through F emerges parallel
3. ray through C reflected along reverse path

diagrammatic image location

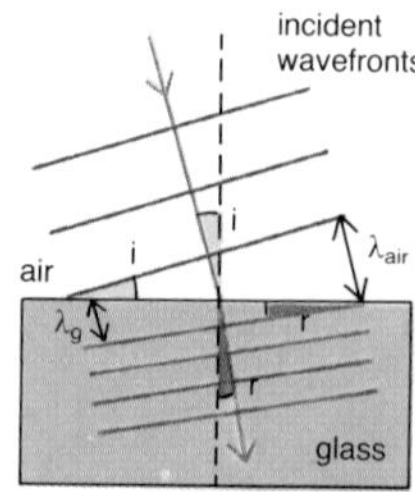

refracted wavefronts

deviation of wavefronts on crossing refracting interface

deviation

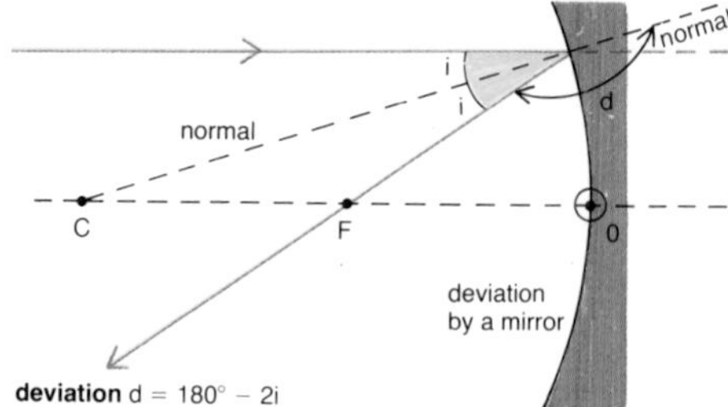

deviation d = 180° − 2i

deviation (*n*) the angular bending of the transmission direction of the path of energy waves (p.52), e.g. by reflection of light at an optical mirror (p.62), or by refraction (↓) of light on crossing the interface (↓) between 2 transparent transmitting media of different optical densities, e.g. air and glass. **deviate** (*v*), **deviated** (*adj*), **deviating** (*adj*).

interface (*n*) a plane or curved surface separating 2 transmitting media, e.g. air/glass, water/glass, air/water for the transmission of light waves. **interfacial** (*adj*).

refraction (*n*) the change in velocity (p.30) and wavelength (p.54) experienced by energy waves (p.52) on crossing the interface (↑) between 2 transmitting media, e.g. light waves passing from air to glass have their velocity reduced, a velocity change at oblique incidence causes deviation (↑) of wavefronts (p.57) towards the normal when the velocity is reduced, and away from the normal when the velocity is increased; no deviation occurs for normal incidence. **refract** (*v*), **refracted** (*adj*), **refracting** (*adj*), **refractive** (*adj*), **refrangible** (*adj*).

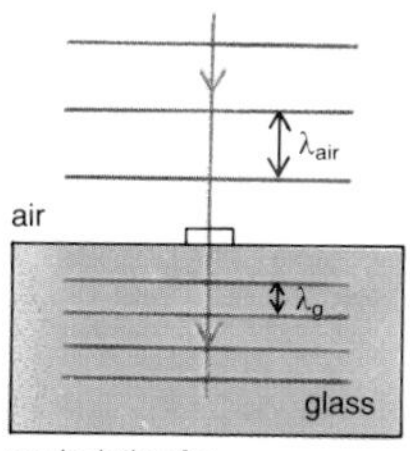

no deviation for normal incidence

refraction
refraction with deviation towards normal

normal

angle of incidence

$\frac{\sin i}{\sin r} = {}_1n_2$

i

medium 1 (air)

interface

angle of refraction

r

incident direction

medium 2 (glass)

incident direction

$\frac{\sin r}{\sin i} = {}_2n_1 = 1/{}_1n_2$

medium 1 (air)

angle of refraction

i

interface

angle of incidence

r

refraction with deviation away from normal

medium 2 (glass)

velocity of light the velocity (p.30) of electromagnetic radiation (p.55) in free space; denoted by c_o for light (p.28); its approximate value is 3.0×10^8 metre/second; this value is reduced in a transparent transmitting (p.52) medium to a characteristic value c for that medium, e.g. for glass $c = 2.0 \times 10^8\,ms^{-1}$, for water $c = 2.3 \times 10^8\,ms^{-1}$; the ratio $c_o \div c$ is the absolute refractive index (↓) of the transmitting medium; the modern accepted value of $c_o = 299\,792\,458 \pm 1.2\,ms^{-1}$.

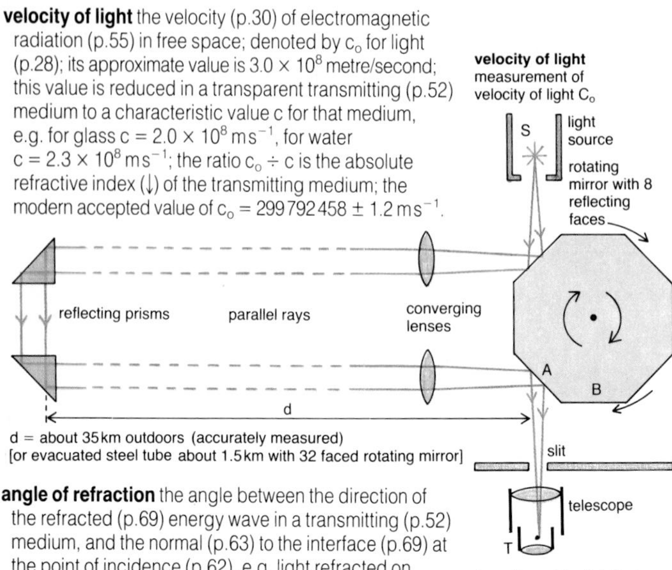

time of travel for light between S and T = $2d/C_o$ seconds

S is seen if B replaces A exactly, making ⅛ revolution in $2d/C_o$ sec.

For drum speed N revs./second:
$N = C_o/8 \times 2d$
so $C_o = 16\,Nd$ m/s

angle of refraction the angle between the direction of the refracted (p.69) energy wave in a transmitting (p.52) medium, and the normal (p.63) to the interface (p.69) at the point of incidence (p.62), e.g. light refracted on entering glass from air.

laws of refraction refer to the refraction (p.69) of energy waves on crossing the interface (p.69) between 2 transmitting (p.52) media, e.g. light (p.28) entering glass: 1. the incident ray, refracted ray and normal (p.63) are in the same plane; 2. the ratio $\sin i \div \sin r$ is constant for 2 transmitting media concerned, where i = angle of incidence (p.63) and r = angle of refraction (↑).

Snell's Law an alternative name for Law 2 of refraction (↑)

refractive index in Law 2 of refraction (↑) for light, the value of the constant ratio $\sin i \div \sin r$ defines the refractive index of medium 2, into which light is entering, relative to medium 1, which the light is leaving; denoted by ${}_1n_2$; the property of reversibility of light ray paths (p.64) gives the ratio $\sin r \div \sin i$ as the refractive index of medium 1 relative to medium 2; denoted by ${}_2n_1$; the relation is denoted by ${}_1n_2 = 1/{}_2n_1$; the wave theory of light (p.90) gives the relation $\sin i \div \sin r$ = velocity of light (↑) in medium 1 ÷ velocity of light in medium 2; its value varies slightly with the wavelength (p.54) of the refracted light; the name refractive index

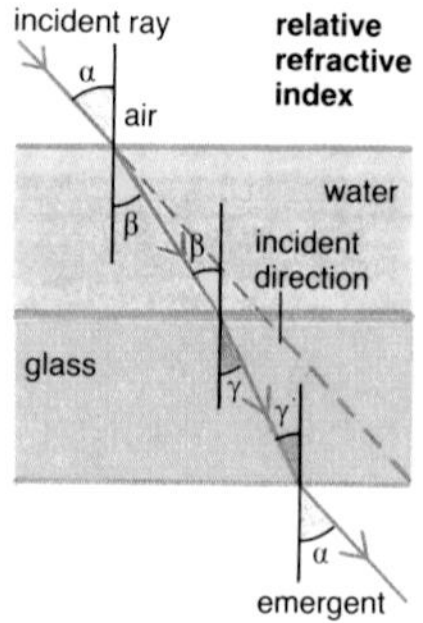

usually refers to the absolute refractive index (↓) for which values are always greater than 1.

absolute refractive index the value of refractive index (↑) of medium 2, into which light is entering, relative to medium 1, when medium 1 is free space or a vacuum, or, for practical purposes, air; refractive indices usually quoted are absolute values.

relative refractive index the value of refractive index (↑) for 2 transparent media; denoted by ${}_1n_2$ for medium 2 relative to medium 1, and ${}_2n_1$ for medium 1 relative to medium 2; values can be related to absolute refractive indices (↑); when medium 2 is glass and medium 1 is water then ${}_an_g/{}_an_w = {}_wn_g$; since ${}_wn_g = 1/{}_gn_w$, ${}_an_w/{}_an_g = {}_gn_w$, e.g. ${}_wn_g = 1.5/1.33 = 1.13$ and ${}_gn_w = 1.33/1.5 = 0.89$.

apparent depth or thickness the apparent displacement of the lower surface of a parallel-sided layer of refracting medium towards the observer; when viewed from air along a normal (p.63) to the surface; the real depth ÷ apparent depth = absolute refractive index (↑); it is the basis of methods for refractive index determination for glass and water.

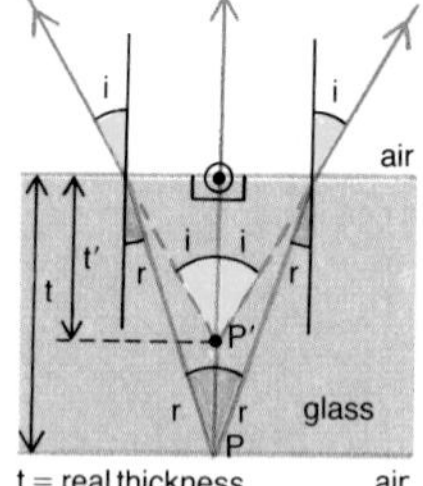

t = real thickness
t' = apparent thickness
apparent displacement of object viewed through refracting layer

internal reflection a phenomenon observed when light (p.28) from within a layer of refracting (p.69) medium emerges into a medium of lower optical density, e.g. from water or glass into air; as the angle of incidence (p.63) on the interface (p.69) increases, the angle of refraction (↑) increases, and the amount of light reflected internally increases; since r>i there is a critical angle c (p.72) for which r = 90° at grazing emergence; for i>c all light is internally reflected.

total internal reflection internal reflection (↑) of all light incident on the interface separating a refracting medium from another of lower optical density; it occurs when the angle of incidence i exceeds the critical angle c.

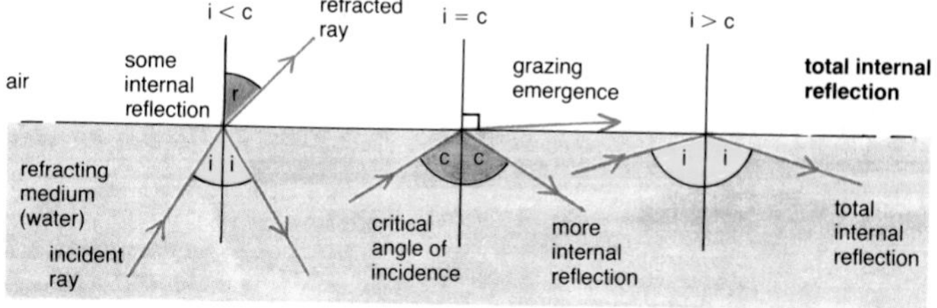

critical angle the angle of incidence for grazing emergence just preceding total internal reflection (p.71); denoted by c; the absolute refractive index (p.71) = 1/sin c; c has a different value for different refracting (p.69) media, e.g. for water $c_w = 48°$ ($\sin^{-1} 1/_an_w$); for glass $c_g = 42°$ ($\sin^{-1} 1/_an_g$).

optical fibre a thin flexible length of transparent material, transmitting (p.52) light by total internal reflection (p.71); fibres have polished surfaces coated with a material of suitable refractive index (p.70); used in bundles to enable an observer to see into otherwise inaccessible places, e.g. the human body's interior, and for the transmission of optical signals without loss of energy after repeated reflections.

fibre-optics system a system using optical fibres (↑) for transmitting images or optical signals; in fibre-optics telecommunication systems electrical signals modulate a laser (p.235) beam which is transmitted through the fibres; optical fibres can transmit far more information than a conventional cable of the same diameter.

prism (*n*) a block of transparent transmitting (p.52) medium of triangular cross-section; usually solid, e.g. glass, quartz; the **refracting angle** is enclosed by the 2 faces meeting in the refracting (p.69) edge of a prism.

deviation by a prism refers to the angular bending of light (p.28) transmitted by a prism, away from the incident direction; usually 2 refractions (p.69) occur; at the refracting faces deviation can occur by total internal reflection (p.71) in reflecting prisms (↓), e.g. in 45° prisms in a periscope and in binocular prisms (↓). **deviate** (*v*), **deviated** (*adj*), **deviating** (*adj*).

reflecting prism a prism (↑) using total internal reflection (p.71) to produce deviation (↑) of light through 90° or 180°, e.g. prism binoculars (p.86).

optical density describes the property of a transparent medium to change the velocity of light (p.70) entering it from another medium with different transmission properties; expressed as greater or less relative to another medium; if the velocity is reduced on crossing the interface (p.69) from medium 1 to medium 2, medium 2 is said to have a greater optical density than medium 1, while the optical density of medium 1 is less than that of medium 2, e.g. for light entering glass or water from air; but for light passing from air through water to glass, water is optically denser than air but less dense than glass.

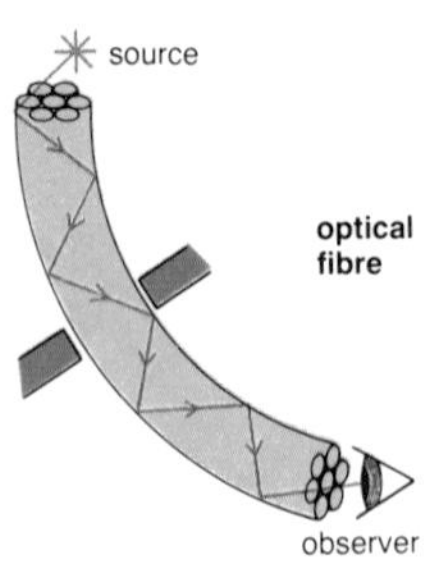

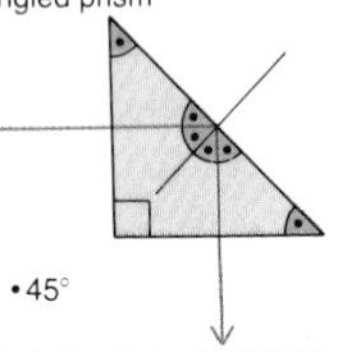

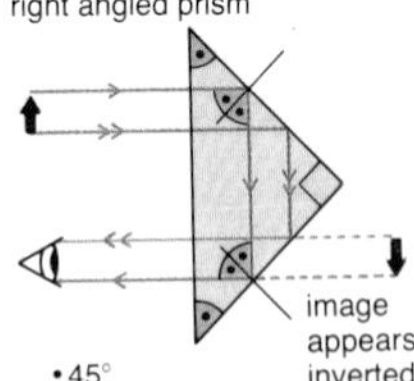

deviation by a prism

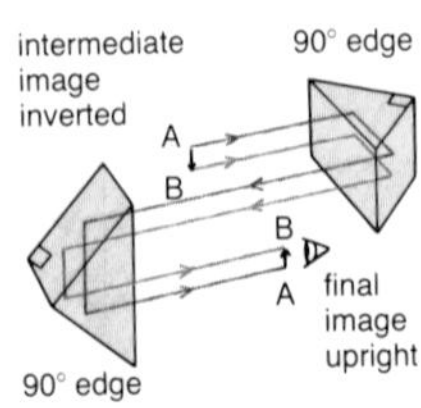

prism binoculars
one pair of prisms for each eye; 90° edges perpendicular to give image inversion

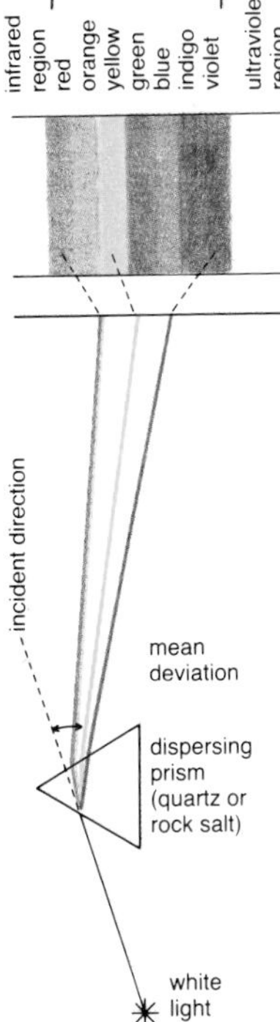

dispersion (*n*) the separation of white light (↓) into its component spectrum colours on refraction (p.69) through a prism (↑); refracting media have different refractive indices (p.70) for different wavelengths (p.54) of light and produce different deviations (p.69) for each colour. **disperse** (*v*), **dispersed** (*adj*), **dispersing** (*adj*).

white light spectrum light from the sun (p.43), or from a lamp, containing all electromagnetic radiations (p.55) in the visible wavelength range; there are seven spectrum colour regions from blue-violet (400 nm) to deep red (800 nm).

optical spectrum the image produced by an optical system, including a dispersing (↑) prism, from which the colour (p.76) and wavelength (p.54) content of light (p.28) emitted from a light source can be observed and analyzed. **spectral** (*adj*).

pure spectrum a spectrum produced on dispersion (↑) of white light (↑) by an optical system including a dispersing prism (↑) and 2 achromatic doublets (p.83), e.g. spectrometer, direct vision spectroscope; the final image has 7 distinguishable colour (p.76) regions formed in the focal plane of the second doublet.

infrared (i.r.) radiation electromagnetic radiation (p.55) detectable beyond the visible red region of the white light spectrum (↑) observed for a quartz or rock salt prism (↑); it has longer wavelength than red light and is less deviated; infrared, heat (p.28) and microwaves (p.101) form a region of the electromagnetic spectrum (p.100), the approximate wavelength (p.54) range is 8×10^{-7} to 10^{-2} m and frequency range 10^{14} to 10^{10} hertz (Hz); they are emitted as radiation from hot body sources, e.g. the sun (p.43), electric heaters and radiators, and detected by a thermopile (p.127) or bolometer; the i.r. spectrum is analysed by infrared spectrometer (p.130); infrared radiation is not scattered (p.98) by fog or mist particles and is used for photography (p.88) under these conditions; it can also produce a photoelectric effect (p.229) with some surfaces.

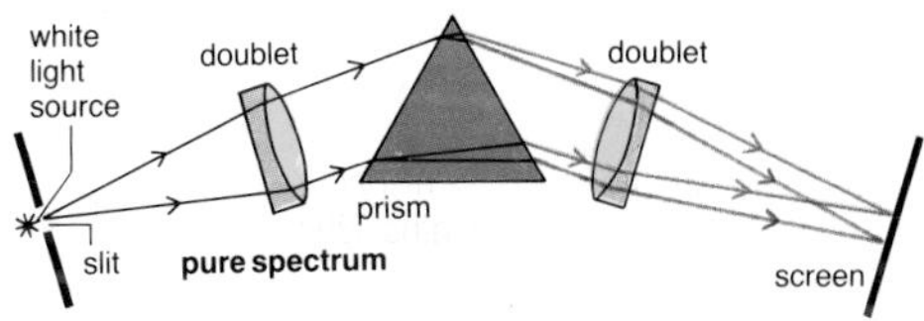

pure spectrum

ultraviolet (u.v.) radiation electromagnetic radiation (p.55) detectable beyond the visible violet region of the white light spectrum (p.73), observed for a quartz or rock salt prism (p.72); it has shorter wavelength than violet light and is more deviated; the ultraviolet forms a region of the electromagnetic spectrum (p.100) approximate wavelength (p.54) range 10^{-8} to 4×10^{-7} m and frequency (p.54) range 10^{17} to 10^{15} hertz (Hz); photon (p.234) energies are approximately 150 eV; it is emitted following quantized electron transitions (p.234) in a gas discharge tube, e.g. mercury vapour lamp, and from the sun (p.43); it is detected by photoelectric cell (p.229) or photographic emulsion, and by its fluorescent effect on certain chemicals and is used in the fluorescent lamp to produce visible light; it causes the skin of the human body to produce vitamin D; it can produce a photoelectric effect with suitable surfaces.

emission spectrum a spectrum (p.73) of radiation emitted from a substance when atoms or molecules change from excited energy levels (p.234) to lower energy levels.

continuous spectrum an emission spectrum (↑) with one wavelength region merging into another without discontinuity, e.g. white light emission spectrum of the sun (p.43) or an incandescent lamp.

monochromatic light luminous energy (p.62) emitted from a light source as wavetrains (p.52) of one observable colour only within a very narrow range of wavelengths; corresponding to radiation with monoenergetic photons (p.234), e.g. the line spectrum (↓) of sodium vapour has 2 wavelengths (p.54) very close together, usually observed as a single line of wavelength 5.893×10^{-7} m (589.3 nm); 2 lines can be resolved as separate by a diffraction grating (p.96) of high resolving power (p.97); a laser (p.235) gives a monochromatic light beam of high energy intensity.

line spectrum an emission spectrum (↑) for light (p.28) emitted from a hot gas or vapours, having characteristic coloured lines of specific wavelength (p.54) and frequency (p.54); it is emitted by atoms in the light source whose electrons are in excited states, e.g. hydrogen spectrum (↓).

sodium line spectrum emission spectrum (↑) of sodium vapour with 2 yellow wavelengths 589.0 and 589.6 nanometer.

emission spectrum
line emission spectrum of sodium

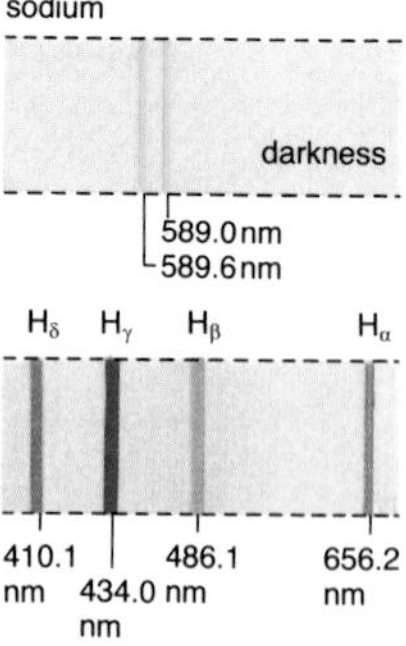

hydrogen spectrum
the visible spectrum of hydrogen contains lines in the 'Balmer series', called the H_α, H_β, H_γ and H_δ lines

hydrogen spectrum the emission spectrum (↑) of hydrogen containing a number of series of lines; the Balmer series is in the visible spectrum, the Lyman series is in the ultraviolet, the Paschen and Brackett series are in the infrared.

band spectrum an emission spectrum (↑) produced when atoms of a heated gas are polyatomic, e.g. diatomic oxygen O_2 and nitrogen N_2; the bands consist of very fine lines whose spacing decreases across each band.

solar spectrum the continuous emission spectrum (↑) of the sun (p.43), including infrared (p.73) and ultraviolet regions (↑), and a very small proportion of X-radiation (p.100); the visible light region is observable on dispersion (p.73) by a prism (p.72) or in a rainbow (p.76).

absorption spectrum
line absorption spectrum of sodium

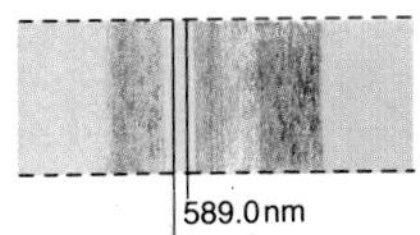

absorption spectrum the spectrum (↑) observed when white light (p.73) is passed through a heated gas; the gas absorbs the characteristic wavelengths of its own line spectrum (↑): the resulting spectrum is the continuous (↑) white light spectrum (p.73) crossed by fine dark lines where characteristic spectral lines have been absorbed, e.g. dark lines at 589.0nm and 589.6nm for sodium vapour (↑); absorption spectra can also be obtained for other regions of the electromagnetic spectrum (p.100), e.g. the Fraunhofer lines (↓), the absorption spectrum lines for certain radio stars.

C D F

656.3 nm 589.3 nm 486.1 nm

Fraunhofer wavelengths

Fraunhofer lines the absorption spectrum (↑) of gases present in the sun, e.g. hydrogen and helium; the solar spectrum (↑) is crossed by fine dark lines identified by letters A to I, ranging from the deep red region to the ultraviolet region (↑): C line (H_α)(656.3nm) in the red region, F line (H_β)(486.1nm) in the blue region and the mean sodium D line (589.3nm) in the yellow region.

dispersive power for an equilateral prism (p.72) of refracting (p.69) medium, this is defined as the angular dispersion (p.73) between blue and red light ÷ mean deviation (p.69) measured for yellow light.

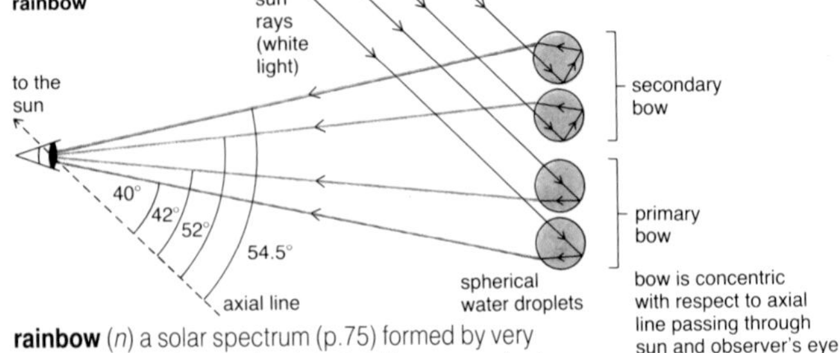

rainbow (*n*) a solar spectrum (p.75) formed by very small spherical water droplets, held in suspension in the Earth's lower atmosphere (p.18) after rain; the droplets deviate (p.69) sunlight by total internal reflection (p.71) and refraction (p.69), producing a semi-circular white light spectrum (p.73) by colour dispersion (p.73) in water; a weaker secondary rainbow, concentric with the primary, is sometimes observed.

colour (*n*) the perception (p.78) experienced by the human eye (p.78) when light-sensitive cone cells on the retina are stimulated by electromagnetic energy (p.28) in the visible wavelength range; it can be described physically (↓) or physiologically (↓); equal stimulation by all colour components of white light (p.73) gives the sensation (↓) of white; the absence of any stimulation gives the sensation of black, e.g. in complete darkness.

physical colour colour (↑) described in terms of its visible light wavelength components; an energy distribution curve shows the energy intensity of different wavelengths.

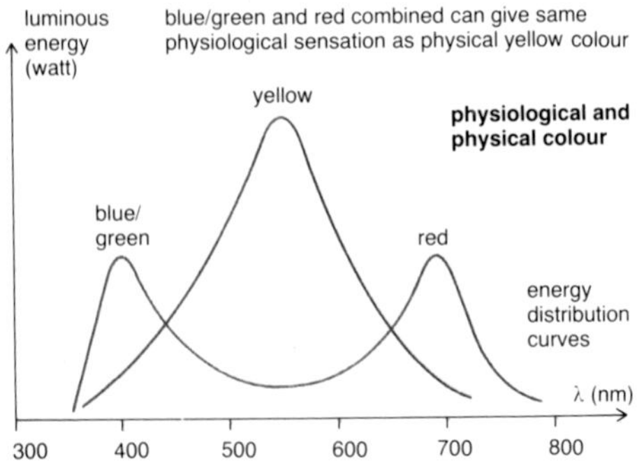

physiological colour colour (↑) described in terms of the perception experienced by the human eye; light-sensitive cells vary in sensitivity with colour wavelengths; perception of any colour can be experienced by combined stimulation of cone cells by 3 basic colours, e.g. yellow can be experienced as a combination of red and blue-green colours (↓).

primary colour sensations cone cells of the retina of the human eye (p.78) are sensitive to red, blue and green colour wavelength ranges; combinations of these 3 colour stimuli can produce all the known colour (↑) sensations.

primary colours red, green and blue light (p.28) cannot be produced by mixing other coloured lights; on mixing together they give the sensation of white.

secondary colours yellow, cyan (blue-green) and magenta (purple) coloured lights are produced by mixing primary coloured (↑) lights.

complementary colours any 2 coloured lights giving white on mixing additively (↑), e.g. yellow and blue.

colour subtraction a pigment absorbing primary colour (↑) from white light reflects the complementary colour (↑), e.g. subtraction of red gives a pigment a cyan (blue-green) colour by reflection; a pigment absorbing red, blue and green appears black; most pigments are impure and reflect more than one colour, e.g. blue paint also reflects indigo and green, yellow paint also reflects green and orange, so blue and yellow paints on mixing reflect predominantly green.

colour filter a transparent sheet of gelatine, glass or plastic, coloured with a dye so that it transmits wavelengths (p.54) of one colour (↑) only, e.g. red filter transmits red light, but absorbs blue and green, from incident white light.

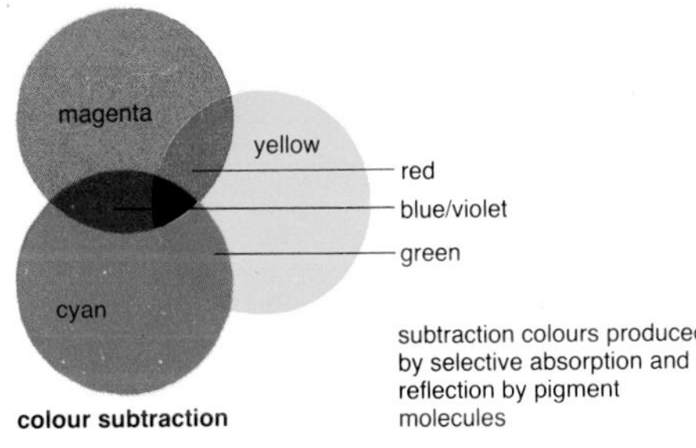

colour subtraction

perception (*n*) the interpretation, by the brain of a human being or other animal, of sensory information about its environment, received as sensations, e.g. visual and auditory perception. **perceive** (*v*), **perceptual** (*adj*).

human eye the organ of visual perception (↑) in human beings; a complex lens system capable of varying its focal length to produce clearly focused images on the retina (↓) of objects situated within its range of accommodation (↓); images are free from chromatic (p.83) and spherical aberration (p.83).

human eye

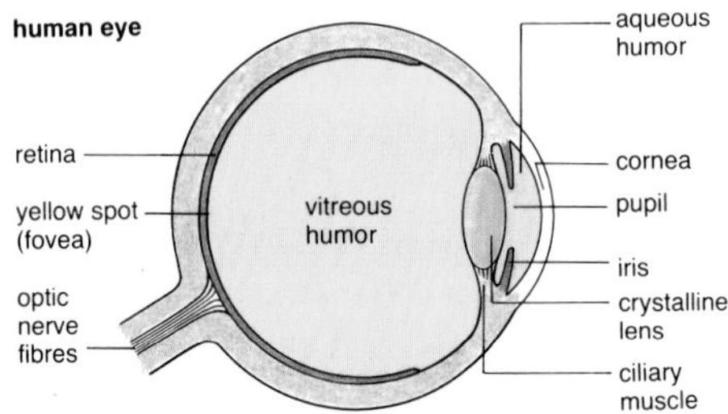

aqueous humor the transparent watery liquid between the cornea (↓) and crystalline lens (↓) of the human eye (↑); it acts as a refracting component of fixed focal length.

vitreous humor a transparent liquid, denser than water, filling the interior of the human eye (↑) between the crystalline lens (↓) and the retina (↓); it acts as a refracting component of fixed focal length.

crystalline lens the central refracting component of the lens system of the human eye (↑); it is a transparent flexible biconvex lens whose surface curvature can be varied by the surrounding ring of ciliary muscles, thus varying the focal length of the whole eye to achieve accommodation (↓); conventionally it is regarded as representing the effective dioptric power (↓) of the whole eye.

cornea (*n*) the transparent front region of the white schlerotica surrounding the human eye (↑); it covers the iris (↓) and the pupil (↓); most of the refraction of light in the human eye (↑) takes place at the cornea.

iris (*n*) the coloured central region at the front of the human eye (↑); it acts as a stop (p.83), varying the diameter of the central pupil (↓).

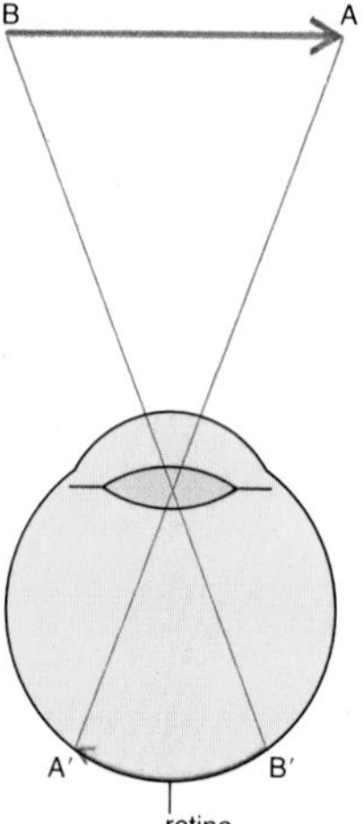

formation of a real inverted image on the retina; crystalline lens representing converging power of the eye
crystalline lens

pupil (*n*) the central hole in the iris (↑) of the human eye, allowing entry of light.

retina (*n*) a layer of light-sensitive cone and rod receptor cells on the back of the human eye (↑); it contains rhodopsin pigment molecules, temporarily and reversibly bleached by the action of light; this bleaching transforms visual stimuli into electrical impulses transmitted by the optic nerve (↓); cone cells are sensitive to 3 primary colour sensations in bright light and rod cells to light at a low level of illumination.

optic nerve a nerve transmitting electrical impulses (p.37) resulting from visual stimulation, from the retina (↑) to the brain.

blind spot a region of the retina (↑) where the optic nerve (↑) fibres connect with the light-sensitive cells; this region is insensitive to light.

fovea (*n*) the region of maximum sensitivity on the retina (↑).

yellow spot an alternative name for fovea (↑).

accommodation (*n*) the ability of the lens system of the human eye (↑) to alter its focus by variation of the focal length of the crystalline lens (↑). **accommodate** (*v*).

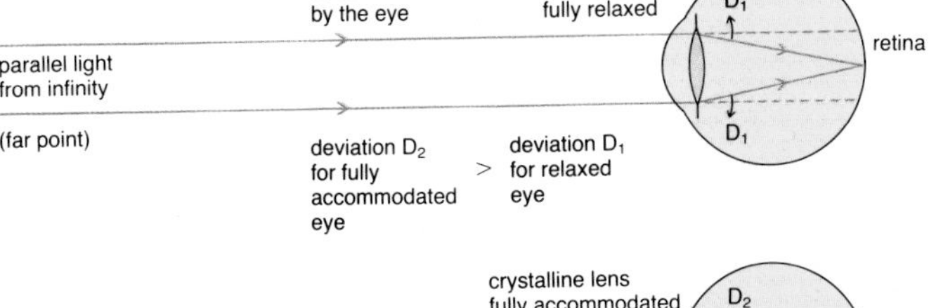

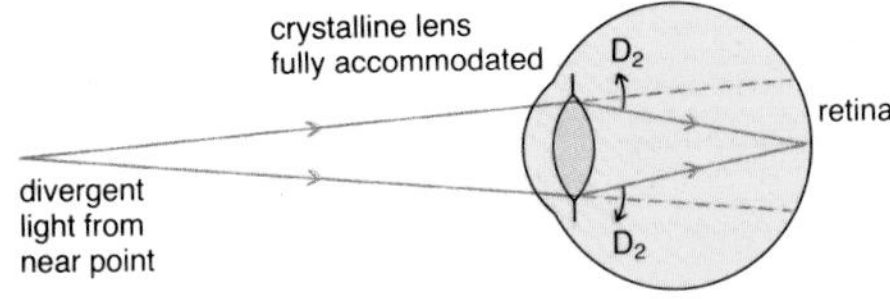

near point the nearest point to the human eye (↑) at which objects can be accommodated (↑) without strain; about 250 mm for a normal eye.

far point the furthest point from the human eye (↑) at which objects can be accommodated (↑) without strain; this is infinity (p.67) for a normal eye.

lens power the reciprocal of its focal length measured in metres and expressed in units of dioptres, e.g. a lens with f = 2 m has power ½ = 0.5 dioptres.

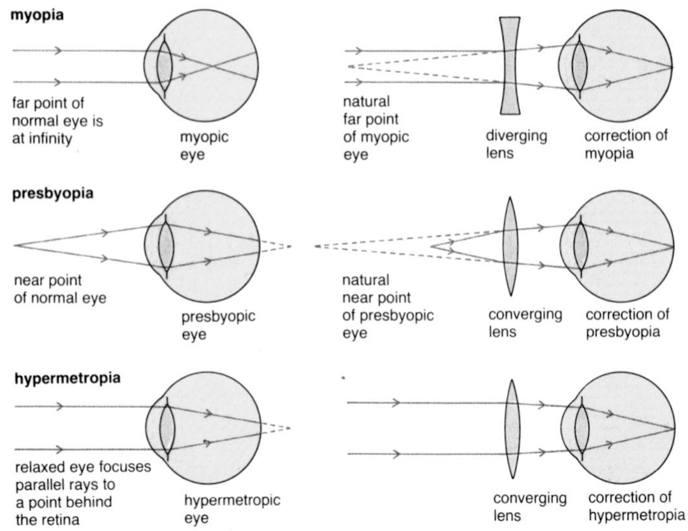

defect of vision a condition of the human eye (p.78) causing images formed on the retina (p.79) to be blurred or unfocused.

short sight a vision defect (↑) in which both near and far points (p.79) are nearer than normal; the eyeball is too long or the eye too strongly converging (↓); it can be corrected by a diverging lens (↓).

myopia (*n*) an alternative name for short sight (↑).

presbyopia (*n*) a vision defect (↑) of people in old age in which the near point (p.79) is further away than normal, due to weakening of the converging (↓) power of the eye and loss in its range of accommodation (p.79); the crystalline lens loses elasticity (p.19) and the ciliary muscles weaken with age; it can be corrected by a converging lens (↓).

long sight a vision defect (↑) in which the near point (p.79) is further away than normal; the eyeball is too short or the eye too weakly converging (↓); parallel rays are focused (p.62) to a point behind the retina (p.79) when the eye is fully relaxed, so ciliary muscles must be constantly used even to focus distant objects and the eye is always under strain; it can be corrected by a converging lens (↓).

hypermetropia (*n*) an alternative name for long sight (↑).

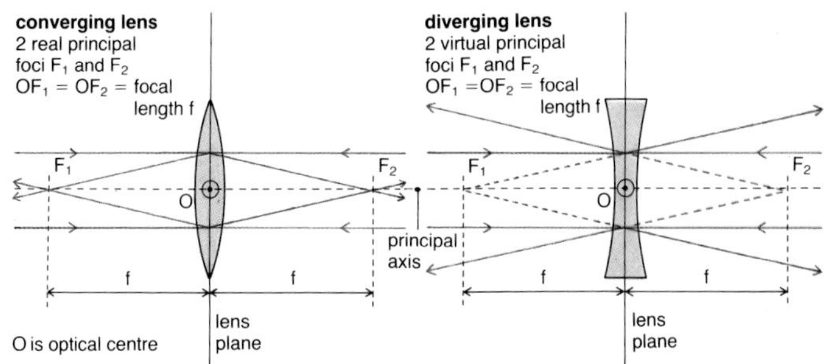

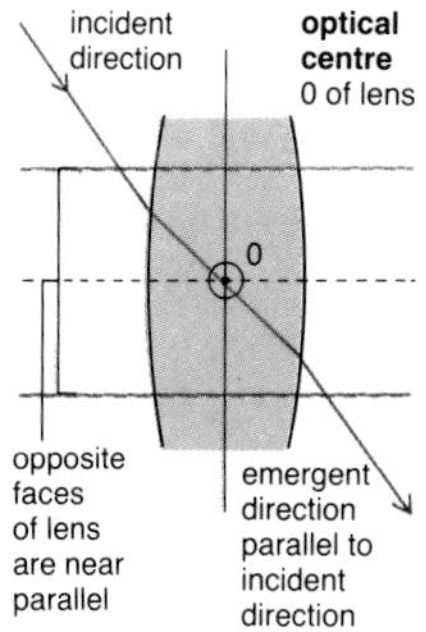

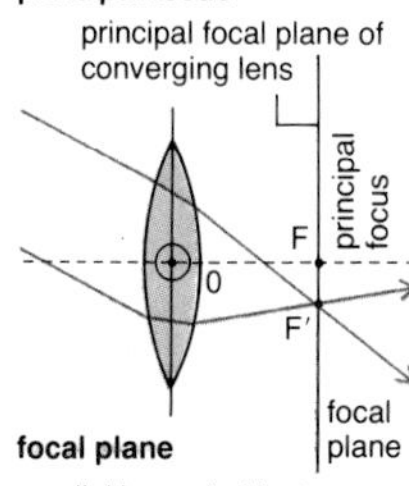

thin lens a glass, plastic or other transparent refracting (p.69) medium bounded by 2 surfaces at least one being spherical; the lens thickness should be small so that its optical centre (↓) is a negligible distance from the surfaces.

converging lens a lens (↑) in which refraction (p.69) at one or both surfaces deviates (p.69) light towards the lens axis (↓).

diverging lens a lens (↑) in which refraction (p.69) at one or both surfaces deviates (p.69) light away from the lens axis (↓).

complex lens system a lens (↑) system of 2 or more components, e.g. human eye (p.78), camera lens (p.89); designed to produce a lens image (p.82) free from aberrations (p.83).

optical centre a point inside the refracting medium of a lens (↑) through which light can pass without deviation (p.69).

lens principal axis an infinitely long straight line passing through the optical centre (↑) of a lens and the centres of curvature of its surfaces.

real principal focus a point on the principal axis (↑) of a converging lens (↑) through which a narrow parallel light beam, incident close and parallel to the principal axis, is converged on refraction by the lens; a converging lens has 2 real principal foci, located on the principal axis equidistant from the optical centre.

virtual principal focus a point on the principal axis (↑) of a diverging lens (↑) away from which a narrow parallel light beam, incident close and parallel to the principal axis, is diverged on refraction by the lens; a diverging lens has 2 virtual principal foci, located on the principal axis, equidistant from the optical centre (↑).

principal focal plane the plane perpendicular to the lens principal axis (p.81) passing through the principal focus (p.81).

lens focal length the distance f between the optical centre (p.81) and the principal focus (p.81), measured along the lens principal axis (p.81).

lens image an image of an object formed by a thin lens (p.81) or a complex lens system (p.81) by refraction; it can be upright or inverted, enlarged or diminished in size; the linear magnification = v/u where v = image distance and u = object distance (p.66); conjugate and self-conjugate foci (p.67) are used in focal length (↑) determination.

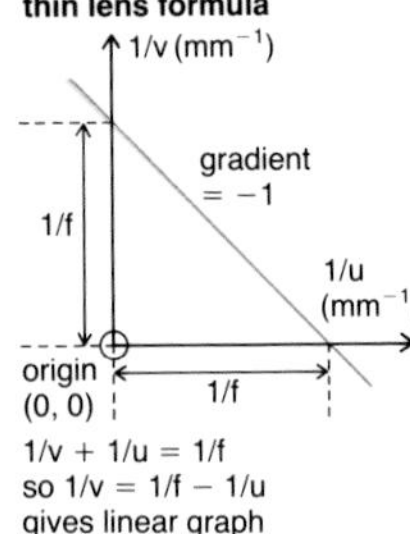

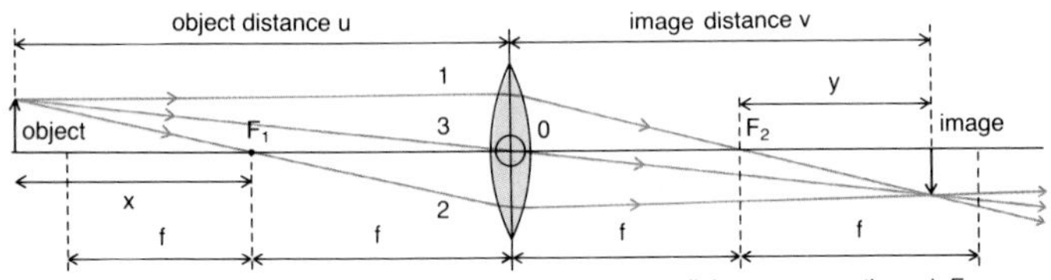

lens image construction

object beyond 2f gives real, inverted, diminished image located between f and 2f. Construction using reverse ray paths with object and image positions interchanged would give a real, inverted, enlarged image located beyond 2f

1. parallel ray emerges through F_2
2. ray through F_1 emerges parallel to axis
3. ray through 0 is undeviated

thin lens formula: $1/v \pm 1/u = 1/f$
Newton's lens formula: $xy = f^2$

lens image construction a diagrammatic method of image location for a thin lens (↑), using a scale drawing and rays of known path: 1. the ray entering the lens close and parallel to the principal axis (p.81) is refracted to emerge through the principal focus (p.81) on the opposite side of the lens: 2. the ray entering the lens through the principle focus on the same side of the lens as the object emerges parallel to the principal axis on the opposite side: 3. the ray entering through the optical centre (p.81) emerges undeviated; 2 refractions at the lens faces are conventionally represented by deviation at the lens plane.

thin lens formula the relation between object distance u, image distance v (p.66) and focal length f (↑) for a converging (p.81) or diverging thin lens (p.81), using the Real is Positive Sign Convention (p.67) it is given by $1/v + 1/u = 1/f$; it may be used to obtain a value for f graphically (p.243) from measurements of u, v and magnification m (p.66).

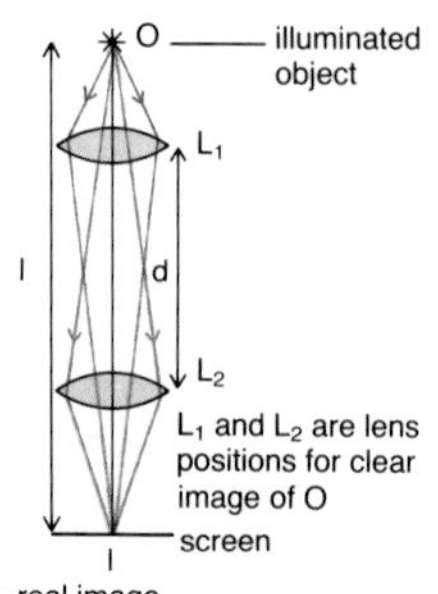

$f = (l^2 - d^2)/4l$

displacement formula

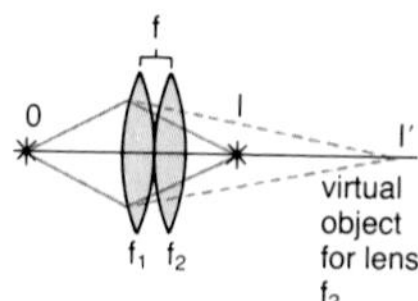

$1/f = 1/f_1 + 1/f_2$

focal length f of 2
thin lenses in contact

spherical aberration

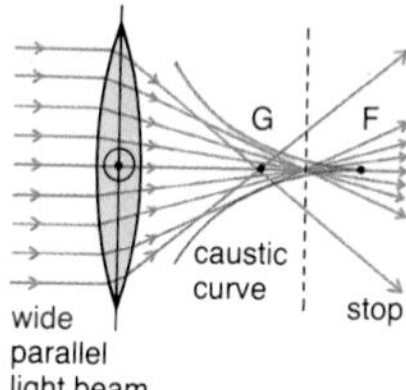

F principal focus
G furthest point from F to which rays are focused

chromatic aberration

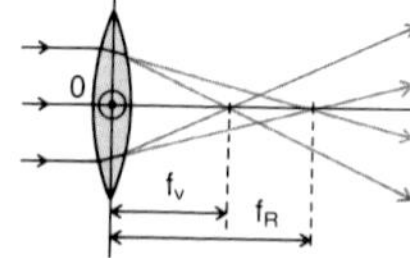

f_v = focal length for blue/violet light
f_R = focal length for red light

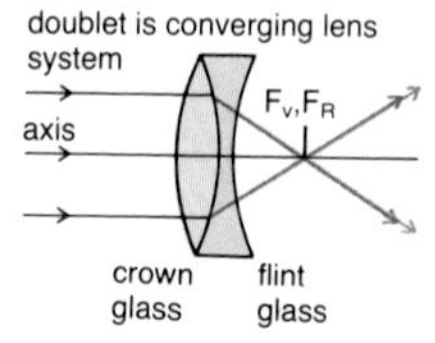

blue/violet and red light are focused to same point F_v, F_R

achromatic doublet

thin lenses in contact the combined focal length f (↑) of 2 thin lenses f_1 and f_2 in contact is given by: $1/f = 1/f_1 + 1/f_2$ on the Real is Positive and Cartesian sign conventions.

thin lens displacement formula for a fixed distance l between an object and a screen, there are 2 positions of a thin lens of focal length f (↑) for which a clearly focused real image (p.63) can be formed on the screen, provided that $l > 4f$; the lens displacement d and fixed distance l are related to the focal length f by: $f = (l^2 - d^2)/4l$, giving a method of determining f using real images.

spherical aberration when a wide parallel beam of light, parallel to the principal axis (p.81), is incident on a lens, the angle of incidence (p.63) on the curved surface increases with distance from the principal axis, and emergent refracted rays lie on a caustic curve (p.65); there is no position for which all refracted rays can be clearly focused on a screen and the resulting blurring of the image is spherical aberration; it can be reduced by using a stop (↓) to admit only the central rays of the incident beam, or by design of the lens surfaces, e.g. telescope objective (p.84); it also occurs in concave and convex mirror images; mirror objectives in telescopes are parabolic to avoid spherical aberration.

stop (*n*) a hole of variable diameter in a folding diaphragm; it restricts light entering a lens system (p.81) to the central rays only of the incident beam, minimizing spherical aberration (↑), e.g. the objective (p.84) lens of a compound microscope (p.85); it also controls the amount of light entering, e.g. the eye pupil (p.79) is a natural stop for the human eye, a camera diaphragm is a variable stop.

chromatic aberration since the focal length of a lens depends on the refractive index (p.70) of its material and hence on the wavelength of light used, the different colour components of white light are focused at different distances from the lens so that there is no position for which all refracted rays can be clearly focused on a screen; the resulting blurring of the image is chromatic aberration.

achromatic doublet a complex lens system (p.81) with 2 thin lenses in contact (↑), whose focal lengths f_1 and f_2 (↑), and dispersive powers ω_1 and ω_2 (p.75) give the system the same focal length f for 2 selected wavelengths; $\omega_1/f_1 = -\omega_2/f_2$ for an achromatic doublet.

visual angle the angle subtended at the human eye (p.78) by the object being viewed.

angular size the ratio of the linear height of an object ÷ distance from the human eye, gives its visual angle (↑) in radians (p.46).

apparent size apparent size is directly proportional to the visual angle (↑); objects of different linear heights, subtending the same visual angle at the eye, have the same angular size (↑).

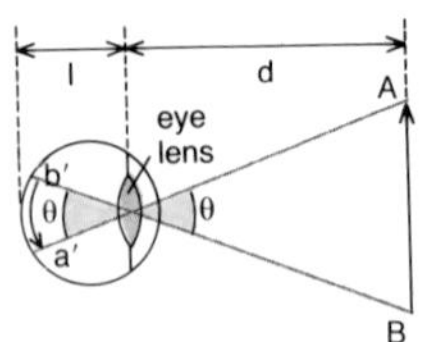

object of height AB subtends visual angle θ at the eye. Eyeball length l and $\theta = a'b'/l$ so image length on eye retina $a'b' = l\theta$

$a'b' \propto \theta$ (radians)

angular size of object $AB = AB/d = \theta$ (radians)

angular size/visual angle

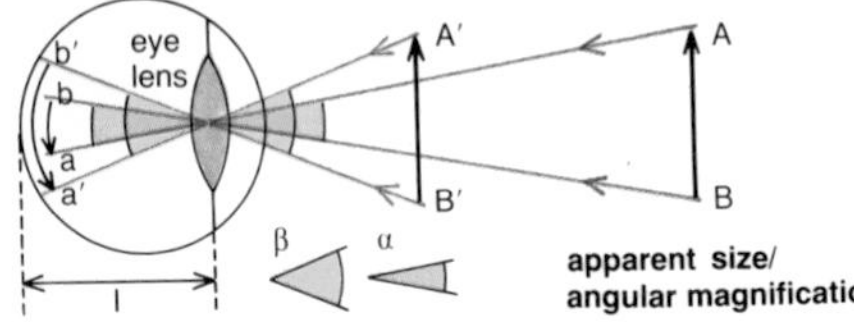

apparent size/ angular magnification

distant object AB forms image of length ab on retina $ab = l\alpha$

same object at A′B′ forms image of length a′b′ on retina $a'b' = l\beta$

since $\beta > \alpha$ then $a'b' > ab$. If AB is seen by eye and A′B′ by optical instrument: angular magnification $= \beta/\alpha$

microscope (*n*) an optical system designed to make visible objects having small angular size (↑) because of their small physical size, e.g. small biological organisms; an image is formed at a greatly increased visual angle (↑). **microscopic** (*adj*).

telescope (*n*) an optical system designed to make visible objects having small angular size (↑) because of their great physical distance from the observer, e.g. stars; an image is formed at greatly increased visual angle (↑). **telescopic** (*adj*).

objective (*n*) the front lens of a microscope (↑) or telescope (↑) receiving incoming light from the field of view (p.68).

eyepiece (*n*) the lens of a microscope (↑) or telescope (↑) nearest to the observer's eye, through which the final image is viewed; usually 2 thin lenses separated by a suitable distance to minimize chromatic and spherical aberration (p.83).

cross-wires (*n*) 2 perpendicular thin wires fixed in the focal plane of an eyepiece (↑) lens; the intersection of the wires gives a reference point for taking measurements.

angular magnification the ratio of angular size (↑) of the image ÷ angular size of the object for a microscope (↑) or telescope (↑); the ratio has no units.

magnifying power an alternative name for angular magnification.

telescope
telescope objective

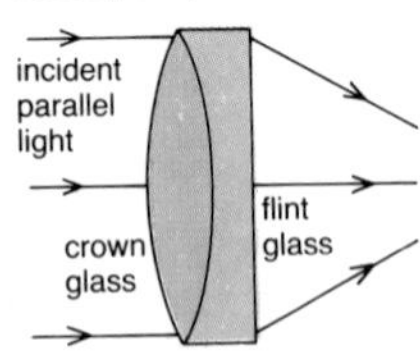

plano-convex achromatic doublet with incidence for minimum spherical aberration

$$\text{angular magnification} = f_o/f_E = \frac{\text{objective diameter}}{\text{eye-ring diameter}}$$

simple microscope a biconvex converging lens (p.81) of short focal length f, used to view the enlarged upright image of a small object, e.g. small print, detailed structure of a plant or insect not visible to the unaided human eye.

magnifying glass an alternative name for a simple microscope (↑).

compound microscope a microscope (↑) formed from 2 lenses with an objective (↑) and eyepiece (↑) both having short focal lengths, but the focal length of the objective f_o being shorter than f_E for the eyepiece, e.g. $f_o = 1$ to 2 cm, $f_E = 2$ to 5 cm; normal adjustment (↓) has the final inverted image at the near point (p.79) of the human eye (p.78) and angular magnification (↑) = linear magnification (p.66) of objective × linear magnification of eyepiece; most magnification is produced by the objective, e.g. × 50 in a × 500 microscope, revealing fine detail of the magnified image under high illumination; the objective diameter should subtend as large an angle as possible at the object to receive the maximum amount of light from every object point; an oil immersion objective (↑) is used for this.

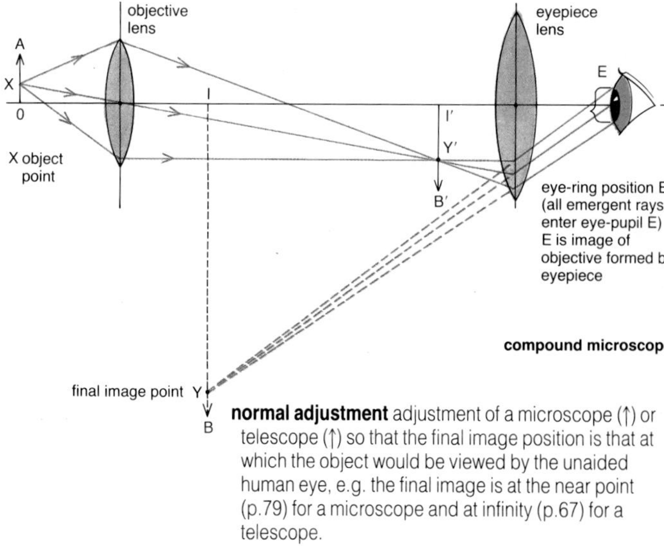

compound microscope

normal adjustment adjustment of a microscope (↑) or telescope (↑) so that the final image position is that at which the object would be viewed by the unaided human eye, e.g. the final image is at the near point (p.79) for a microscope and at infinity (p.67) for a telescope.

astronomical telescope a 2 lens telescope (p.84) with the objective (p.84) having focal length f_o = 50 to 100cm, and the eyepiece (p.84) having shorter focal length f_E = 5 to 10cm; normal adjustment (p.85) has the final inverted image at infinity (p.67) and angular magnification (p.84) = f_o/f_E; the objective aperture should be as large as possible to collect the maximum amount of light from the faintly illuminated night sky, e.g. 20–40cm for research instruments; a large value of f_o, e.g. 20m for research instruments gives high angular magnification.

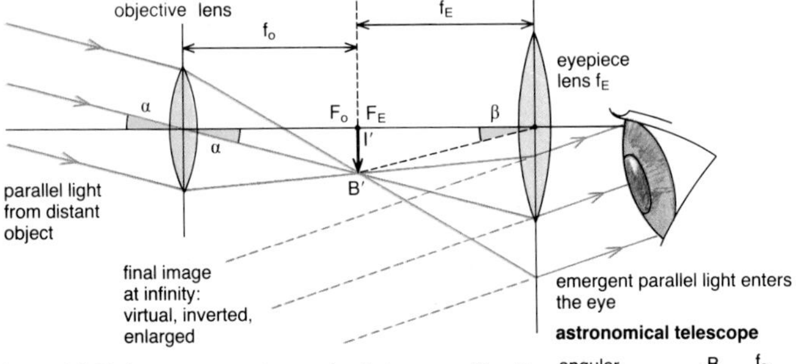

astronomical telescope

angular magnification $= \frac{\beta}{\alpha} = \frac{f_o}{f_E}$

terrestrial telescope an astronomical telescope (↑) with an intermediate erecting lens so that the final image is viewed upright; this instrument is very long and for most practical purposes has been replaced by prism binoculars (↓).

prism binoculars 2 astronomical telescopes (↑), with the separation between objective (p.84) and eyepiece (p.84) much shortened by a pair of 45° reflecting prisms (p.72) placed between the lenses to erect the final image.

field glasses alternative name for prism binoculars (↑).

eye-ring the eye-pupil (p.79) acts as a stop (p.83) and should be placed near the eyepiece of a microscope or telescope (↑) to receive the maximum amount of light from the objective; the image of the objective formed by the eyepiece gives the eye-ring position.

telescope aperture the area of a refracting (↓) or reflecting (↓) telescope surface receiving incoming light from the field of view (p.68); usually represented by the objective lens diameter for a refracting telescope and by the mirror diameter for a reflecting telescope; the resolution (↓) of a telescope increases with its aperture.

parallel light from astronomical bodies
plane mirror
F
Newton position

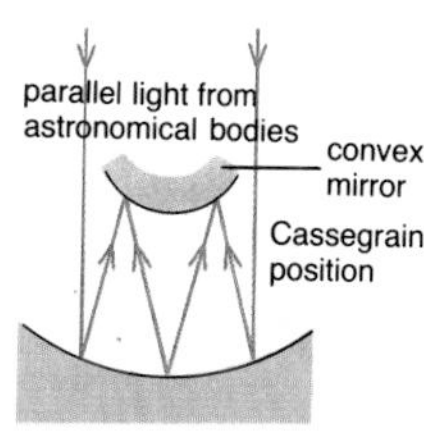

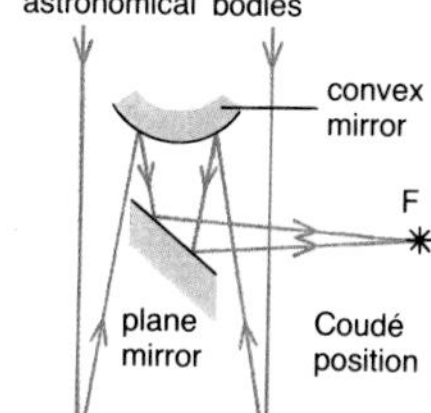

parabolic
reflecting telescope

resolution (*n*) the ability of an optical instrument, e.g. microscope (p.84) or telescope (p.84) to produce clear definition of detail in the final image; a numerical value for a particular optical instrument or diffraction grating (p.96) is its resolving power (p.97). **resolve** (*v*), **resolving** (*adj*).

refracting telescope an astronomical telescope (↑) using a lens system.

reflecting telescope an astronomical telescope (↑) using an optical mirror (p.62) system, e.g. Palomar telescope; free from chromatic aberration (p.83).

radiotelescope (*n*) a telescope (p.84) for receiving radio waves (p.101) in the wavelength (p.54) range 10 m to 10 cm or less, from distant galaxies and astronomical bodies, e.g. quasars, pulsars, studied in radio-astronomy; the parabolic dish reflector can be fully steerable, e.g. Jodrell Bank, UK, with dish diameter approximately 80 m; or it can be a fixed natural bowl in the Earth, e.g. Arecibo, Puerto Rico, USA, with a dish diameter approximately 300 m; an interferometer instrument combines the signals from several different dishes or aerial arrays, e.g. Merlin, Jodrell Bank, UK, which has a variable separation range of 11 km–134 km; the greater the separation of the dishes the greater the resolution (↑) of the interferometer instrument.

projector (*n*) an optical system designed to project an enlarged image of a highly illuminated photographic (p.88) slide, or other translucent object, on to a vertical screen; light source is a quartz iodine lamp; the condensing lens converging the maximum amount of white light (p.73) on to the slide; the objective (p.84) lens is an achromatic doublet (p.83) forming an image of the slide on the screen. **projection** (*n*), **project** (*v*), **projected** (*adj*).

slide projector

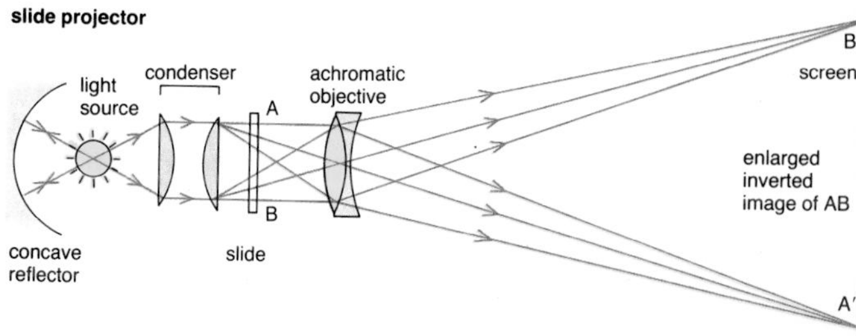

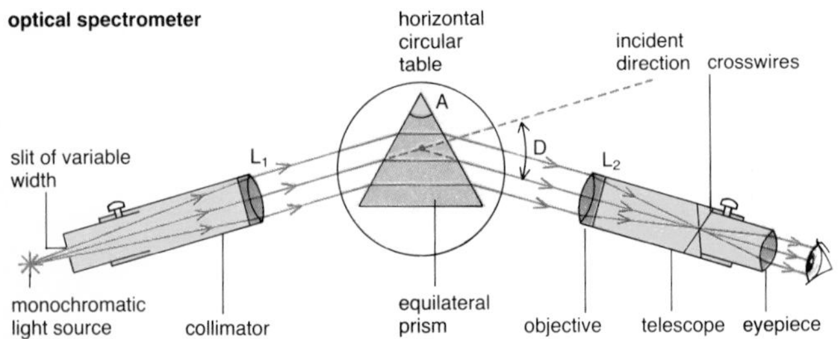

optical spectrometer (*n*) an optical system, consisting of a collimator (↓) and an astronomical telescope (p.86) capable of independent rotation about a common vertical axis, around which a horizontal circular table can also rotate; the angle of rotation of the circular table or of the telescope can be measured; angles of diffraction (p.61) can be measured with a diffraction grating (p.96) mounted vertically on the table, and the wavelength of light (p.28) present in the source can be determined; the refracting angle A (p.72) and angle of minimum deviation D (p.72) can be measured for a prism and the refractive index (p.71) of the prism glass determined.

collimator (*n*) an optical system for producing a parallel light beam including different wavelengths (p.54); a light source illuminates a slit of variable width, whose distance from an achromatic doublet (p.83) can be varied until it is at the focal length distance for red and blue wavelengths; it is used in a spectrometer (↑). **collimate** (*v*).

photograph (*n*) a permanent record on film of the image formed in a photographic camera (↓); the film is a light-sensitive emulsion of silver bromide particles in gelatine, affected chemically on exposure to light so that a latent image is formed; a reducing solution reduces particles to silver, developing the latent image, which is then fixed chemically by dissolving all remaining silver bromide particles in sodium thiosulphate (hypo) solution; the film has the greatest density of silver particles per cm^2, and hence the greatest blackening, where most light has reached it, giving a photographic negative reversed into a positive black and white photograph by contact printing; colour photography requires a colour sensitive film. **photography** (*n*), **photographic** (*adj*).

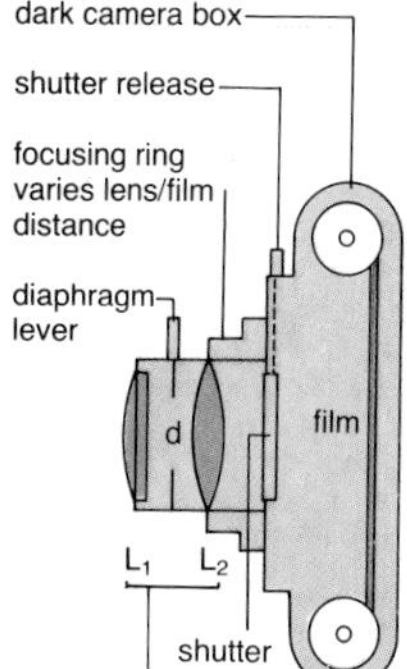

photographic camera

camera a light sensitive film and camera lens (↓) are at opposite ends of a light-excluding box and their separation can be adjusted so that clear images of objects, situated at different distances in front of the lens, can be formed on the film; a variable diaphragm (↓) controls the intensity of illumination of the image and a shutter (↓) controls the exposure time (↓) of the photograph (↑).

camera lens a complex lens system (p.81) with several components designed to minimize chromatic and spherical aberration (p.83); for colour photography (↑) the system is achromatic (p.83) throughout the white light spectrum (p.73).

diaphragm (*n*) a variable diameter stop (p.83) inside a camera (↑).

shutter (*n*) a mechanically operated device admitting light to the camera (↑) during the exposure time (↓).

exposure time the time for which light is admitted to the camera (↑), and is incident on the film, during photography; together with the f-number (↓) it controls the total luminous energy affecting the film and hence the image brightness.

f-number the ratio f/d of the focal length f of the lens in an optical system to the diameter d of the diaphragm (↑), e.g. f-4 for d = f/4. f-8 for d = f/8; f/d is also called the relative aperture; typical f-number settings for a camera are 2.8, 4, 5.6, 8, 11, 16, 22; the larger the f-number the smaller the aperture.

telephoto lens a 2 lens system equivalent in action to a single lens of long focal length; parallel light from infinity (p.67) is converged towards the principal focus (p.81) of the converging lens; the intermediate image acts as a virtual object (p.67) to the diverging lens, forming a real final image; it is used in television and photographic camera (↑) to give good enlargement of images of distant objects.

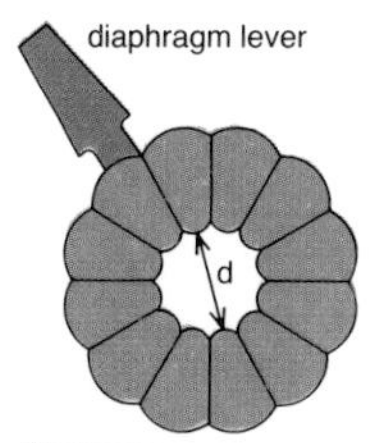

diaphragm
(variable diameter stop)

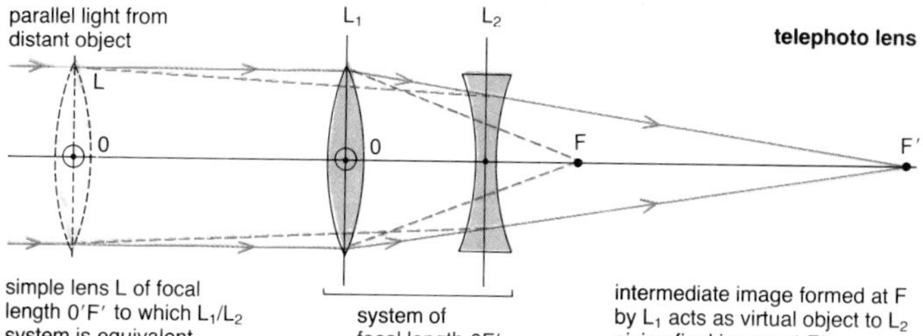

telephoto lens

wave theory of light
1. reflected wavefronts
2. incident wavefronts

A. reflected wavefront
B. incident wavefront
C. refracted wavefront

= i

= r

wave theory of reflection and refraction

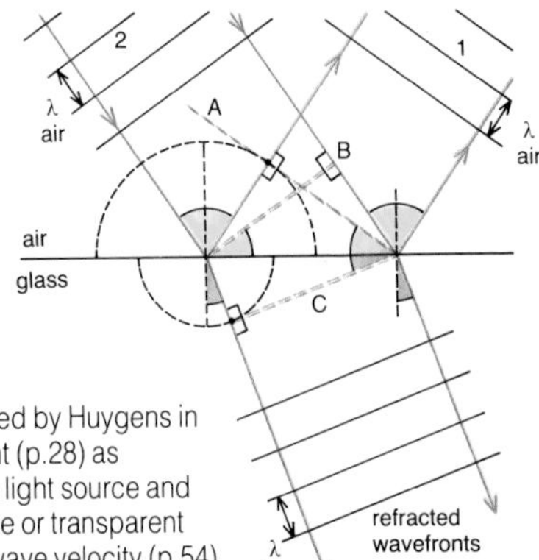

wave theory of light a theory proposed by Huygens in the late 17th century, describing light (p.28) as luminous energy (p.62) emitted by a light source and transmitted (p.52) through free space or transparent media as transverse waves (p.54); wave velocity (p.54) has its maximum value in free space and reduced values in different transparent media; wavefronts (p.57) are established by the combined effects of secondary wavelets (↓); the theory can account for light reflection (p.62), refraction (p.69), diffraction (p.61) interference (p.58) and polarization (p.61), but not for the photoelectric effect (p.229); practical evidence supporting Huygens' theory was not available until the early 19th century, with Young's (↓) experimental determination of the wavelength (p.54) of light.

secondary wavelets Huygens' wave theory of light (↑) describes a light source as a vibrating or oscillating (p.49) source from which transverse light waves are emitted in all directions in space around the source; as waves touch points in their path, these points vibrate transversely, emitting secondary wavelets in all directions in space around the points; common tangents to the secondary wavelets form the wavefronts, spherical near the surface and planar at greater distances; light rays (p.62) are perpendicular to wavefronts.

Huygens' principle the wave theory of light based on the idea of secondary wavelets (↑) spreading from each point as a wavefront such that the envelope of these wavelets forms the new wavefront.

optical interference describes interference (p.58) occurring between wavetrains (p.52) with wavelengths (p.54) in the visible region of the electromagnetic spectrum (p.100); certain practical conditions must be fulfilled for observable (↓) interference.

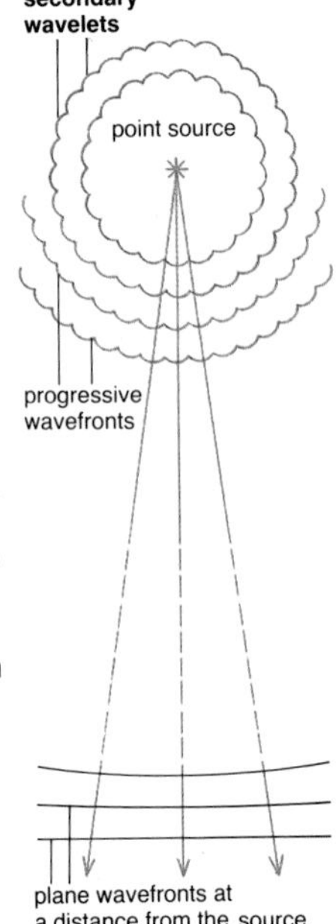

observable optical interference a permanent observable optical interference (↑) pattern in the region of crossing of 2 wavetrains requires: 1. monochromatic light (p.74); 2. coherent sources (↓); 3. 2 sources very close together and small in size compared with other dimensions of the apparatus, e.g Young's experiment (↓).

coherent light sources light sources emitting wavetrains (p.52) of the same frequency (p.54) and wavelength (p.54), always in phase (p.53) or with a constant phase difference (p.59); in practice 2 coherent sources are usually derived from the same source, e.g Young's experiment (↓).

Young's experiment practical determination of the wavelength (p.54) of monochromatic light (p.74), made in 1801 using the wave theory of light (↑) as a basis for calculation of the result; 2 narrow parallel slits, close and parallel to the central axis of the apparatus, are illuminated by monochromatic light and act as coherent sources (↑) of wavetrains, producing observable optical interference (↑) in their region of crossing; the permanent interference pattern is observed in any plane perpendicular to the axis of the apparatus in the region of crossing, as a series of narrow parallel alternate dark and bright monochromatic bands or fringes parallel to the slits and centred on the axis of the apparatus; the wavelength λ of monochromatic light is calculated from the fringe spacing y, source separation a and distance D between the plane of the slits and the plane of the fringes; the wavelength $\lambda = ay/D$; luminous energy (p.62) is redistributed between dark and bright interference fringes (p.92).

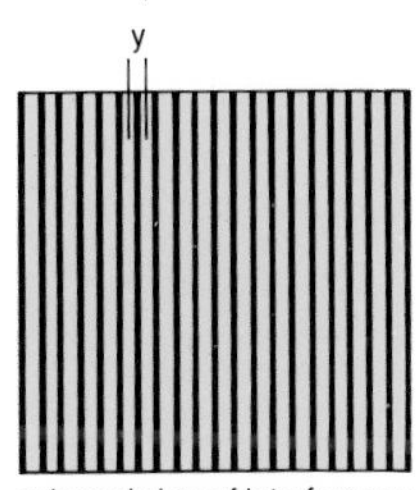

enlarged view of interference fringes in a plane perpendicular to the axis of the system in the region of crossing

y = fringe spacing

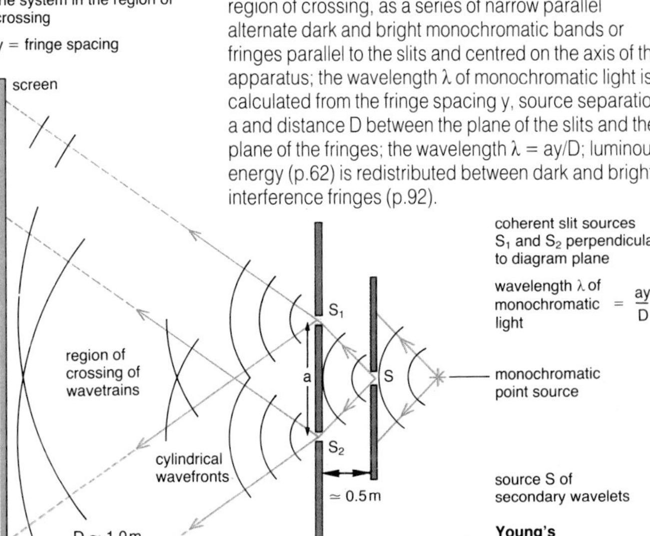

Young's experiment

path difference the difference in physical path length travelled in air or free space by interfering (p.58) wavetrains (p.52); for monochromatic light (p.74) of wavelength λ: a path difference nλ, where n = an integral number, e.g. 0, 1, 2, 3, etc., gives the condition for constructive interference (p.58) and bright interference fringes (↓) in the observable optical interference pattern of Young's experiment (p.91), and for diffraction orders (p.97) observed with a diffraction grating (p.96); a path difference (n + ½)λ gives the condition for destructive interference (p.58) and dark interference fringes in Young's experiment.

interference fringes observable optical interference pattern in Young's experiment (p.91); each bright or dark band represents a series of points in the interference pattern, having constant path difference (↑) between wavetrains arriving from 2 coherent sources, e.g. for a bright fringe on the axis of the apparatus and at the centre of the interference pattern the path difference = 0; for the first bright fringes equidistant from this axial fringe the path difference = nλ, where n = 1; for the first dark fringes adjacent to the axial fringe the path difference = (n + ½)λ, where n = 0; the fringe pattern is symmetrical with respect to the axial fringe.

Lloyd's mirror an alternative method of obtaining 2 coherent sources (p.91) derived from the same monochromatic light source for the calculation of the wavelength of light as in Young's experiment (p.91); a monochromatic source placed close to an optical plane mirror (p.63) forms a virtual image (p.63) in the mirror equidistant from the mirror plane; the source and image are coherent and an observable optical interference (p.91) pattern of alternate dark and bright interference fringes (↑) is seen, with a dark fringe on the axis of the apparatus.

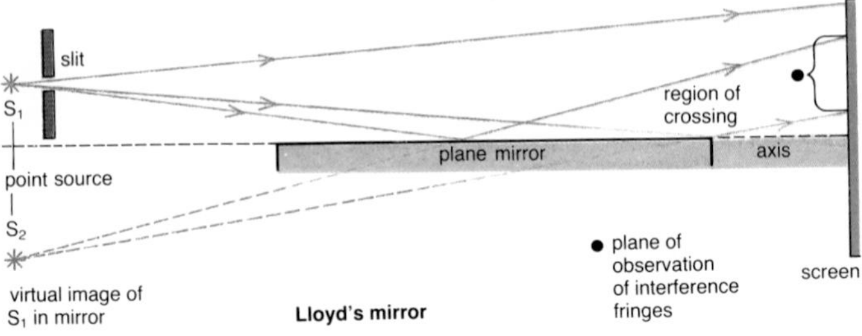

Lloyd's mirror

phase change
in thin film interference

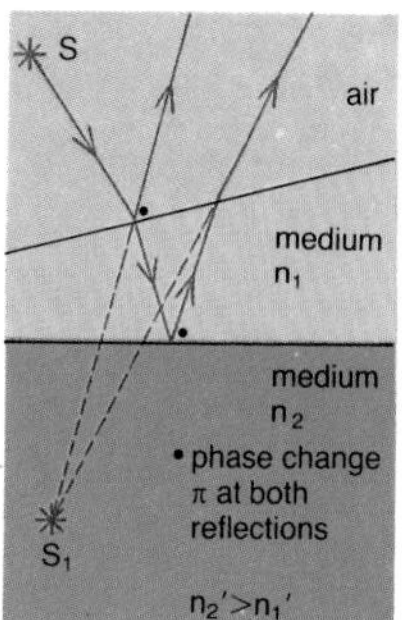

phase change on reflection when a wavetrain (p.52) is reflected at the interface (p.69) between 2 media with different transmission properties, there is a phase change (p.59) of π radians (p.46) if the wave velocity (p.54) is greater in the first medium than in the second, e.g. reflection of light at the air-glass interface in Lloyd's mirror (↑) experiment; a π phase change also occurs on reflection of a sound compression wave (p.56) at a rigid boundary, e.g. closed pipe (p.104).

optical path when the path of a light wavetrain (p.52) is through media with different transmission properties, the wave velocity (p.54) is different in different media, e.g. air, oil, water, glass; for calculation purposes the distance travelled in a medium of refractive index n′ (p.71) is converted to the optical path, or the distance which would be travelled in free space or a vacuum in the same time t seconds, so that the optical path travelled in a vacuum in time t = n′ × physical path length travelled in a medium in time t.

thin film a thin layer of refracting (p.69) medium whose thickness is not more than several light wavelengths (p.54), e.g. approximately 10^{-8}m; soap film, oil film on water.

thin film interference refers to optical interference (p.90) between wavetrains (p.52) reflected from the upper and lower surfaces of a thin film (↑); the interference will be constructive (p.58) or destructive (p.58), depending on the value of the path difference and on the wavelength λ of the illuminating light.

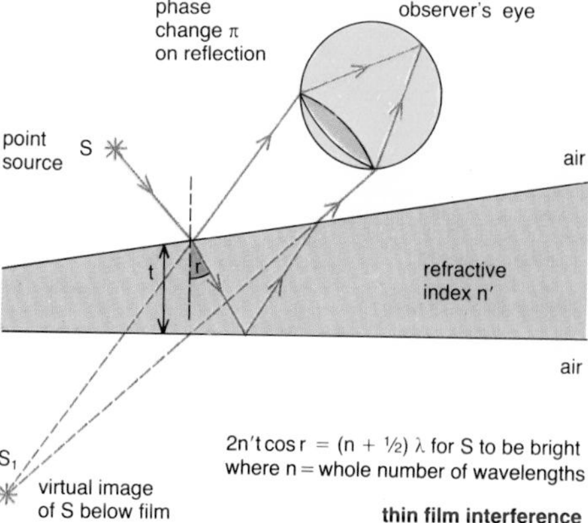

thin film interference

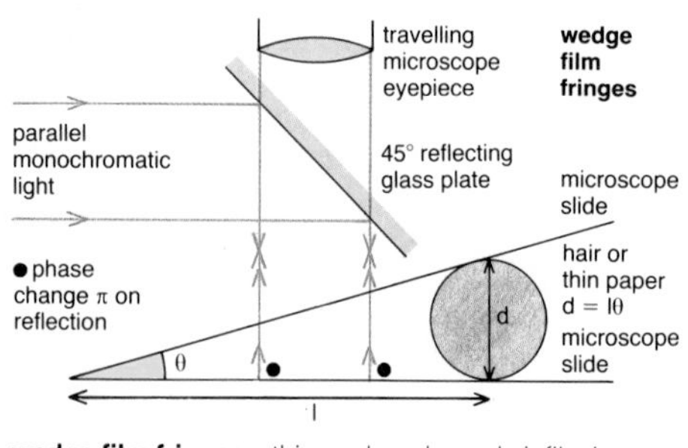

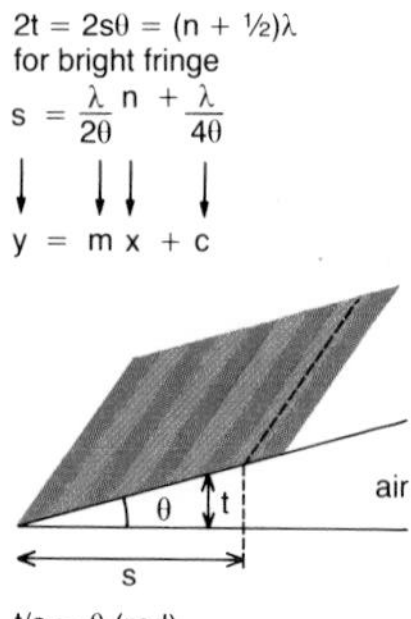

wedge film fringes a thin wedge-shaped air film is formed by placing a hair or a thin piece of paper between 2 microscope slides; thin film interference (p.93) occurs between wavefronts of monochromatic light reflected from upper and lower surfaces of the air film giving parallel, equally spaced bright interference bands parallel to the wedge tip.

Newton's rings a plano-concave air film is formed between the lower surface of a plano-convex lens and the upper surface of an optically flat (↓) glass plate; over the central region of the thin air film, interference (p.92) occurs between wavetrains of monochromatic light reflected from the upper and lower surfaces of the air film, thus giving rise to a series of alternate bright and dark rings concentric with the point of contact between the lens and the glass plate.

optically flat surface a glass or metal surface polished so that surface irregularities are smoothed out to within tolerance limits less than the wavelength of light, e.g. less than 10^{-7}m, and do not disturb interference patterns, e.g. Newton's rings (↑).

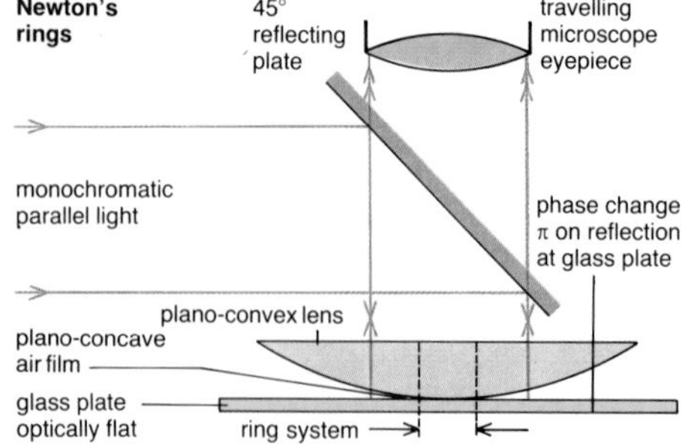

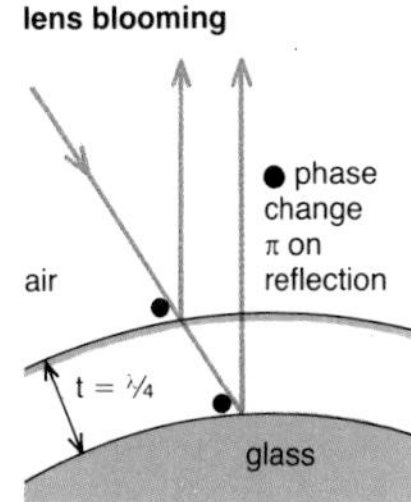

lens blooming surfaces of a glass lens are coated with a thin film of transparent magnesium fluoride, refractive index 1.2 (p.71), to a precise thickness $t = \lambda/4$, where λ is the wavelength central in the range to be used, e.g. about 5.5×10^{-7} m for visible light; phase changes of π occur on reflection (p.62) at both interfaces, and for a path difference (p.92) $2t = \lambda/2$, reflected wavetrains (p.52) interfere destructively (p.58) and reflections from the lens surface are annulled, giving clearer final images from microscope (p.84), telescope (p.84) and camera (p.89) lenses.

dielectric mirror thin film multilayers each of thickness λ/2, giving path difference λ (p.92) for constructive interference (p.58) and 100% reflections.

diffraction of light diffraction (p.61) phenomena observed with visible light wavelengths (p.54); the wave theory of light (p.90) explains these effects as optical interference (p.90) occurring between secondary wavelets (p.90) in the same wavefront, e.g. diffraction patterns (↓) due to a straight edge and narrow slit.

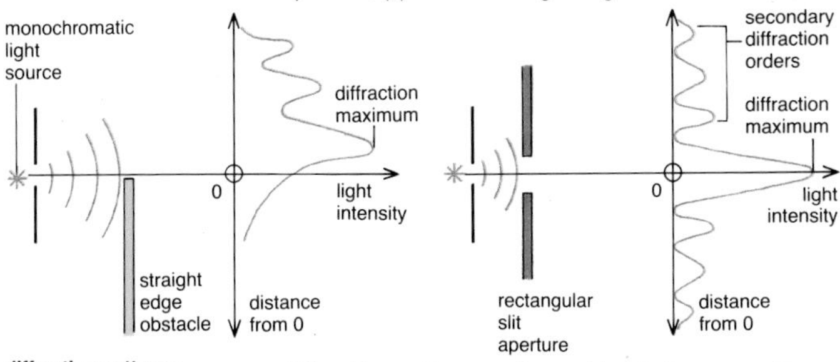

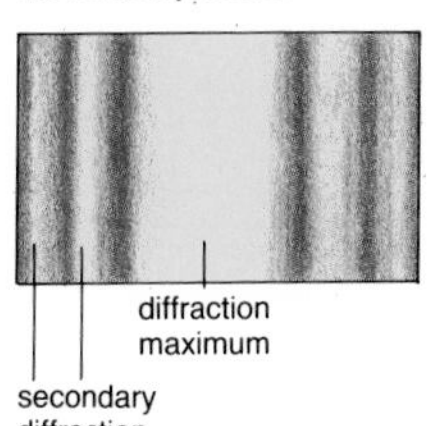

single slit diffraction patterns

diffraction pattern an observable interference pattern (p.91) resulting from diffraction of light (↑) by an obstacle obstructing part of the wavefront; there is a redistribution of luminous energy in the region of interference and the pattern has alternate regions of brightness and darkness with a central bright region, e.g. diffraction at a slit gives parallel bands of varying energy intensity, symmetrically placed with respect to the central diffraction maximum (↓).

diffraction maximum the central region of maximum energy intensity in a diffraction pattern (↑).

secondary diffraction orders regions of energy intensity less than the central diffraction maximum (↑) and symmetrically placed around it.

diffraction grating a series of narrow, parallel, closely spaced lines ruled originally on a glass plate about 3 cm × 2 cm; the rulings act as coherent sources (p.91) of cylindrical wavefronts (p.57) when illuminated normally with plane wavefronts of monochromatic light (p.74); wavefronts of wavelength λ (p.54) are diffracted (p.61) in all directions on a semicircle around each set of coherent point sources; those travelling in direction θ with a path difference (p.92) nλ, where n = an integral number, e.g. 0, 1, 2, 3, etc., combine to produce a bright diffraction image of the illuminated slit source of monochromatic wavefronts; for values of θ corresponding to increasing values of n, pairs of symmetrically placed bright images or diffraction orders (↓) are formed around the central axial image or zero order (n = 0); values of θ are measured by a spectrometer; for constructive interference (p.58) and the formation of a diffraction maximum (p.95): d sin θ = nλ; so if n, d and θ are known, λ can be calculated.

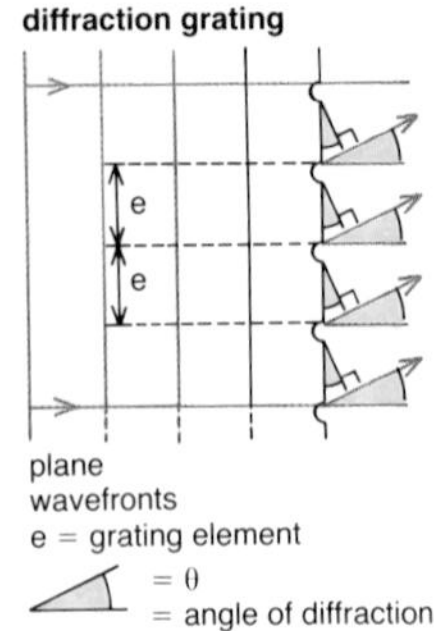

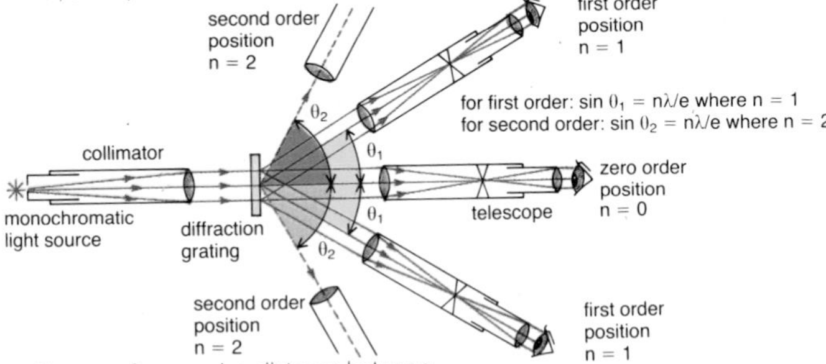

grating spacing spacing distance between corresponding points in consecutive rulings on a diffraction grating (↑).

grating interval alternative name for grating spacing (↑).

transmission grating a diffraction grating which transmits light through the slits and absorbs it at all other points.

reflection grating a mirror ruled as for a diffraction grating such that light is reflected from the equally spaced lines and absorbed at other points; reflection gratings have the advantage that they can be used for wavelengths that would normally be absorbed by glass, e.g. infrared and ultraviolet; by arranging for the grating to be concave the reflected orders may be focused without the use of a lens.

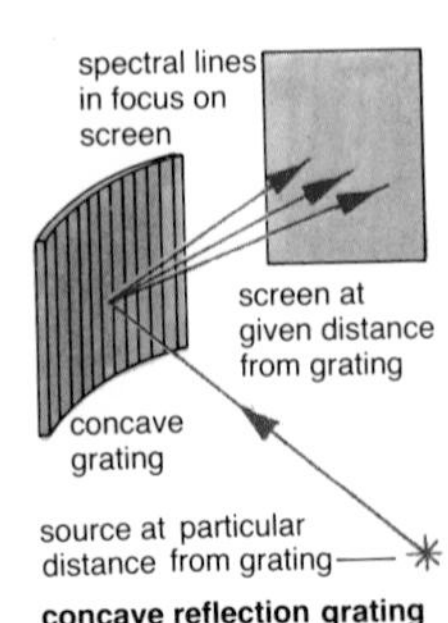

concave reflection grating

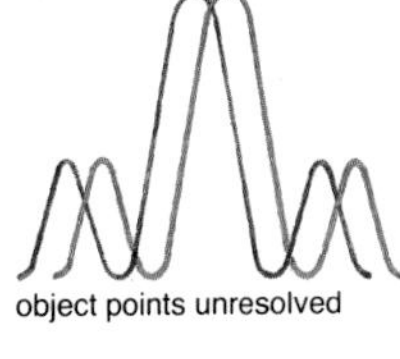

object points unresolved

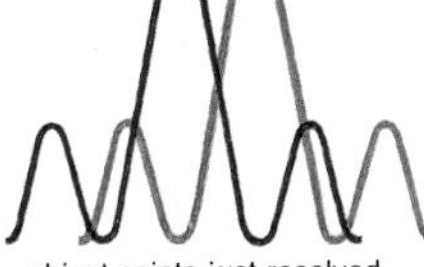

object points just resolved

diffraction orders refers to diffraction images formed by a diffraction grating (↑) illuminated by plane wavefronts (p.57) of monochromatic light (p.74); symmetrically placed with respect to the central axial zero order (n = 0) and identified by increasing numbers 1, 2, 3, etc.

resolving power represents the ability of an optical system to resolve as separate 2 closely spaced points on its final image, e.g. human eye (p.78) can resolve as separate 2 points subtending a visual angle (p.84) not less than about 1 minute (1/60°) of arc; the resolving power depends on the dimensions of the optical system and on the wavelength λ of the light used.

holography a way of producing stereoscopic images without a camera or lens; monochromatic, coherent, collimated laser (p.235) light is split into 2 beams; one beam is diffracted off the subject on to a photographic plate and the other is directed directly on to the plate; the two beams form an interference pattern (p.58) on the plate called a hologram; once the hologram is developed the original subject may be recreated by placing it in a beam of coherent light; the hologram behaves as a diffraction grating (↑) producing two beams of diffracted radiation, one giving a real image and the other a stereoscopic, virtual image; unlike an ordinary photograph, the hologram records information about the phase (p.53) of the light as well as its intensity.

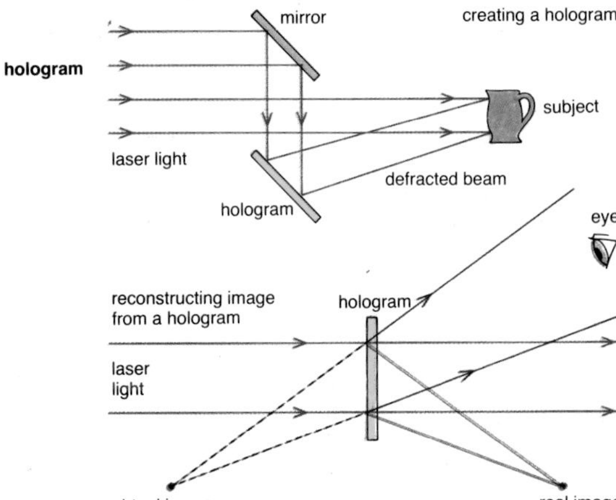

unpolarized wave a transverse wave (p.54) whose vibrations are not restricted to one plane, but take place in all possible directions perpendicular to the direction of energy transmission.

plane polarized wave a transverse wave (p.54) whose vibrations have been restricted to one plane by some means of polarization (p.61), e.g. passage of visible light through a polarizing crystal (↓) such as tourmaline, polaroid (↓).

polarized light
1. vibrations in all directions
2. vertical components
3. horizontal components

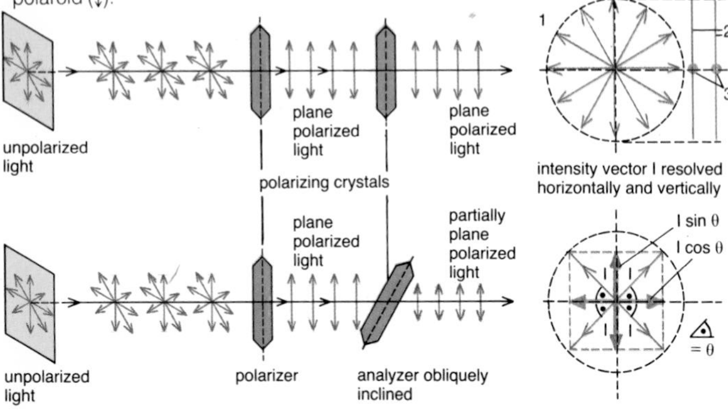

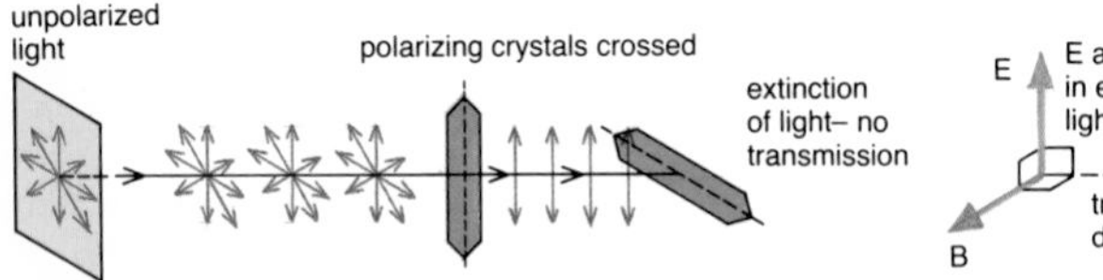

partially plane polarized wave a transverse wave (p.54) whose vibrations have been partially restricted to one plane, by the partial removal of 1 of the 2 sets of perpendicular vibrations into which the unpolarized wave (↑) can be resolved, e.g. scattering (↓) or reflection (↓) of visible light.

colloidal solution a suspension of solid particles of diameter 10^{-3}mm or less in a fluid.

scattering (*n*) an effect of the interaction between electromagnetic radiation (p.55) and molecules or colloidal (↑) particles of the transmitting medium; observed with visible light, u.v. radiation (p.74) and X-rays (p.100); electric charges (p.165) in molecules are set into vibration (p.49) by the electric field

scattering
light both red and partially plane polarized due to scattering through long length of atmosphere

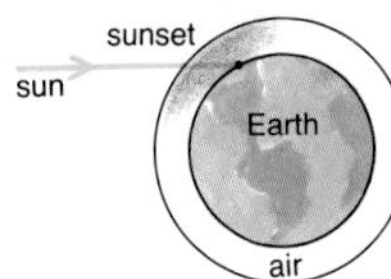

polarization by scattering

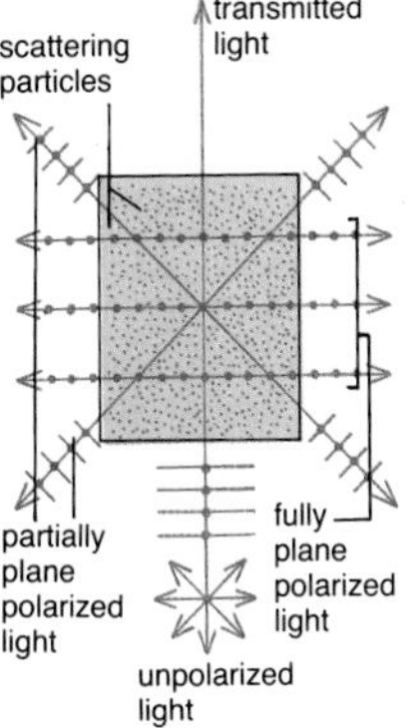

component of an unpolarized (↑) beam of radiation; these vibrating electric charges re-radiate the energy, scattering it in all directions around the molecules, perpendicular to the direction of vibrations; air molecules and colloidal (↑) dust particles in the Earth's atmosphere (p.18) preferentially scatter short wavelength (p.54) blue light from the white light spectrum (p.73) of sunlight, giving the sky a blue colour; at sunrise and sunset, light travels a greater distance through the Earth's atmosphere, giving a red appearance since blue light has been removed by scattering; red and yellow light of longer wavelengths can penetrate fog and mist whereas white light is scattered. **scatter** (*v*), **scattered** (*adj*), **scattering** (*adj*).

polarization by scattering an unpolarized light wave (↑) incident on colloidal particles (↑) in suspension in water is scattered (↑); light perpendicular to the incident direction is plane polarized.

polarizing crystal certain crystalline (p.12) substances have the property of transmitting the vibrations (p.49) of an unpolarized (↑) light beam in 1 plane only, e.g. calcite; the beam emerging from the crystal is plane polarized (↑); 2 similar polarizing crystals can be used to test a light beam for its degree of polarization by placing them in the crossed position with their transmission directions perpendicular.

polarizer (*n*) refers to a polarizing crystal (↑) or medium, e.g. polaroid (↓), whose function is to produce a plane polarized (↑) light beam.

analyzer (*n*) a second polarizing crystal (↑) or medium, e.g. polaroid (↓), used with a polarizer (↑) to test the degree of polarization of a light beam.

polaroid (*n*) a polarizing medium now replacing large mineral crystals as a polarizer (↑) and analyzer (↑) of light; very small crystals of herapathite (an iodine compound of quinine sulphate), in colloidal suspension (↑) in nitrocellulose, are forced to pass under pressure through a narrow horizontal slit; the crystals align themselves parallel to the slit and the resulting polaroid medium consists of a suspension of parallel crystals acting as a single polarizing crystal (↑); used in sunglasses to remove 50% of the incident light by transmitting only plane polarized light (↑), and to further reduce glare due to partially plane polarized light incident from the sky and reflected from the surroundings. **polaroid** (*adj*).

polarization by reflection when an unpolarized (↑) light beam is reflected from a glass, water or other reflecting surface, the reflected light is partially plane polarized (↑); when light is incident at the specific polarizing angle for that medium, the reflected light is fully horizontally plane polarized (↑).

polarizing angle the angle of incidence ϕ on a surface for which reflected light is fully horizontally plane polarized (↑); the reflected and transmitted beams are perpendicular and the relation refractive index $n = \tan \phi$ is known as Brewster's Law.

electromagnetic spectrum the continuous spectrum (p.74) of all the electromagnetic radiations (p.55) found in nature or generated by laboratory sources of electromagnetic energy (p.28); the wavelengths (p.54) range from approximately 10^{-13}m for gamma-radiation (↓) to approximately 2 km for long radio waves (↓); the spectrum is divided into approximate regions according to wavelength, frequency (p.54), source of electromagnetic radiation and distinguishing properties; all electromagnetic radiations have certain wave properties in common, e.g. straight line transmission (p.52) at wave velocity $3 \times 10^8 \text{m s}^{-1}$ in free space; attenuation (p.57) according to an inverse square law; reflection (p.62); refraction (p.69); diffraction (p.61); interference (p.58) and polarization (p.98); each kind of electromagnetic radiation demonstrates these properties under circumstances appropriate to its wavelength.

gamma (γ) **rays** electromagnetic radiations forming a region of the electromagnetic spectrum (↑); the wavelength (p.54) range is approximately 10^{-13} to 10^{-10}m; with a frequency (p.54) range of approximately 10^{20} to 10^{18} hertz (Hz); photon (p.234) energies are above 50 keV (p.228); γ-rays are emitted on disintegration of radioactive nuclei (p.240) and detected by a Geiger-Muller counter (p.238).

X-rays (*n*) electromagnetic radiations forming a region of the electromagnetic spectrum (↑); the wavelength (p.54) range is approximately 10^{-10} to 10^{-8}m; the frequency (p.54) range is 10^{18} to 10^{16} hertz (Hz); photon (p.234) energies are 30 to 50 keV; X-rays are emitted on bombardment of a tungsten target by high energy electrons in an X-ray tube (p.236) and detected by Geiger-Muller counter (p.238) or by photographic emulsion

polarization by reflection

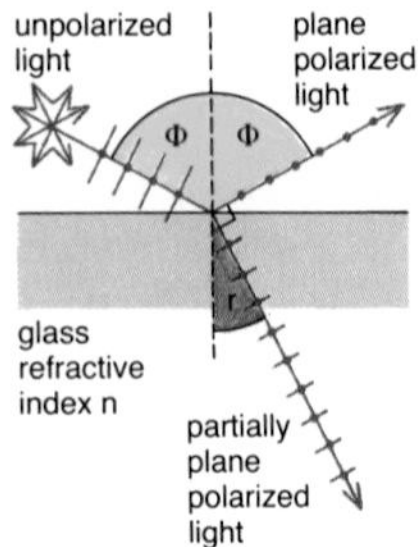

$n = \sin \Phi / \sin r$
$= \sin \Phi / \cos (90 - r)$
$= \sin \Phi / \cos \Phi = \tan \Phi$

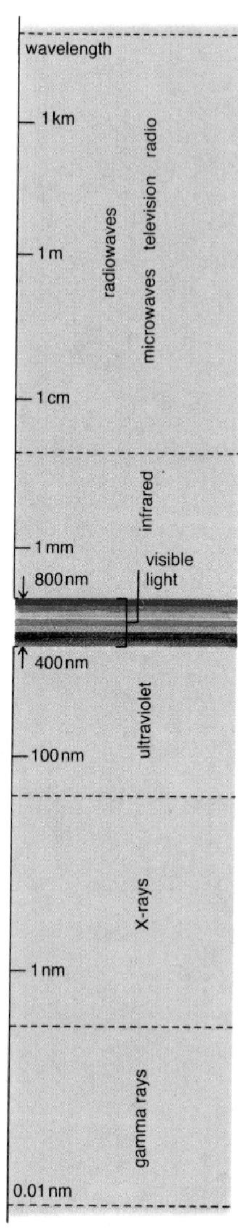

microwaves (*n*) includes electromagnetic radiations of approximate wavelength (p.54) range 10^{-3} to 10^{-1} m, forming part of the electromagnetic spectrum (↑) between infrared radiation (p.28) and radio waves (↓); used for microwave cooking and in radar (↓) and telecommunications (↓) systems; microwaves may be generated by a Gunn diode (p.225) and may be detected (p.221) using a diode detector.

radar (*n*) an acronym for RAdio Direction And Ranging; includes electromagnetic radiations of centimetre wavelengths (p.54) found in the microwave (↑) region of the electromagnetic spectrum (↑); they are generated by a high frequency oscillator as continuous wavetrains (p.52) or as pulsed radiation; the direction and distance of a target are found by measuring the time elapsing between transmission (p.52) of a pulse or wavetrain and the detection of its reflection or echo (p.107) by the receiving aerial; the velocity (p.30) of a moving target can be found using the Doppler effect (p.110), observed in the frequency (p.54) change between transmitted and reflected waves.

radio waves electromagnetic radiations forming a region of the electromagnetic spectrum (↑) with a wavelength (p.54) range of 10^{-2} m to 2 km and a frequency (p.54) range of 10^{10} to 10^{5} hertz (Hz); they are generated by oscillatory circuits and detected by the receiving aerials of acceptor circuits; radio waves are used in telecommunications (↓) over very long distances, including transmissions (p.52) round the Earth's surface by reflection at the E and F ionized layers of the ionosphere of the Earth's atmosphere (p.18); also used in radioastronomy (↓) for investigation of the nature of astronomical bodies and systems in remote parts of the universe.

telecommunications (*n*) transmission of information over long distances as modulated electromagnetic signals, e.g. radio, microwaves or light; orbiting satellites (p.44) can be used to relay signals.

radioastronomy (*n*) the study of radio waves (↑) emitted by galaxies and astronomical bodies in remote parts of the universe, e.g. quasars (p.44), pulsars (p.43), which can be transmitted through the Earth's atmosphere (p.18) and ionosphere; in the wavelength (p.54) range 10 cm to 10 m but typically 21 cm, a characteristic wavelength of un-ionized hydrogen; they are detected by radiotelescopes (p.87).

velocity of sound the velocity (p.30) of a longitudinal compression wave (p.56) in a transmitting medium, e.g. in a solid, velocity v (ms^{-1}) = $\sqrt{E/\rho}$, where E = Young's modulus of elasticity (Nm^{-2}) (p.20) and ρ = density of the medium (kgm^{-3}), for steel v = 5060 ms^{-1}; in a liquid v = $\sqrt{p/\rho}$, where p = bulk modulus of elasticity (p.19) of the liquid, for water v = 1430 ms^{-1}; in a gas v = $\sqrt{\gamma p/\rho}$, where γp = adiabatic bulk modulus of elasticity (p.19) of the gas; also v = $\sqrt{\gamma r T}$, where r = gas constant (J/kg/K) and T = Kelvin temperature (p.116), for air v = 330 ms^{-1} at 0°C (273K).

free vibration the vibration (p.49) of a mechanical system at its natural vibration frequency (↓).

natural vibration frequency a mechanical system, having elasticity (p.19) and inertia (p.36), vibrates freely (↑) after displacement (p.52) and release, e.g. plucked string, or air in a closed or open pipe (p.104); its characteristic natural vibration frequency (p.54) depends on its dimensions and inertia and on the elastic properties of the system.

forced vibration the vibration (p.49) of a mechanical system at frequencies other than its natural vibration frequency (↑).

periodic driving force an harmonic (p.48) force applied to produce forced vibration (↑) of a mechanical system, e.g. electromagnetic vibrator driven by a.c. mains.

response (*n*) the vibration amplitude (p.49) of a mechanical system, vibrating when a periodic driving force (↑) is applied; as with an electric current (p.152) in an electromagnetic tuned circuit (p.221). **respond** (*v*).

frequency/response curve a graph of vibration amplitude (p.49) against applied frequency (p.54) for an oscillating system.

resonance (*n*) maximum response (↑) of a mechanical system undergoing forced vibration (↑), when frequency (p.54) of the periodic driving force (↑) coincides with the natural vibration frequency f_o (↑) of the system; the frequency/response curve (↑) shows a resonance peak at f_o, whose sharpness depends on the amount of damping (↓) in the system; resonance is demonstrated in acoustical systems such as the resonance tube (↓), in an electromagnetic tuned circuit (p.221) and in the optical resonance of line absorption spectra (p.75), e.g. Fraunhofer lines (p.75); a system absorbs maximum energy from the driving force at f_o. **resonate** (*v*), **resonant** (*adj*), **resonating** (*adj*).

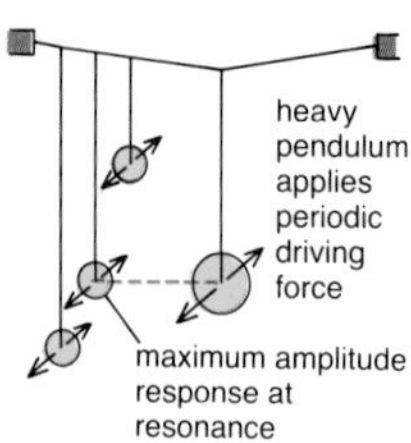

forced vibration and resonance

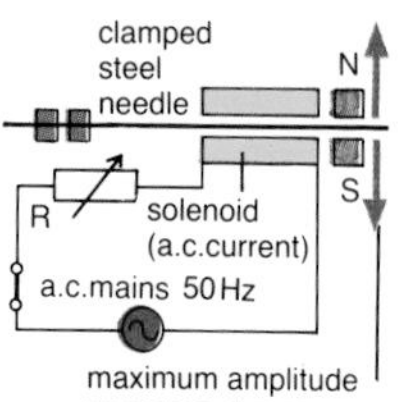

electromagnetic vibrator

periodic driving force
phase relation between periodic driving force and driven system; in phase at f_o

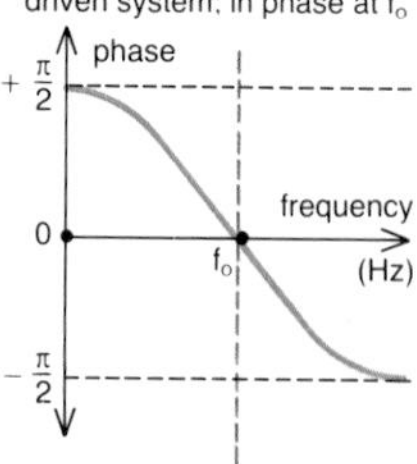

frequency/response curve

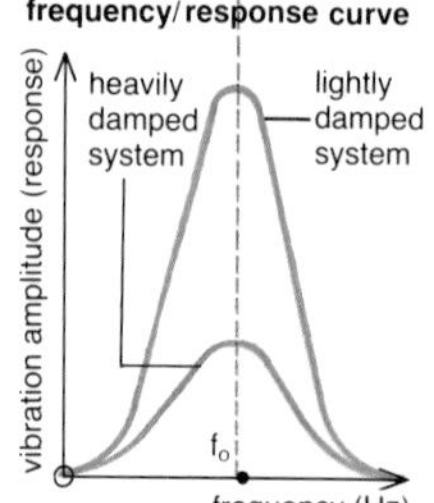

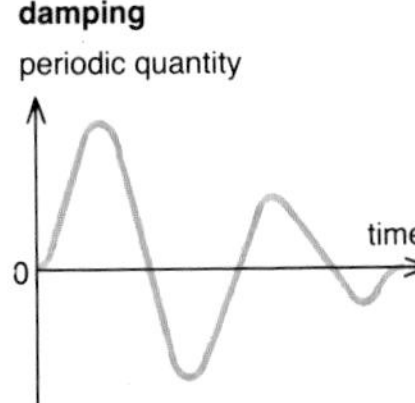

damping (*n*) describes the response (↑) of a system undergoing forced vibration (↑) when inhibited by features of the system reducing the effectiveness with which it absorbs energy from the periodic driving force, e.g. friction (p.21); the amplitude of response is reduced at all frequencies and the frequency/response curve (↑) is flattened at the resonance (↑) peak f_o; the vibration amplitude of a system undergoing free vibration (↑) is also reduced by damping, e.g. a simple pendulum, oscillating spring (p.51), coil oscillating in a magnetic field in a moving coil galvanometer, electromagnetic tuned circuit (p.221) with resistance. **damp** (*v*), **damped** (*adj*), **damping** (*adj*).

resonance tube a hollow cylindrical glass tube, closed at one end by water, whose level can be varied so that the length l of the enclosed air column is variable; a vibrating (p.49) tuning fork, held over the open end, acts as the periodic driving force (↑), and the air column responds with weak longitudinal forced vibrations (↑) as l is varied, emitting a loud note of the tuning fork frequency as the air column resonates (↑). *see page 104*.

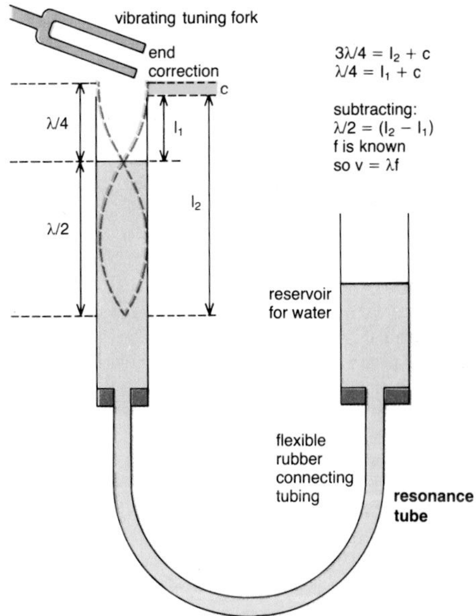

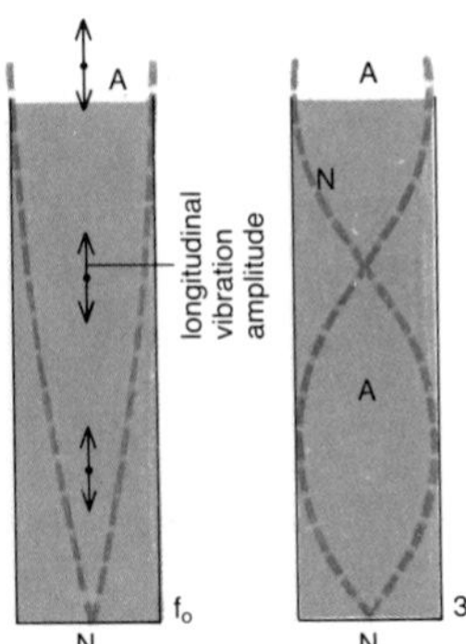

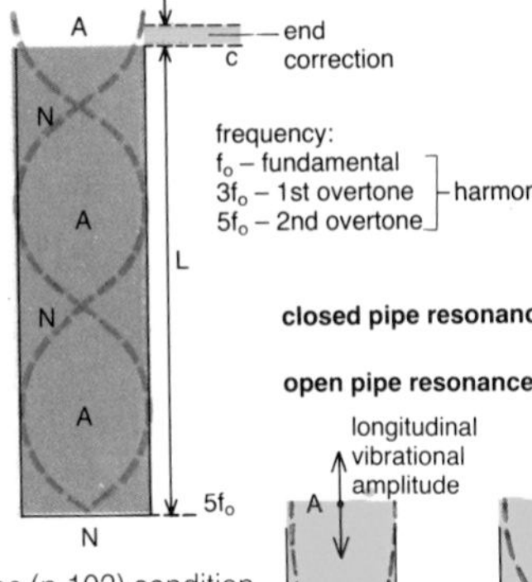

closed pipe resonance

open pipe resonance

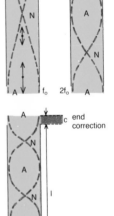

longitudinal
vibrational
amplitude
A
N
A
f_o
$2f_o$
A
N
A
N
A

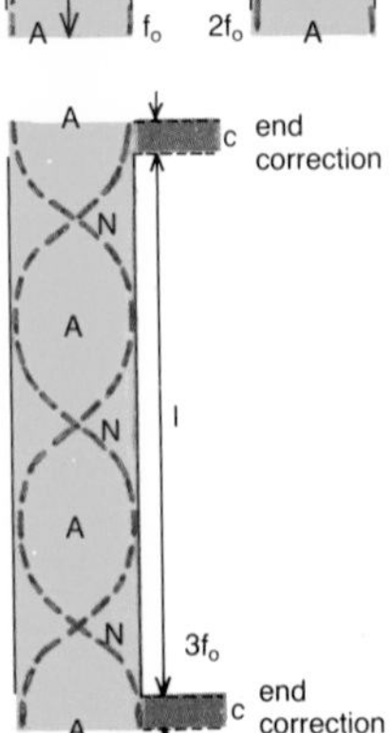

closed pipe resonance the resonance (p.102) condition in the enclosed air column in a cylindrical pipe closed at one end; the closed end defines the node (p.60) position and the open end defines the antinode (p.60) position, with various possible vibration nodes (↓) between them, giving odd-numbered harmonics (↓), all present in the natural pipe note.

open pipe resonance the resonance (p.102) condition in the enclosed air column in a cylindrical pipe open at both ends; both open ends define antinode (p.60) positions; possible vibration modes (↓) between them give all harmonics (↓) in the natural pipe note.

pipe end correction at the non-rigid boundary between free air and air enclosed in a pipe (↑), the antinode (p.60) of the stationary wave (p.61) is located a small distance c (mm) from the open end of the pipe; $c \simeq 0.6r$ where r = pipe radius; open pipe has two end corrections.

vibration mode a possible stationary wave (p.61) system for a closed or open pipe (↑), a bowed or plucked string, or other vibrating system, given the limiting node and antinode positions.

harmonic (*n*) an alternative name for vibration mode (↑); a musical note produced by a particular vibration mode; notes played on musical instruments contain a range of harmonics, giving the instrument the characteristic tone quality (p.106); frequencies (p.54) of harmonics are simply numerically related.

fundamental (*n*) the simplest and dominant vibration mode (↑) for a particular vibrating system; first harmonic of frequency f_o.

overtone (*n*) refers to vibration modes (↑) other than the fundamental (↑); harmonics (↑) of frequency above f_o.

sonometer (*n*) a thin steel wire held in tension (p.60), stretched across 2 fixed bridges and 1 movable bridge supported on a hollow wooden box; used to investigate Mersenne's Laws (↓).

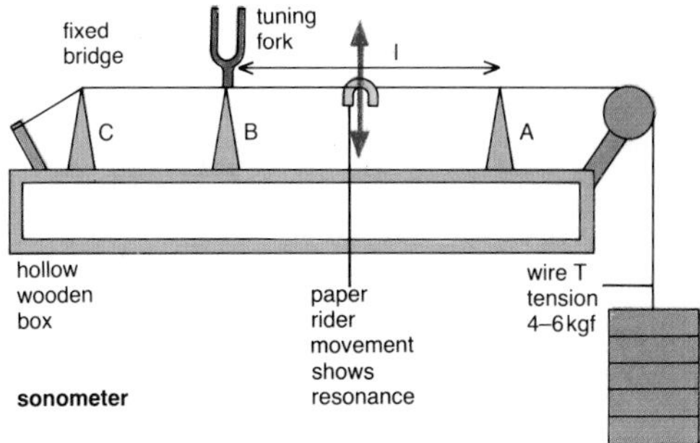

sonometer

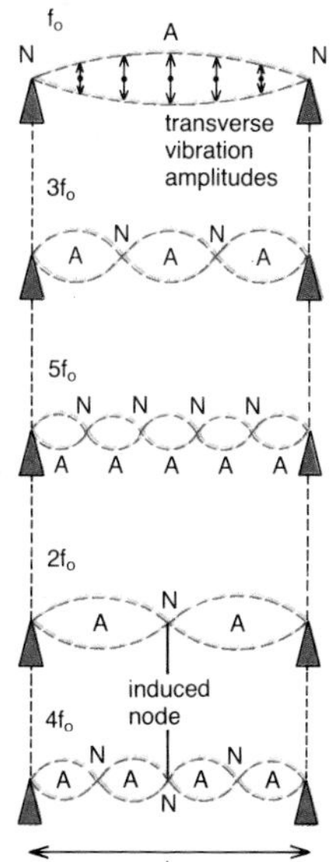

tensioned wire harmonics

Mersenne's laws stated for a wire or string of length l (m) and mass/metre m ($kg\,m^{-1}$), under tension T (p.60) (newton), vibrating with fundamental (p.104) frequency f_o, e.g. sonometer (↑): 1. $f_o \propto 1/l$ for constant tension T; 2. $f_o \propto \sqrt{T}$ for constant length l; 3. $f_o \propto \sqrt{1/m}$ for wires of constant length l under constant tension T but having different mass/metre m.

tensioned wire harmonics refers to possible vibration modes (↑) of a wire or string in tension (p.60), its fixed ends defining node (p.60) positions, e.g. Melde's experiment (p.60).

pure note a musical note (p.106) containing the fundamental (p.104) only, without overtones (↑).

frequency range of audibility a range of frequencies (p.54) to which the normal human ear is sensitive; from approximately 20 Hz threshold (↓) to 20 000 Hz limit for a young person, though the upper limit can be as high as 25 000 Hz; the upper limit is much reduced with age, e.g. 10 000 Hz; animal hearing, e.g. dogs, bats, whales, dolphins can greatly exceed the human upper limit and function in the ultrasonic (p.109) range.

audio-frequencies (*n*) frequencies within the range of audibility (↑) of the human ear.

threshold (*n*) the point or boundary at which a phenomenon begins or becomes observable, e.g. hearing and feeling thresholds (p.107) for the human ear.

musical note a periodic (p.48), continuous and sustained vibration (p.49) within the frequency range of audibility (p.105) of the human ear, and giving a pleasurable sensation and perception (p.78) of hearing; produced by the human voice or by a musical instrument; it is not usually a pure note (p.105) but contains numerous overtones (p.105) of varying intensities, giving the recognizable characteristic tone quality (↓) of the instrument.

pitch (*n*) the position of a note in a musical scale, a note of high pitch has a high frequency (p.54) and a note of low pitch has a low frequency, e.g. c‴ (1024 Hz) has a higher pitch than c″ (512 Hz), and c′ (256 Hz) has a lower pitch than c″.

tone quality musical notes (↑) of the same pitch (↑) and loudness (↓) sound identifiably different when played by different musical instruments or sung by different voices; the different tone quality depends on the presence of numerous overtones (p.105) of varying intensities (↓) in addition to the fundamental (p.104) and it is characteristic of the instrument or voice; the different harmonic (p.104) content gives the note a different complex waveform (p.53) as shown by a cathode ray oscilloscope (p.232).

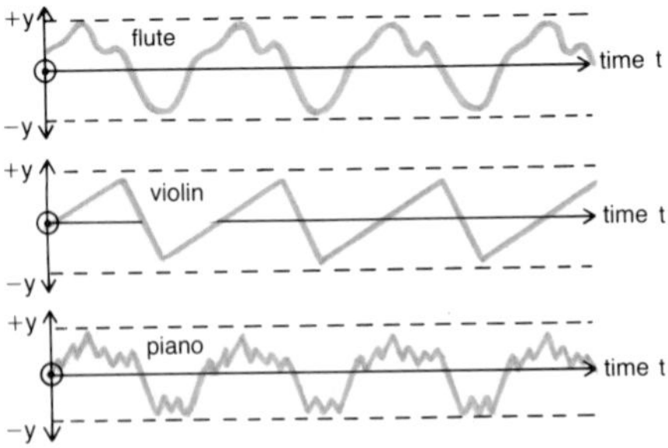

noise[1] (*n*) a sound (p.28) of unacceptable loudness (↓) and quality; it usually contains vibrations of many frequencies (p.54) combined unharmoniously to give unmusical notes (↑) or sounds. **noisy** (*adj*).

noise[2] an unwanted signal in, for example, an electronic system.

white noise describes noise (↑) containing all frequencies (p.54) within the range of audibility (p.105) at approximately equal levels of loudness (↓).

pitch
notes in a musical scale (e.g. piano keyboard)

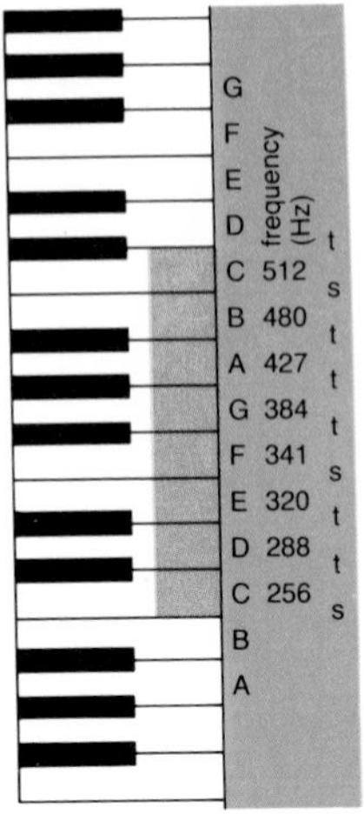

octave
s semitone
t tone

octave (*n*) a musical interval spanning 8 successive notes of a diatonic, or 13 of a chromatic, scale; for notes an octave apart the upper is double the frequency of the lower.

loudness (*n*) an observer's sensation of how loud a sound is; a high intensity level (decibels) (↓) gives a high loudness level (phons).

hearing threshold a threshold (p.105) reference level for sound intensity (↓) detectable by the normal human ear; its value is 10^{-12} watt/m^2 (Wm^{-2}) for a note of frequency (p.54) 1000 Hz; this is made the zero level to which other sound intensity levels can be related.

intensity of sound the rate of sound energy transfer/m^2 across a plane normal to the direction of propagation.

bel (*n*) a unit of relative sound intensity; if a sound source power changes from P_1 to P_2, so that the sound energy intensity changes from I_1 to I_2 at distance r from the source the number of bels (B) = $\log_{10}(P_2/P_1) = \log_{10}(I_2/I_1)$, e.g. when $P_1 = 0.1$ W and $P_2 = 0.2$ W; $\log_{10}(0.2/0.1) = \log_{10} 2 = 0.3$ B; for most practical purposes the smaller decibel (↓) unit is used.

decibel (*n*) a unit of relative sound intensity; subdivision of the bel (↑); 1 decibel (dB) = 0.1 bel (B): the number of decibels (dB) = $10 \log_{10}(P_2/P_1) = 10 \log_{10}(I_2/I_1)$, e.g. when $P_1 = 20$ mW and $P_2 = 200$ mW, $10 \log_{10}(200/20) = 10$ dB; when $P_1 = 100$ mW and $P_2 = 20$ mW, $10 \log_{10}(20/100) = -7$ dB; a negative (−) sign indicates a decrease in source power output; the decibel is also used as a unit of change in electrical power in amplifying circuits.

echo (*n*) a sound heard indirectly after reflection.

echo sounding a method of determining depth by measuring the time taken for a pulse of high frequency sound to reach the sea bed or a submerged object and for the echo to return. **echo sounder** (*n*).

sonar (*n*) alternative name for echo sounder; acronym standing for SOund NAvigation Ranging.

sound refraction the refraction (p.69) of sound waves (p.56) on crossing an interface (p.69) between transmitting media in which the velocity of sound v (p.102) has different values.

sound diffraction diffraction (p.61) of sound waves (p.56) by obstacles whose dimensions are comparable with the sound wavelengths (p.54), 0.5 m to 1 m, e.g. outdoor sounds can be heard round the corners of buildings by edge diffraction.

echo sounding

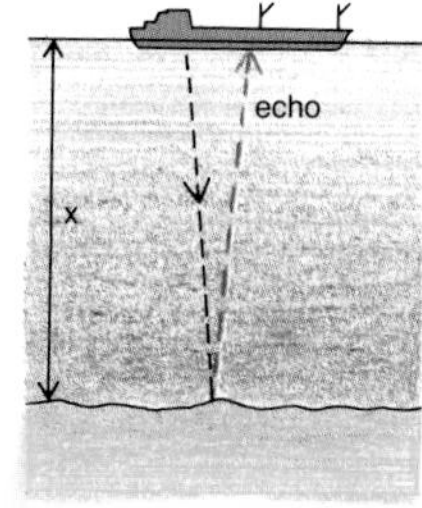

If time elapsing between sound being sent and echo being received is t (s) and velocity of sound in water is v (m/s) then depth of sea x = ½vt; this distance is given by the echo sounder

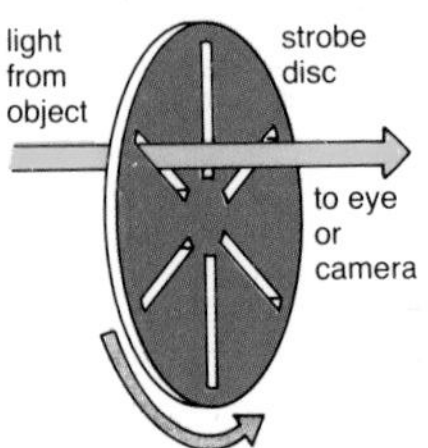

disc rotated at constant speed by motor

stroboscope

signal generator an electromagnetic oscillatory circuit (p.221), which can be tuned to produce an output signal of required amplitude (p.49) frequency (p.54) and waveform; the term is usually applied to sine-wave generators.

stroboscope (*n*) abbreviation 'strobe', an experimental arrangement for measuring the frequency (p.54) of a vibrating mechanical system, e.g. Melde's experiment (p.60); a flashing xenon lamp, electrically maintained at a variable frequency, illuminates the system intermittently; the flashing frequency is varied until it coincides with the vibration frequency of the system, when the system appears stationary; an alternative form of stroboscope is a rotating disc containing a number of equally spaced slots; by varying the speed of rotation the frequency can be found at which the system appears stationary. **stroboscopic** (*adj*).

stroboscopic photography a method of taking a photograph of a moving object with a flashing strobe lamp or through a rotating strobe disc such that several images of the object are obtained on the photo; widely spaced images show fast movement of the object and closely spaced images show slow movement.

beats (*n*) periodic (p.48) variation in sound intensity (p.107) heard when 2 notes of nearly equal frequencies (p.54) f_1 and f_2 are sounded together; the 2 wavetrains (p.52) interfere (p.58) when superimposed, giving a resultant waveform (p.53) whose maximum amplitude (p.49) gives the maximum sound intensity of each beat; the beat frequency $f = f_2 - f_1$, where $f_2 > f_1$; beats can account for certain difference tones heard in musical instruments. **beat** (*v*), **beating** (*adj*).

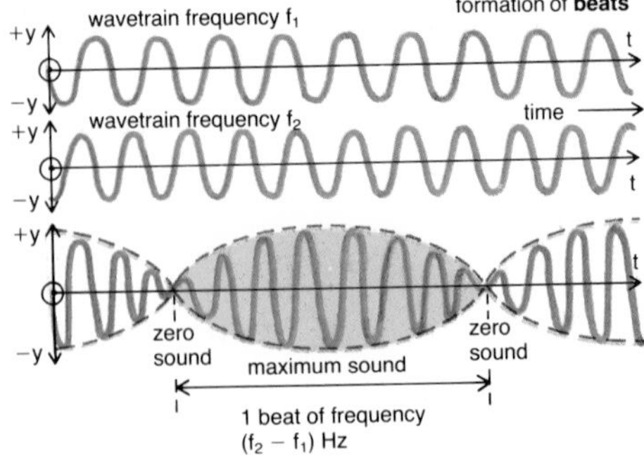

formation of **beats**

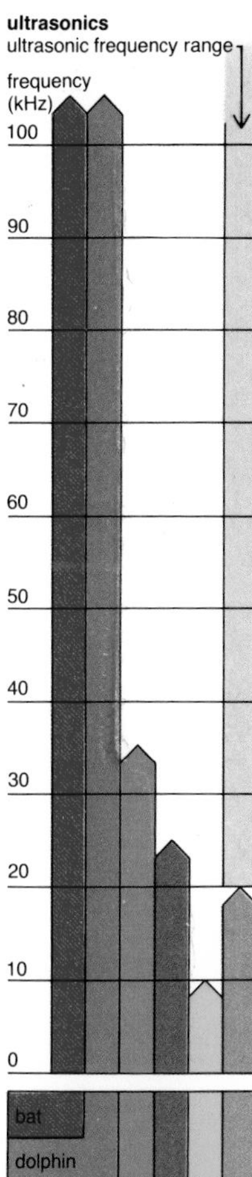

acoustics (*n*) the study of sounds; the science of designing auditoria; the features of an auditorium making it good for listening to, or recording, music or speech. **acoustical** (*adj*).

reverberation (*n*) a sound heard repeatedly after multiple reflections within a room, auditorium or enclosed space; the continuation of sound in an auditorium after the sound source, e.g. voice or musical instrument, has ceased to emit energy; it ceases after the reverberation time (↓). **reverberate** (*v*), **reverberating** (*adj*).

reverberation time the time interval T (seconds) for which a sound continues to reverberate (↑) within a room or auditorium after the sound source has ceased to emit energy; acoustical design (↑) determines the value of T, given by Sabine's Law (↓); good acoustics for orchestral music require T between 1 and 2.5 s; speech requires T between 0.5 and 1.5 s for words to be clearly heard as separate.

absorptive power[2] describes the effectiveness of a surface in absorbing sound, it is expressed relative to 1 square foot of open window, which would absorb all sound incident upon its plane; denoted by a; for polished wood, a poor sound absorber, $a = 0.01$ to 0.03; for thick carpet, a good sound absorber, $a = 0.3$ to 0.6; the value of a varies with frequency (p.54) of the absorbed sound within specified ranges.

Sabine's Law the reverberation time T (↑) of an auditorium depends on its volume V (m^3), its surface area A (m^2) and the absorptive powers a (↑) of its surfaces; the relation is $T = kV/aA$, where k = a constant characteristic of the acoustics (↑) of that particular room; it is used in acoustical design of auditoria.

ultrasonics (*n*) the study of compression waves (p.56) of frequency (p.54) above 20 000 Hz and beyond the range of audibility of the human ear; reflection properties of these short-wavelength (p.54) compression waves are used in echo sounding (p.107) to determine water depth at sea, in detection of internal cracks in metal castings and in ultrasonic scanning for the internal medical investigation of the human body; ultrasonic waves are generated when a high frequency alternating e.m.f. (p.149), applied between opposite faces of a quartz crystal, causes the crystal to expand and contract alternately. **ultrasound** (*n*).

piezoelectric effect a quartz crystal under stress (p.19) acquires equal and opposite electric charges (p.165) on opposite faces and an electric p.d. (p.154) is developed between them; the application of a compression wave (p.56) causes generation of an alternating e.m.f. (p.149); piezoelectric crystals are used in microphones and pick-up heads of record players. **piezoelectricity** (*n*).

Doppler effect the apparent change in the frequency (p.54) of a wavetrain (p.52) emitted by a source moving relative to the observer; the frequency is apparently increased as the source approaches, and decreased as it recedes from the observer; this effect is due to modification of the wavelength (p.54) as approaching wavefronts (p.57) move closer together and as receding wavefronts move further apart; the apparent frequency f′ is related to true frequency f: $f' = Vf/(V \pm U_s)$, where V = velocity of wave (p.54) and U_s = source velocity (p.30); the Doppler effect is most commonly observed for sound waves (p.56), e.g. a train whistle or jet aircraft passing an observer; it is assumed to be the cause of the red shift (↓) in light emitted from distant galaxies; when the observer is moving relative to a stationary source, the frequency is apparently

Doppler effect

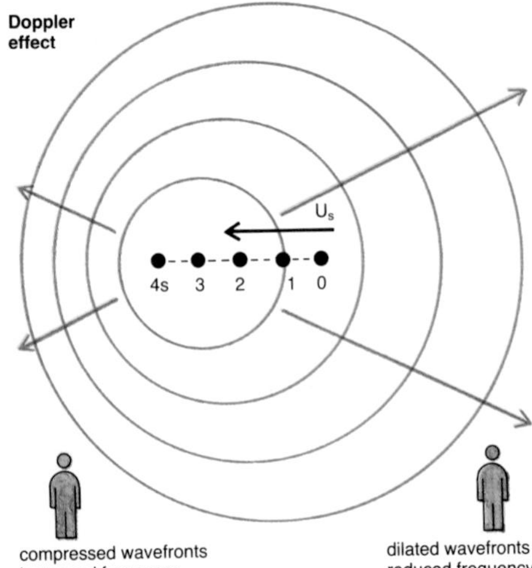

increased as the observer approaches the source and decreased as he recedes from it; this effect is due to modification of wave velocity relative to the observer; $f' = (V \pm U_o)f/V$, where U_o = observer's velocity; for source and observer both moving: $f' = (V \pm U_o)\,f/(V \pm V_2 s)$.

red shift characteristic lines in the emission and absorption spectra (p.75) of certain stars are displaced towards the longer wavelength (p.54) red end of the spectrum, when compared with the same emission or absorption line spectra (p.75) obtained in the laboratory, e.g. star Delta Leporis has 2 dark lines, K and H, identified as absorption spectral lines of calcium, 3933.664 Å and 3968.470 Å (1 Angstrom unit Å = 10^{-10} metre) wavelength; the K line shift of 1.298 Å towards the towards the red end is interpreted as a Doppler shift (↑), giving the recession velocity of the star as 99 km s^{-1}; some approaching stars show blue shifts; the reddening of emitted light from distant galaxies is assumed to be a Doppler effect as galaxies recede in an expanding universe; optical observations with Mount Palomar 200-inch telescope (p.86) gave recession velocities up to 60 000 km s^{-1} from red shift measurements; red shift effect of wavelength lengthening is observed in radio waves (p.101) in radio-astronomy (p.101), giving recession velocities according to Hubble's Law (↓) of the order 10^8 m s^{-1} for certain luminous quasi-stellar radio sources (quasars); other quasars have several different red shifts, and red shift calculations of distance do not agree with the observed brightness of quasars; thus red shift as a Doppler effect may need reinterpretation in some circumstances.

Hubble's Law this states that the recessional velocity of external galaxies is proportional to their distance; the constant of proportionality, the Hubble constant, H is about 75 km/sec/megaparsec or $2.4 \times 10^{-18} s^{-1}$, where 1 parsec $\simeq 30 \times 10^{12}$ km (1 Mpc = 10^6 parsec).

Hubble's Law

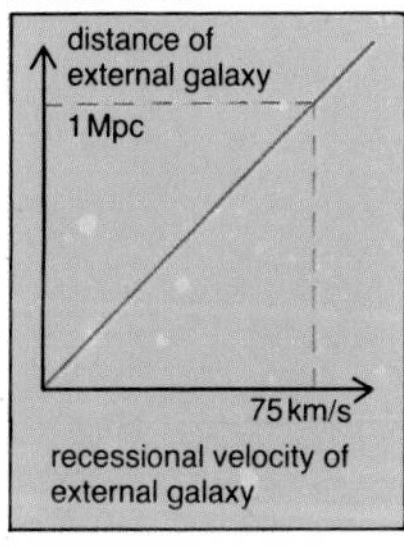

gravitational red shift according to the theory of relativity a photon leaving the sun or other star loses energy by doing work against the gravitational field; the frequency of the light is reduced and the wavelength shifted towards the red end of the spectrum; a gravitational red shift of approximately 10^{-9} m has been measured for the solar spectrum, in agreement with the theory.

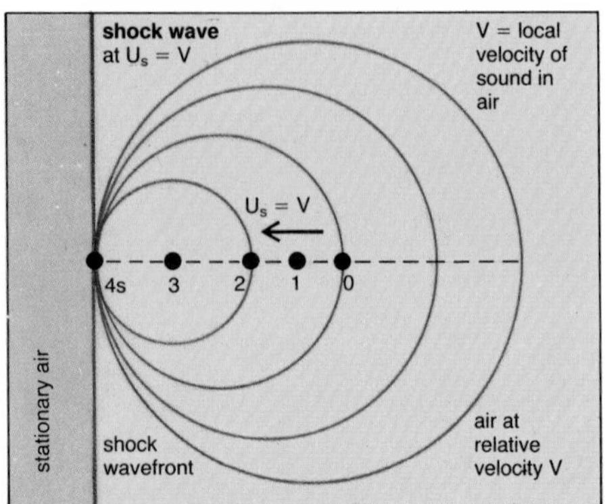

shock wave a common wavefront (p.57) formed by all compression waves (p.56) emitted from an energy source moving at a velocity equal to, or greater than, the local velocity of sound in air (p.102); air ahead of the shock is stationary and air behind the shock has a relative velocity equal to that of the source; shocks form on aircraft flying supersonically, e.g. Concorde; the shape of the shock wavefront surface is conical.

Mach cone the shape of the shock wavefront (↑) surface attached to an aircraft flying supersonically is a cone of semi-vertical angle θ, where $\sin\theta$ = velocity of aircraft ÷ local velocity of sound = U_s/V; supersonic aircraft speeds can be given as Mach numbers, e.g. Mach 1 ($U_s = V$), Mach 2 ($U_s = 2V$), Mach 3 ($U_s = 3V$).

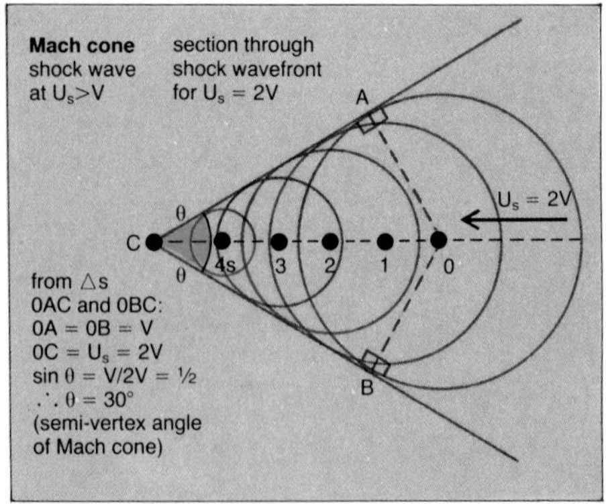

calorimeter

calorimetry (*n*) measurement of the quantity of heat energy (p.28) involved in a heat exchange process, e.g. determination of specific heat capacity (↓) or specific latent heats (p.16) of a substance. **calorimetric** (*adj*).

calorimeter (*n*) an apparatus in which calorimetric (↑) experiments are carried out, e.g. a copper vessel for mixture methods using solids and liquids; a continuous flow calorimeter (p.115) for electrical heating methods for liquids.

specific heat capacity the quantity of heat energy (p.28) required to raise the temperature (p.116) of 1 kilogram (kg) of a specified substance by 1 degree Kelvin (1 K) (p.145); units are joule/kg/K ($J\,kg^{-1}\,K^{-1}$), e.g. for water value is 4200 $J\,kg^{-1}\,K^{-1}$ or 4.2 $kJ\,kg^{-1}\,K^{-1}$ where 1 kJ = 1000 J; denoted by s.

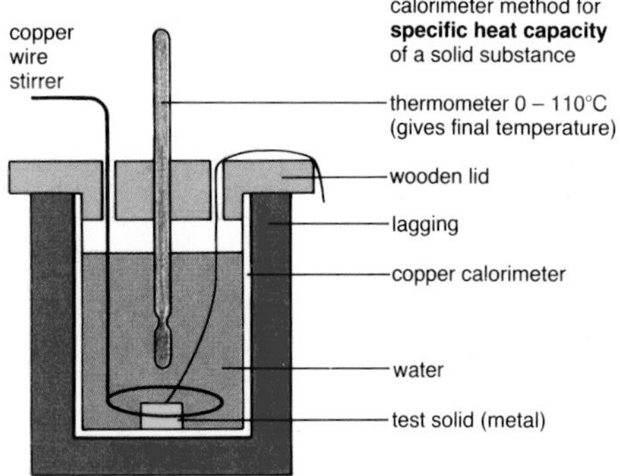

calorimeter method for **specific heat capacity** of a solid substance

heat capacity the quantity of heat energy (p.28) required to raise the temperature (p.116) of a body (p.12) of mass m kg by 1 degree Kelvin (1 K) (p.145); units measured in joule/K ($J\,K^{-1}$); value is m s, where s = specific heat capacity (↑) of the body's substance.

thermal capacity an alternative name for heat capacity (↑).

food calorie the unit of calorific value (p.114) for foodstuffs used as the body's fuel; it is based on the former heat calorie as a unit of heat energy; 1 heat calorie = 4.2 joule (J); 1 food calorie (C) = 1000 calories = 1 kilo-calorie (Kcal); for fats the calorific value is 9 $Kcal\,g^{-1}$, for carbohydrates and proteins 4 $Kcal\,g^{-1}$.

calorific value the quantity of heat energy (p.28) released on combustion of 1 kilogram of fuel measured in units of joule/kg ($J\,kg^{-1}$); used to express the effectiveness of a fuel as an energy source, e.g. for fossil fuels, coal and oil; for foodstuffs it is expressed in food calories (p.113).

lagging (*n*) material which is a poor heat conductor (p.124), used as thermal insulation (↓).

thermal insulation a means of preventing the escape of heat from a physical system, e.g. by lagging (↑), by surrounding with a vacuum jacket as in a thermos flask (p.126), by surrounding with an air layer in an enclosure to limit convection (p.124), as in a specific heat capacity measurement by cooling (↓) experiment.

heat insulation alternative name for thermal insulation (↑).

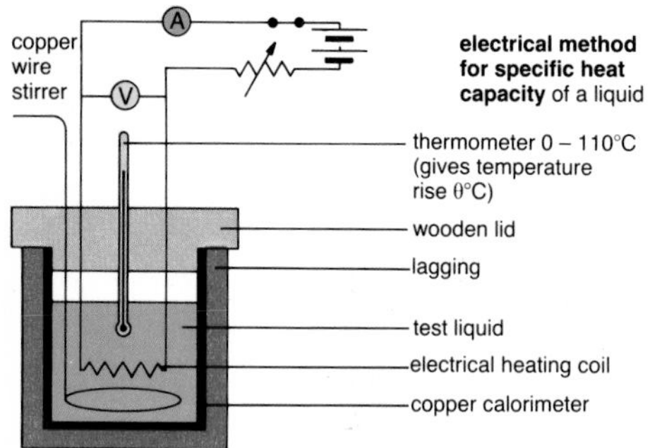

electrical method for specific heat capacity of a liquid

electrical method for specific heat a calorimetric (p.113) method commonly used to determine the specific heat capacity s (p.113) of a liquid; a mass m (kg) of a specified liquid is heated electrically for time t (seconds) in a lagged (↑) copper calorimeter (p.113) of known thermal capacity C (joule K^{-1}); an electric potential difference V (volt) (p.154) across the heating coil and electric current I (ampere) (p.152) are maintained constant, and the temperature (p.116) rise θ (°C) in time t (s) is measured; electrical energy provided in time $t = VIt$ (joule); assuming no heat losses, the heat energy absorbed by calorimeter and liquid $= (ms + C)\theta$; $VIt = (ms + C)\theta$, so $s(J\,kg^{-1}\,K^{-1})$ can be calculated.

rate of cooling the rate of fall of temperature (p.116) of a body (p.12) or physical system, e.g. calorimeter (p.113), losing heat to its surroundings; temperature fall/second is denoted by $d\theta/dt$, units are $K\,s^{-1}$; $K\,min^{-1}$.

rate of heat loss the rate of loss of heat energy (p.28) from a body (p.12) or physical system, e.g. calorimeter, to its surroundings; measured in units of joule/second (Js^{-1}); denoted by dQ/dt; dQ/dt = m s dθ/dt, where m = body mass (kg); s = specific heat capacity (p.113) of body or substance ($Jkg^{-1}K^{-1}$); dθ/dt = rate of cooling (↑) (Ks^{-1}; $Kmin^{-1}$).

excess temperature the temperature (p.116) difference between a body (p.12) or physical system, e.g. calorimeter (p.113) and cooler surroundings.

Newton's Law of Cooling the rate of heat loss (↑) from a body or physical system is directly proportional to its excess temperature (↑); $dQ/dt \propto (\theta - \theta_R)$ where θ = temperature of body or system and θ_R = temperature of surroundings; it is approximately true in still air for excess temperatures up to about 30°C (30K); more strictly true for all excess temperatures in conditions of forced convection (p.126); the relation $dQ/dt \propto \theta^{5/4}$ applies for excess temperatures θ between 50°C (50K) and 300°C (300K); Newton's Law in equation form is: $dQ/dt = kS(\theta - \theta_R)$, where S = surface area losing heat (m^2) and k = constant ($Js^{-1}m^{-2}K^{-1}$), depending on the type of surface, e.g colour, texture, polish.

continuous flow calorimeter a calorimeter (p.113) system for use with liquids and gases, in which fluid flowing through under constant pressure (p.39) at a measured rate m (kgs^{-1}), is heated by a steady measured supply of electrical power VI (watt) (p.159); steady state (p.124) inlet and outflow liquid temperatures θ_1 and θ_2°C are measured and the specific heat capacity s (p.113) of the fluid can be calculated; its advantages are that the heat capacity (p.113) of the calorimeter need not be known and that compensation can be made for unavoidable heat losses (↑) h (Js^{-1}) by repeating the experiment between the same inlet and outflow temperatures, so that cooling conditions are identical.

continuous flow calorimeter

Callendar and Barnes calorimeter:

$$\begin{cases} VI \text{ (watt)} = ms(\theta_2 - \theta_1) + h\,(Js^{-1}) \\ V'I' = m's(\theta_2 - \theta_1) + h \end{cases}$$

so $(VI - V'I') = (m - m')s(\theta_2 - \theta_1)$ and s can be calculated

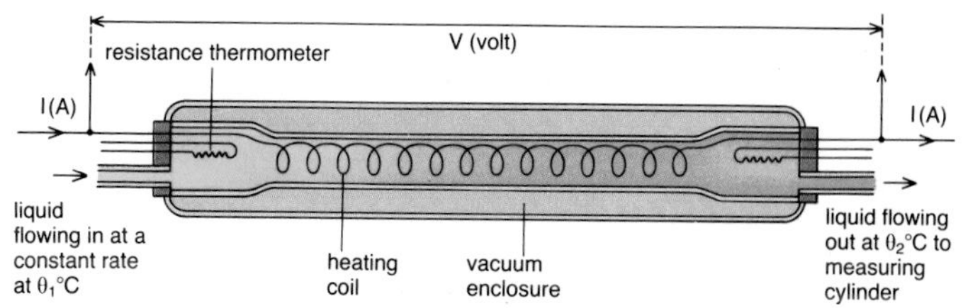

temperature (*n*) a measure of the degree of hotness or coldness of a body (p.12) or substance on a particular scale of temperature (↓); the definition of temperature (p.118) is made on a scale based on a specific thermometric property (↓) of a thermometric substance (↓).

thermometry (*n*) temperature (↑) measurement. **thermometric** (*adj*).

thermometer (*n*) a device or instrument for measuring temperature (↑); it incorporates the thermometric substance (↓) and a scale of temperature (↓) based on a specific thermometric property (↓) of the substance.

thermometric property a property of matter varying with temperature (↑) in a way that is reproducible, measurable and preferably linear, e.g. pressure of a fixed mass of ideal gas (↓) at constant volume; electrical resistance (p.155) of a conductor within a specified temperature range; volume of a fixed mass of liquid.

thermometric substance a substance possessing a thermometric property (↑), e.g. a fixed mass of hydrogen gas behaving ideally, as in the standard constant volume gas thermometer (↓); platinum metal made into a platinum coil, as in a platinum resistance thermometer (p.118); mercury or alcohol in a liquid-in-glass thermometer (p.119).

scale of temperature a scale based on a specified thermometric property (↑) of a specified thermometric substance (↑), e.g. constant volume gas scale (↓) based on the standard constant volume hydrogen gas thermometer (↓); platinum scale (↓) based on the platinum resistance thermometer (p.118); the scale must have 2 or more fixed points (p.118) with the fundamental interval (p.118) between them subdivided into degrees of temperature interval (↓).

temperature interval the difference in temperature (↑) between equal subdivisions of a scale of temperature (↑); measured in degrees Kelvin (K) (p.145).

constant volume gas scale a scale of temperature (↑) based on the linear variation in gas pressure (p.136) of a fixed mass of hydrogen gas at constant volume, with variation in temperature, under conditions of ideal behaviour (p.138); it is realized practically in the standard constant volume hydrogen gas thermometer (↓); the definition (p.118) of temperature θ_g on the gas scale is given by $\theta_g/100 = (p_\theta - p_0)/(p_{100} - p_0)$, where p_0, p_{100} and p_θ are gas pressures at constant volume

scale of temperature

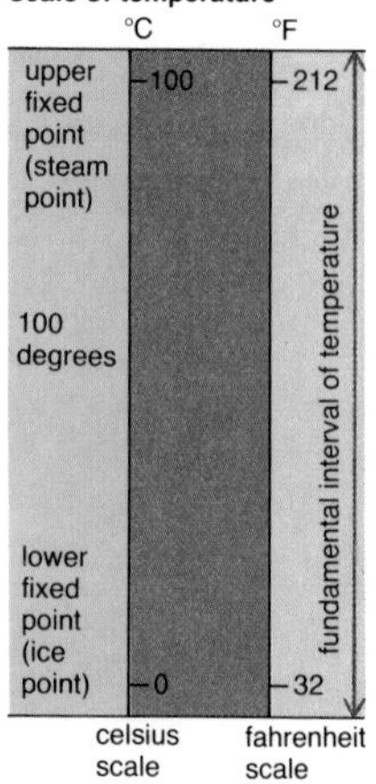

$\theta_C = (\theta_F - 32) \times 5/9$

$\theta_F = (\theta_C \times 9/5) + 32$

for temperatures 0°C(273 K), 100°C(373 K) and θ_g°C(θ_g + 273 K); gas scale is now replaced by the Kelvin thermodynamic scale (p.145), with which its temperatures coincide.

ideal gas scale under conditions of ideal behaviour (p.138), at very low gas pressure (p.136), all gases have the same pressure coefficient 1/273 K^{-1} (p.136), and an equal mass of any gas could replace hydrogen in the standard constant volume gas thermometer (↓); the ideal gas scale is the constant volume gas scale (↑) operating at low gas pressures.

standard constant volume gas thermometer a practical means of measuring temperatures (↑) on the constant volume gas scale (↑); it comprises a long narrow platinum-iridium bulb, containing a fixed mass of hydrogen gas, connected by glass tubing to a mercury barometer (p.39) and mercury reservoir; an index marker fixes the constant gas volume at which gas pressure (p.136) measurements are made by observing the difference in mercury levels; its range is −250°C to +500°C using hydrogen; this can be extended to +1500°C using nitrogen, and to below −250°C using helium.

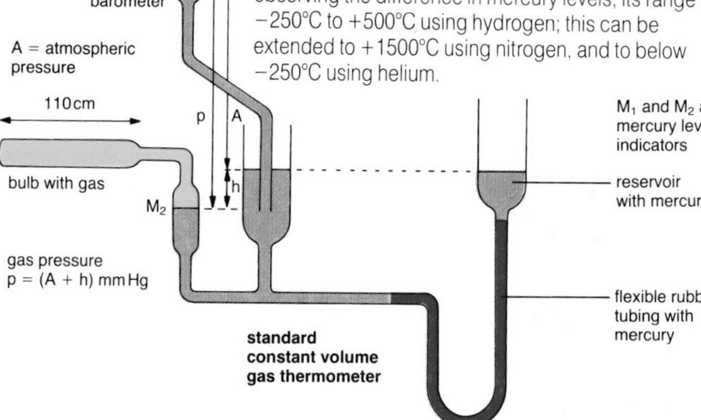

standard constant volume gas thermometer

platinum resistance scale a scale of temperature (↑) based on the linear variation in the electrical resistance (p.155) of a platinum coil with variation in temperature within a specified temperature range; it is realized practically in the platinum resistance thermometer (p.118); the definition of temperature θ_p on the platinum scale is given by $\theta_p/100 = (R_\theta - R_0)/(R_{100} - R_0)$, where R_0, R_{100} and R_θ are resistance values for temperatures 0°C(273 K), 100°C(373 K) and θ_p°C(θ_p + 273 K).

platinum resistance thermometer a practical means of measuring temperature (p.116) on the platinum resistance scale (p.117); it comprises a platinum coil wound non-inductively (p.156), the coil leads running parallel to a pair of dummy leads undergoing the same changes of temperature and resistance; coil leads and dummy leads are connected into opposite sides of a Wheatstone network (p.161) arranged practically as a Callendar and Griffiths Bridge; the platinum coil is the unknown resistance whose value is measured at 2 fixed points (↓) and at the unknown temperature θ_p; a length of uniform slide-wire is included to obtain a very precise bridge balance point; temperature is measurable to ±0.01C(K); its range is −200°C to +1200°C; it is accurate but slow-reading and has small thermal capacity (p.113).

fixed point a constant temperature (p.116), easily and accurately reproducible under laboratory conditions, e.g. boiling and freezing points (p.16) of pure water at standard atmospheric pressure 760 mm Hg (p.40); at least 2 fixed points are required to establish a scale of temperature (p.116); arbitrary values can be given to the chosen upper and lower fixed points (↓), but their values are conventionally agreed in defining particular scales of temperatures, e.g. the Celsius or Centigrade scale (↓).

upper fixed point the higher of 2 fixed point (↑) temperatures defining a scale of temperature (p.116), e.g. the boiling point of pure water at 760 mm Hg is given the value 100 on the Celsius scale (↓) and 212 on the Fahrenheit scale (↓).

lower fixed point the lower of 2 fixed point (↑) temperatures defining a scale of temperature (p.116), e.g. the freezing point of pure water at 760 mm Hg is given the value 0 (zero) on the Celsius scale (↓) and 32 on the Fahrenheit scale (↓).

fundamental interval the temperature difference between upper and lower fixed points (↑) on a temperature scale; it is subdivided into an arbitrary number of degrees of equal temperature interval (p.116), e.g. 100 on the Celsius scale (↓); 180 on the Fahrenheit scale (↓).

definition of temperature as made on the Celsius scale (↓) it is based on a specific thermometric property (p.116) X of a specific thermometric substance (p.116); the property values are measurable at 2 fixed points (↑) and at the unknown temperature θ_x to be defined;

platinum resistance thermometer

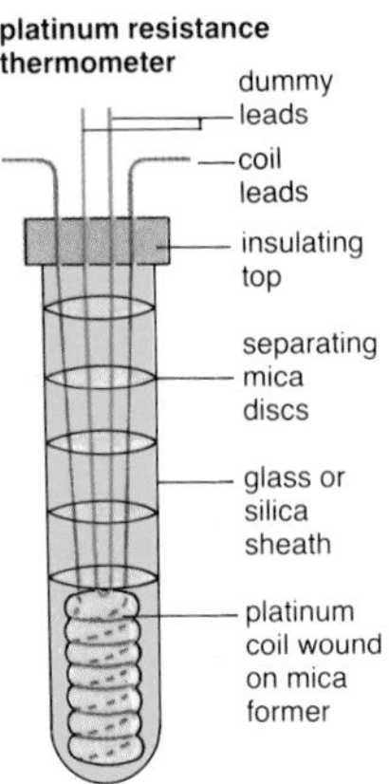

Callendar and Griffiths Bridge

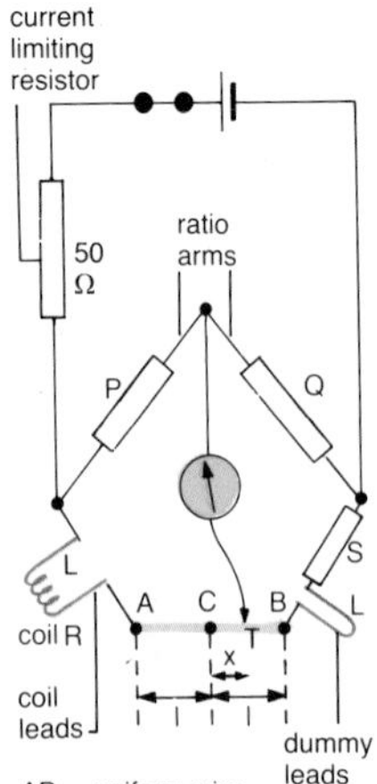

AB = uniform wire length 2l, centre C, resistance r ohm/mm

For bridge balance at T:

$$\frac{P}{Q} = \frac{R + L + r(l + x)}{S + L + r(l - x)}$$

for P/Q = 1: R = S − 2rx

Celsius temperature scale

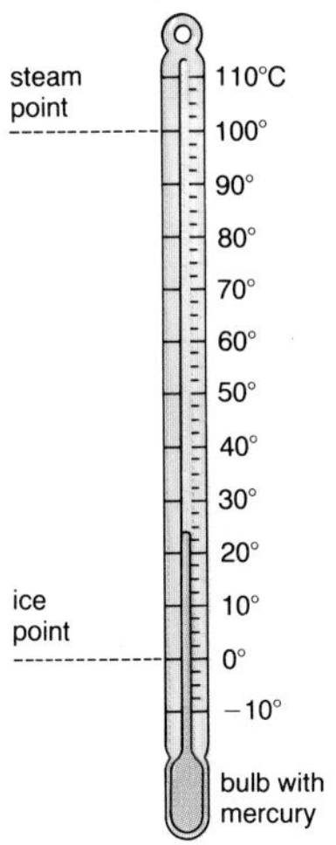

mercury-in-glass thermometer: range 0–110°C

property values at 100, 0 and θ_x are X_{100}, X_0 and X_θ; temperature θ_x is defined as $\theta_x/100 = (X_\theta - X_0)/(X_{100} - X_0)$; scales based on different thermometric properties agree when there is a linear relation between the property variation on one scale and temperature measured on the other, e.g. platinum resistance scale (p.117) and gas scale.

degree of temperature a temperature interval (p.116) which is a subdivision of the fundamental interval (↑) of a specific temperature scale; 1 Fahrenheit degree (°F) = 5⁄9 Celsius degree (°C), representing equal changes in temperature on different scales.

Celsius temperature scale a scale of temperature for which the upper and lower fixed points (↑) are 100 and 0 (zero); the fundamental interval (↑) has 100 equal degrees of temperature (↑); the temperature interval (p.116) is defined in units of 1 Kelvin (K); a position on the scale above the lower fixed point is denoted as θ°C (θ + 273K).

Centigrade temperature scale an alternative common name for the Celsius scale (↑).

Fahrenheit temperature scale a scale of temperature for which the upper and lower fixed points (↑) are 212 and 32; the fundamental interval (↑) has 180 equal degrees of temperature (↑); $(\theta_F - 32)/180 = \theta_C/100$ relates Fahrenheit scale temperature θ_F to Celsius scale (↑) temperature θ_C.

mercury-in-glass thermometer a thermometer (p.116) using mercury as its thermometric substance (p.116) and the linear variation in volume of a fixed mass of liquid with temperature as the thermometric property (p.116); its range is from the boiling point of mercury 358°C to the freezing point of mercury −39°C; it has limited accuracy and range of measurement compared with the standard constant volume gas thermometer (p.117), the platinum resistance thermometer (↑) and the thermoelectric thermometer (p.120), but it is direct and rapid reading and has a low thermal capacity (p.113).

Bourdon gauge

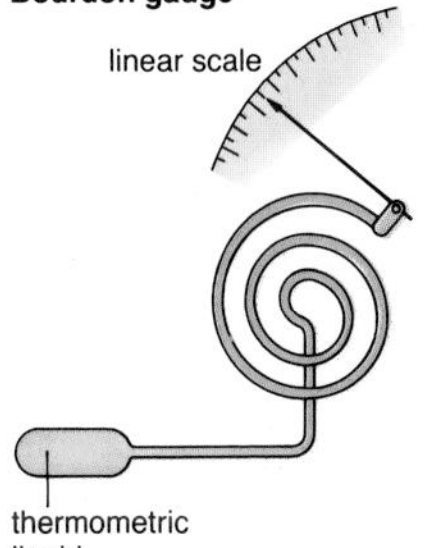

Bourdon gauge the gauge coil is a flattened spiral of several turns of fine metal tubing; it is sensitive to changes of pressure from inside, tending to unwind with fluid pressure increase, and conveys the fluid movement to a pointer moving over a linear scale; it is used generally for measuring pressure and in the liquid-in-metal and gas-in-metal (p.120) thermometer.

liquid-in-metal thermometer a thermometer (p.116) using mercury or xylene as its thermometric substance (p.116) and the linear variation of a fixed mass of liquid with temperature as its thermometric property (p.116); a Bourdon gauge (p.119), connected to the thermometer bulb and stem, detects the volume change and conveys the liquid movement to a pointer moving over a linear scale; its range is from the freezing point of mercury −39°C to its boiling point 358°C; for xylene it is from −40°C to 400°C.

gas-in-metal thermometer an industrial thermometer (p.116) similar in construction to the liquid-in-metal thermometer but using gas, e.g. nitrogen, as its thermometric substance (p.116), and the pressure of the fixed mass of gas at constant volume, indicated by the Bourdon gauge (p.119), as its thermometric property (p.116); its range using nitrogen is −51°C to 538°C.

industrial thermoelectric thermometer

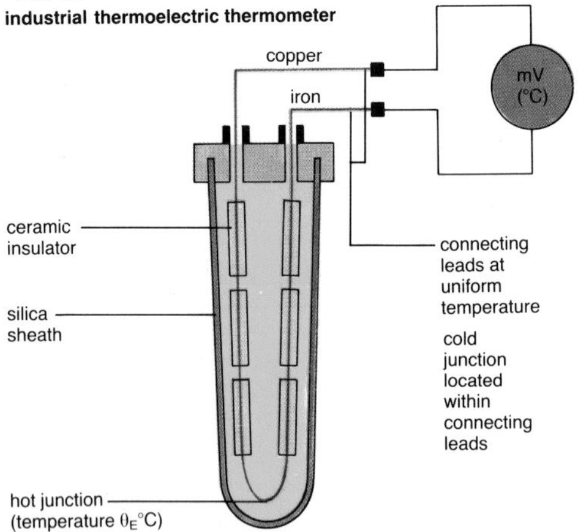

thermoelectric thermometer a thermometer (p.116) using the thermoelectric effect (p.163) as its thermometric property (p.116) and 2 metals of a thermocouple (p.163) as its thermometric substances (p.116); thermocouple metals are chosen with neutral temperature θ_N (p.164) outside the range of measurement, e.g. copper-iron couple from −200°C to 300°C; 40% rhodium-in-platinum alloy with 20% rhodium-in-platinum alloy from −200°C to +1800°C;

thermocouple characteristic curve

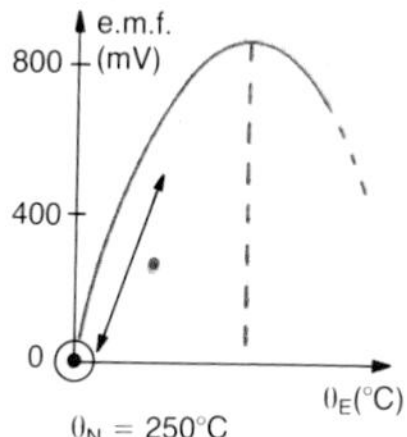

• linear part of thermocouple characteristic

tungsten-molybdenum couple above 1800°C; for moderate accuracy ± 0.1°C the thermoelectric e.m.f. (p.163) is measured by a sensitive millivoltmeter calibrated directly in °C (K), giving a rapid-reading instrument; for greater accuracy ± 0.01°C e.m.f. measurement is made by a modified potentiometer (p.162) having a high resistance in series with the wire, reference to standard thermocouple tables giving the value of temperature; if the e.m.f. measuring instrument is situated a long distance from the hot junction, e.g. for measurement of a nuclear reactor (p.241) core temperature, compensating leads (↓) are introduced to extend the length of the thermocouple; if required, the hot junction can be made very small with negligible thermal capacity; for the initial linear part of the thermocouple characteristic curve, temperature θ_E on the Celsius scale (p.119) is defined by $\theta_E/100 = (E_\theta - E_0)/(E_{100} - E_0)$, where E_0, E_{100} and E_θ are thermoelectric e.m.f. values for temperatures 0°C (273K), 100°C (373K) and θ_E°C = (θ_E + 273K); for the cold junction at 0°C, E_0 = 0(zero) and $\theta_E/100 = E_\theta/E_{100}$.

compensating leads refers to leads used to extend the length of thermocouple leads in a thermoelectric thermometer (↑); the leads should have approximately the same thermoelectric characteristic curve as the thermocouple metals, though they are made from less expensive metals.

thermoelectric e.m.f

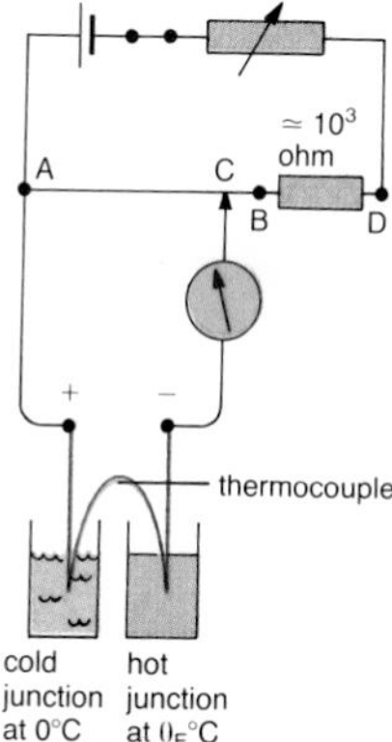

modified potentiometer circuit for measurement of thermoelectric e.m.f.

magnetic thermometer a thermometer (p.116) used for measurement of very low temperatures, e.g. below 1K, approaching the absolute zero of temperature (p.137); its thermometric substance (p.116) is a paramagnetic salt and its thermometric property (p.116) is the variation of its magnetic susceptibility K with Kelvin temperature T (p.145) according to the Curie relation (p.188) $K \propto 1/T$; the scale is non-linear.

radiation pyrometer thermometer (p.116) for high temperature measurement; its thermometric property (p.116) is the variation in emissive power e (p.128) of a heat and light (p.28) energy source with temperature, within a very narrow wavelength range; the Lummer-Pringsheim curves (p.129) are characteristic curves for a blackbody radiator (p.128) at specified temperatures; a pyrometer is calibrated by comparision with a standard blackbody, and can then be used to measure temperature from about 500°C to about 3000°C, e.g. optical pyrometer; total radiation pyrometer.

International Practical Temperature Scale a practical means of reproducing standard constant volume gas thermometer (p.117) readings for certain fixed point (p.118) temperatures used in thermometry (p.116) ; it refers to a series of freezing and boiling points (p.16) at standard atmospheric pressure 760mm Hg (p.40) for substances of a specified high level of purity; these are measured by a gas thermometer and corrected for low pressure, giving a series of practical fixed points in degrees Celsius: 1. Oxygen Point; boiling point of oxygen: −182.970°C. 2. Triple Point of Water (p.146): 0.010°C (this replaces the Ice Point). Ice Point; freezing point of water: 0.000°C (now only used for limited accuracy calibrations, e.g. mercury-in-glass thermometer (p.119)). 3. Steam Point; boiling point of water: 100.000°C. 4. Sulphur Point; boiling point of sulphur: 444.600°C. Antimony Point; freezing point of antimony: 630.500°C (this is not now used, as supercooling (p.16) by up to 20 or 30K is common and makes accurate measurement difficult). 5. Silver Point; freezing point of silver: 960.800°C. 6. Gold Point; freezing point of gold: 1063.00°C.

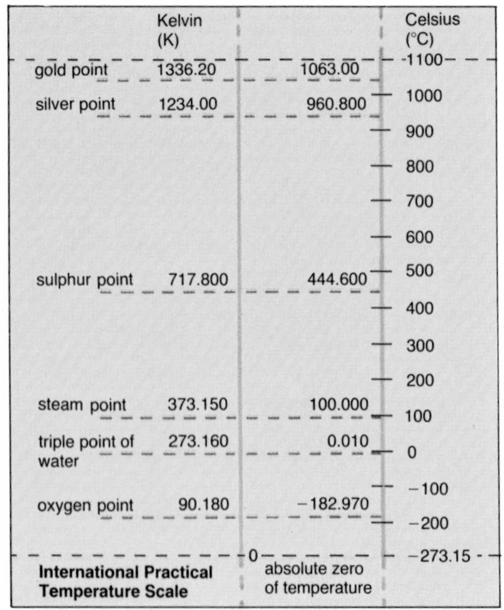

heat transference heat energy (p.28) transmission (p.52) between 2 places having different temperatures (p.116), the natural energy flow being from high to low temperature; heat is transferred down a temperature gradient (↓); the 2nd law of thermodynamics (p.145) shows that heat cannot be transferred from one place at a given temperature to another place at a higher temperature unless work is done; heat transference through material media can be by conduction (↓), convection (p.124) and radiation (p.124): through free space or a vacuum it is by radiation only; for liquids and non-metallic solids, with a crystal lattice structure, heat conduction is by transfer of molecular vibration energy on contact; for liquids and gases heat transfer is mainly by convection. **transfer** (*n*), **transfer** (*v*).

heat transference

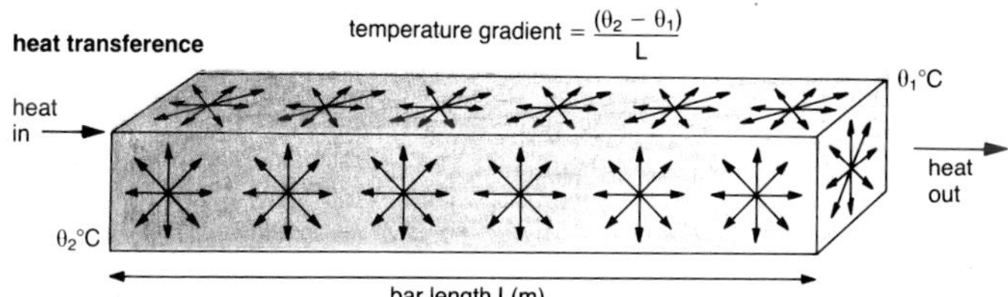

heat transference by conduction through a solid bar

temperature gradient a temperature (p.116) difference $(\theta_A - \theta_B)$K between 2 points A and B ÷ their distance L (m) apart: $(\theta_A - \theta_B)/L(Km^{-1})$, or $d\theta/dL$, in units of Km^{-1}.

heat conduction heat transference (↑), from one place to another at a lower temperature, by the transfer on contact of molecular vibrational energy and the transfer of kinetic energy by electrons (p.7) through a transmitting medium; the vibration amplitude (p.49) increases with temperature, causing an increase in volume (p.133) of the crystal (p.12) lattice and thermal expansion (p.131); in metals, where there is a large supply of free electrons, heat conduction is predominantly due to the movement of electrons through the material; being light, electrons move through the material quickly, whereas molecular vibrational energy is transmitted rather slowly; thus metals are good conductors of both heat and electricity and non-metals poor conductors of both. **conductor** (*n*), **conduct** (*v*), **conducting** (*adj*).

convection (*n*) heat transference (p.123) by movement of fluid (p.15) molecules of higher kinetic energy (p.27) to another place where molecules have lower kinetic energy; it occurs in transmitting media in liquid (p.15) and gaseous states (p.15) by the formation of convection currents (p.125), e.g. in a domestic hot water system, in the formation of land and sea breezes. **convector** (*n*), **convect** (*v*), **convecting** (*adj*).

heat radiation heat transference (p.123), from one place to another at a lower temperature, as electromagnetic energy (p.28) in the heat (p.28) and infrared radiation (p.73) wavelength range, approximately 8×10^{-7} m to 8×10^{-5} m; it can occur through free space or a vacuum and through certain transparent materials, e.g. air, glass; the sun's (p.43) energy reaches the Earth by radiation. **radiator** (*n*), **radiate** (*v*), **radiating** (*adj*).

heat conductor a medium through which heat transference (p.123) occurs by conduction (p.123); most metals have high thermal conductivity (↓), e.g. silver, copper, aluminium; materials having low thermal conductivity include glass, cork, asbestos, liquids (p.15) and gases.

heat insulator a conductor (↑) having low thermal conductivity (↓), e.g. cork, asbestos, expanded polystyrene, magnesia mineral fibre insulation; it is used for thermal insulation (p.114).

heat conduction rate the rate of heat transference (p.123) by conduction (p.123) dQ/dt (joule s^{-1}); for a uniform sample of heat conducting (↑) material of cross-sectional area A (m^2), e.g. cylindrical copper bar or circular glass disc, in a region of a temperature gradient $d\theta/dl$ (Km^{-1}); under steady state (↓) and linear flow (↓) conditions $dQ/dt = KA\, d\theta/dl$ (Js^{-1}), where K = thermal conductivity (↓) of heat conducting material.

steady state condition a condition achieved in a physical system when the rate at which energy (p.26) enters the system equals the rate at which energy leaves it; for a uniform sample of heat-conducting (↑) material, e.g. cylindrical copper bar, circular glass disc, heat energy (p.28) per second entering any part of the sample = heat energy per second leaving that part of the sample, and no heat is absorbed by the sample to give any further temperature (p.116) rise; all parts of the sample are at a constant maximum temperature; the steady state condition is necessary for a uniform heat conduction rate (↑) $dQ/dt = KA\, d\theta/dl$ (Js^{-1}).

convection

domestic hot water system: water circulates by convection

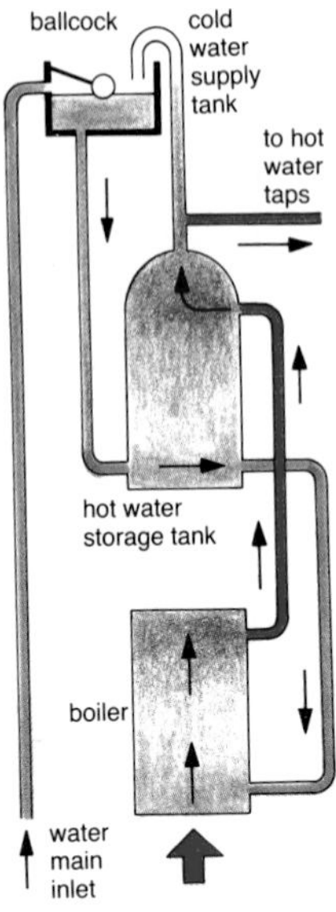

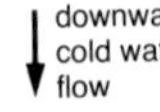

linear flow condition a condition achieved in a sample of heat conducting (↑) material, e.g. cylindrical copper bar or circular glass disc, when the temperature gradient (p.123) $d\theta/dl$ (Km^{-1}) across the sample is uniform; heat flow is parallel to the sample axis, and lagging (p.114) prevents heat escape through the exposed surfaces; it is a necessary condition for a uniform heat conduction rate (↑) $dQ/dt = KA\, d\theta/dl$ (Js^{-1}); for an unlagged bar in the steady state condition (↑) $d\theta/dl$ (Km^{-1}) is non-uniform.

thermal conductivity the proportionality constant K in the heat conduction rate (↑) equation $dQ/dt = KA\, d\theta/dl$ (Js^{-1}); units of K are watt/m/K ($Wm^{-1}K^{-1}$); the value of K indicates whether a specified material is a good or bad heat conductor (↑), e.g. for copper $K = 385$, for glass $K = 0.8$, for cork $K = 0.05$, for air $K = 0.02$ ($Wm^{-1}K^{-1}$).

thermal conductivity of a sample

linear heat flow under steady state conditions

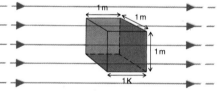

1m
1m
1m
1K

convection current the movement of a fluid (p.15) during heat transference (p.123) by convection (↑); fluid near the heat source, e.g. convector, hot body or surface, expands and rises as its density (p.7) decreases; colder, denser fluid moves downwards to replace it; a circulating fluid current is formed, transferring heat energy (p.28) by movement of fluid molecules, e.g. heating of air in a room, cooling of a hot body or surface.

natural convection the formation of convection currents (↑) around a hot body or surface, causing loss of heat; the cooling rate (Ks^{-1}; $Kmin^{-1}$) depends on the excess temperature (p.115), the exposed area and the type of surface, e.g. colour, texture, polish.

convection currents in water

potassium permanganate crystals slowly dissolve

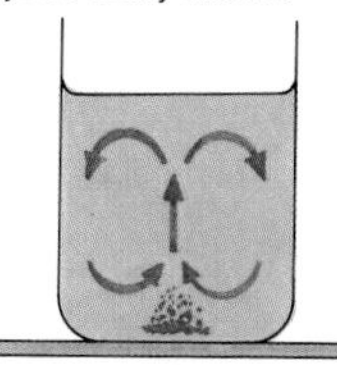

gentle heat

convection currents in air

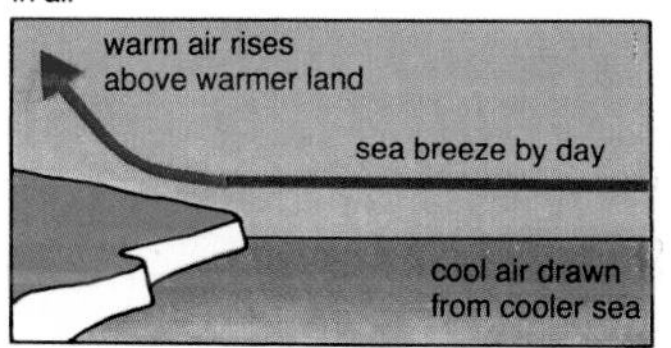

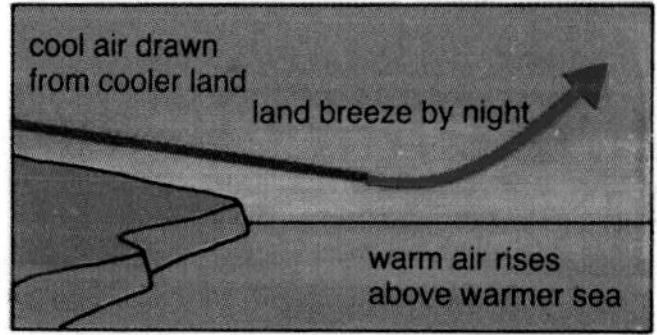

forced convection the cooling of a hot body or surface by a draught from an external source, in addition to natural convection effects (p.125); the cooling rate (Ks^{-1}; $K min^{-1}$) depends on the draught speed ($m s^{-1}$) as well as on the factors affecting natural convection; Newton's law of cooling (p.115) applies to forced convection.

Dewar vessel a double-walled glass vessel, with the space between its walls evacuated to prevent heat loss by convection (p.124); its walls are silvered, so that heat energy (p.28) radiated (p.124) across the vacuum from the vessel contents is repeatedly internally reflected between the walls, preventing heat loss by radiation; physical contact between the vessel walls and outer container is limited to the stopper and support, minimizing heat loss by conduction; it is used as a container for liquid gases, e.g. boiling liquid air at −196°C (77 K).

thermos flask a Dewar vessel (↑) of small, portable size.

greenhouse effect the way in which planetary atmospheres are heated by trapping long wavelength infrared radiation; short wavelength infrared and light radiation from the sun is transmitted through the atmosphere, is absorbed and re-radiated as long wavelength infrared radiation for which the atmosphere is not completely transparent; this causes the atmosphere to heat up to a higher temperature than otherwise would be the case; on Earth, water vapour and carbon dioxide in the atmosphere are responsible for the greenhouse effect; on Venus which has a dense atmosphere of carbon dioxide the surface temperature reaches 750 K due to the greenhouse effect.

greenhouse (*n*) an outdoor glass enclosure maintained above the environmental temperature by the greenhouse effect (↑).

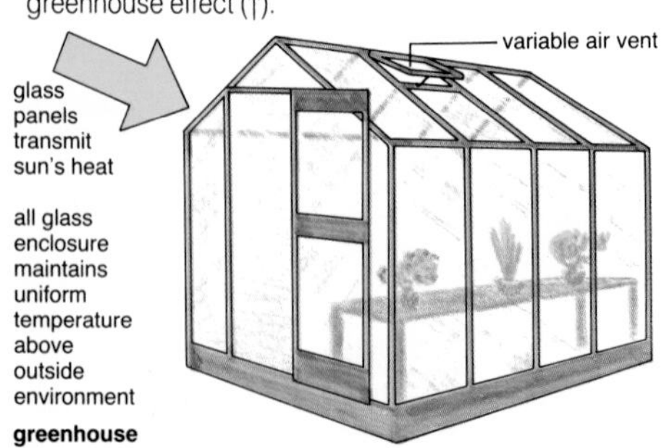

greenhouse

Dewar vessel
in outer container

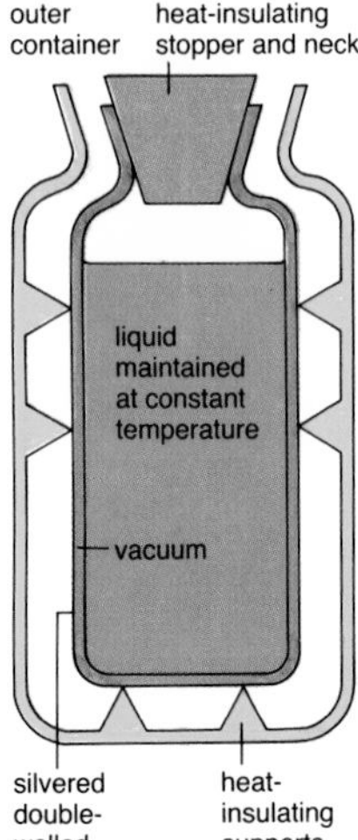

heat absorption bodies and substances receive heat (p.28) from other bodies and from surroundings at higher temperatures (p.116) by conduction (p.123), convection and radiation (p.124); heat received passes through the body's surface and becomes increased internal energy (p.13) of molecules within the body, whose temperature rises; the ability of a surface to absorb heat depends on its colour, texture and polish, e.g. a dull, rough, black surface absorbs heat better than a smooth, highly polished surface. **absorb** (*v*).

heat reflection refers to reflection by a surface of some of the heat energy (p.28) incident upon it; heat not absorbed (↑) or transmitted (p.52) is reflected; a good absorbing surface is a poor reflector, and a poor absorber is a good reflector.

thermopile (*n*) a detector of heat (p.28) and infrared (p.73) radiation; many thermocouples (p.163) are connected in series with their blackened hot junctions adjacent, to absorb (↑) the maximum amount of radiation; the combined thermoelectric e.m.f.s (p.163) produce an electric current detectable by a sensitive galvanometer.

blackbody absorber describes an object that absorbs (↑) all heat, infrared (p.73) and light (p.28) radiation incident (p.62) upon it, e.g. a hollow blackened cavity; a practical approximation is a surface blackened with unburnt carbon (soot) from a carbon-burning flame.

solar panel a device used for absorbing heat from the sun; it often consists of a hollow black metal container inside a glass-fronted panel; heat radiation is absorbed by the black surface and is conducted (p.124) through the metal: the inside of the panel is kept warm by the greenhouse effect (↑); water circulated through the metal container takes away the heat for use in domestic hot-water systems and for heating swimming pools.

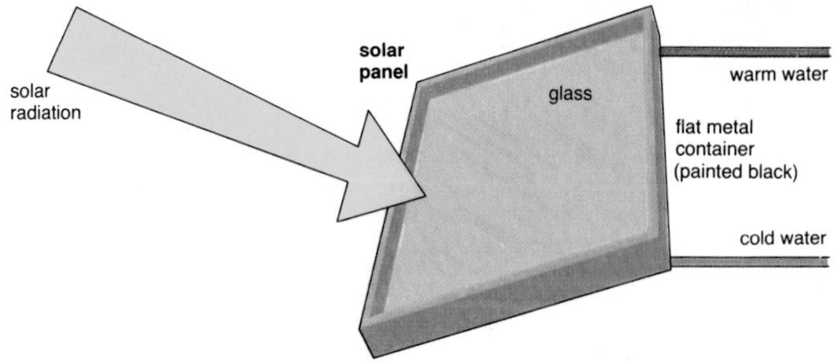

blackbody radiator describes an object that radiates (p.124) heat, infrared (p.73) and light (p.28) radiation of wavelengths and quantities of energy appropriate to its temperature (p.116), with the restriction imposed by the minimum wavelength of the radiation energy spectrum, given by the characteristic Lummer-Pringsheim curve (↓); a blackbody radiates energy according to Stefan's Law (↓).

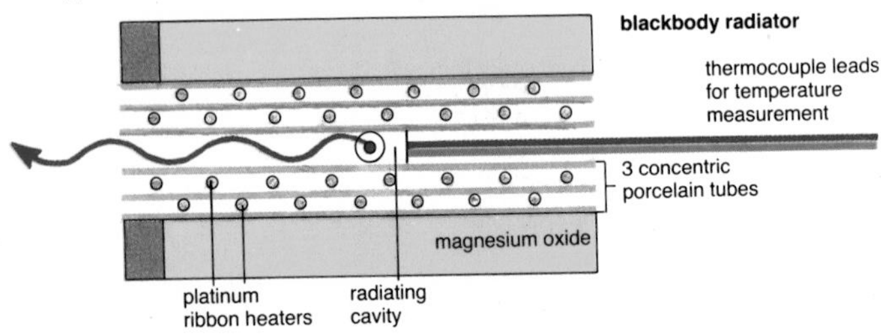

blackbody radiator

full radiator an alternative name for a blackbody radiator (↑).

imperfect black surface a surface radiating with the properties of a blackbody radiator (↑), but with energy at all wavelengths proportionally reduced.

coloured surface spectrum a heated surface radiates (p.124) the same colour wavelengths (p.54) as it absorbs (p.127) from incident white light (p.73); the characteristic Lummer-Pringsheim curve (↓) is modified, having only certain emitted wavelengths.

absorptive power[1] for absorption (p.127) of heat, infrared (p.73) and light (p.28) radiation energy by a specified surface: energy absorbed/s/m^2 ($Js^{-1}m^{-2}$) ÷ energy received/s/m^2 ($Js^{-1}m^{-2}$) is the ratio a_λ for wavelength λ, ratio has no units; for a blackbody radiator (↑) a_λ has maximum value = 1.

emissive power for emission of heat, infrared (p.73) and light (p.28) radiation energy from a specified surface at a given temperature (p.116): the energy emitted/s/m^2 ($Js^{-1}m^{-2}$) within a narrow wavelength range $d\lambda$ centred on λ is denoted by e_λ; it has maximum value E_λ for a blackbody radiator (↑); the total radiation energy E ($Js^{-1}m^{-2}$) emitted by the surface = area under the graph of the Lummer-Pringsheim (↓) curve for the specified surface temperature and is given by $E = \int_{\lambda_1}^{\lambda_2} e_\lambda . d\lambda = \sigma T^4$ (Wm^{-2}), where T = Kelvin temperature (p.145) and σ = Stefan's constant (↓).

Kirchoff's Law the emissive power e_λ (↑) is directly proportional to the absorptive power a_λ (↑) for the same specified surface at a given temperature, within a narrow wavelength range $d\lambda$ centred on λ;e_λ/a_λ = constant, and for a blackbody radiator (↑) a_λ = maximum value 1, so e_λ has maximum value E_λ (Wm^{-2}) for that temperature; a good radiation emitter of certain wavelengths is a good absorber but a bad reflector of those wavelengths, e.g. the line emission spectrum (p.74) of heated sodium vapour has 2 characteristic yellow lines of wavelength 589.0 nm and 589.6 nm; its line absorption spectrum (p.75) has dark lines at these specific wavelengths.

emissivity the ratio of the emissive power e_λ (↑) for a specified surface at a given temperature (p.116) ÷ emissive power E_λ for a blackbody radiator (↑) at the same temperature within the same narrow wavelength range $d\lambda$ centred on λ; denoted by $e = e_\lambda/E_\lambda$; it has no units; for a blackbody radiator e = 1.

Stefan's Law the total energy E (Wm^{-2}) emitted by a blackbody (↑) at Kelvin temperature T is given by $E = \sigma T^4$, where σ = Stefan's constant (↓).

Stefan's constant a constant of proportionality in Stefan's Law (↑); denoted by σ; its value is 5.75×10^{-8} watt/m²/K⁴ ($Wm^{-2}K^{-4}$).

Lummer-Pringsheim curves sets of graphs showing the distribution of heat, infrared (p.73) and light (p.28) radiation energy in the emission spectrum (p.74) of a blackbody radiator (↑) at temperatures (p.116) up to 2000°C; the relative energy intensity (p.57) is plotted against wavelength to give the characteristic curve for a specified temperature; the characteristic features of any specific curve are: 1. the position of the most probable wavelength λ_m corresponding to maximum energy intensity E_m, related to Kelvin temperature T (K) (p.145) by Wien's law: $\lambda_m T$ = constant; 2. the position of the threshold (p.105) wavelength λ_T of energy emission, decreasing at higher temperatures, so that hotter bodies emit increasing amounts of shorter wavelength radiation, e.g. visible red, white light spectrum colours, blue and ultraviolet; 3. total area under the characteristic curve, representing the total radiation energy present in the emission spectrum, increasing with Kelvin temperature T (K) according to Stefan's law (↑); Planck's formula forms the starting point of the quantum theory (p.234), which describes the experimental results for all wavelengths.

the distribution of energy in blackbody radiation spectrum at different temperatures

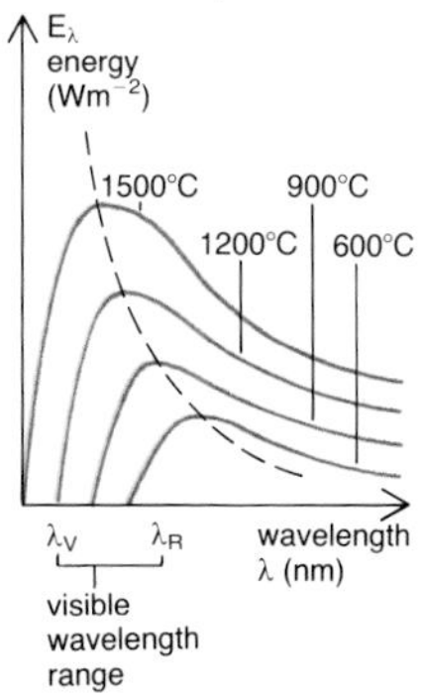

Lummer-Pringsheim curves

Planck's blackbody radiation curves alternative name for Lummer-Pringsheim (p.129) curves; Planck's formula describes the curves for all wavelengths and temperatures.

Prevost's theory of exchange this suggests that a blackbody radiator (p.128) at Kelvin temperature T_1 (K) (p.145), radiating heat and infrared (p.73) energy to another body or to surroundings at lower temperature T_2 (K), emits energy at a rate σT_1^4 (Wm^{-2}) and receives energy at a rate σT_2^4 (Wm^{-2}); net radiation energy lost $= \sigma(T_1^4 - T_2^4)$, according to Stefan's law (p.129); when $T_1 = T_2$, a temperature equilibrium state exists and energy radiated by the body = energy received from its surroundings.

infrared spectrometer an instrument similar in function to the optical spectrometer (p.88), but with components changed or modified for use at infrared (p.73) and heat (p.28) radiation wavelengths; energy from a blackbody radiator (p.128) source, whose temperature is measured by a total radiation pyrometer (p.121), passes through a slit wavelength selector and is collimated (p.88) by a parabolic mirror (p.64); the parallel radiation beam is dispersed (p.73) by a quartz or rock salt prism, and focused (p.62) by a parabolic mirror on to a thermopile (p.127) detector replacing an optical telescope; glass lenses and prisms are not transparent to infrared and heat.

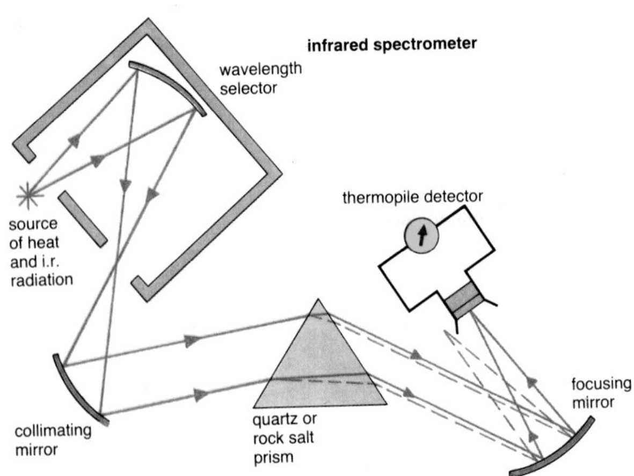

infrared spectrometer

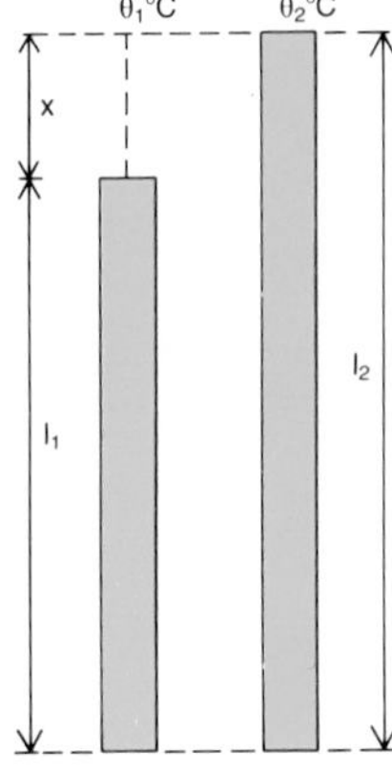

$x = l_1\lambda\theta$ where $\theta = (\theta_2 - \theta_1)$
$l_2 = l_1 + l_1\lambda\theta = l_1(1 + \lambda\theta)$
mean linear expansivity

thermal expansion (*n*) the increase in dimensions (p.12) of a physical system due to heating which increases its internal energy (p.13); molecules become further apart, due to their increased kinetic energy. **expand** (*v*), **expanding** (*adj*), **expanded** (*adj*).

thermal contraction contraction effects due to cooling which reduces the internal energy of a physical system.

linear expansion refers to the thermal expansion (↑) for solids considered in 1 dimension only, because other dimensions are relatively small, e.g. expansion of metal bar or rod, railway lines.

mean linear expansivity a measure of linear expansion (↑); the change in length e (m) per unit of original length l_o (m) at 0°C (273 K) per unit of temperature interval (K) (p.116); it is denoted by λ; with units of K^{-1}: $\lambda = e/l_o\theta$, where θ = temperature change (K); for lengths l_1 and l_2 (m) at temperatures θ_1 and θ_2 (°C), the approximate relation is: $\lambda = (l_2 - l_1)/l_1\theta$, where $\theta = (\theta_2 - \theta_1)$ (K), and $l_2 = l_1(1 + \lambda\theta)$; values of λ are very small, e.g. the average for metals is $18 \times 10^{-6}K^{-1}$, for glass $8 \times 10^{-6}K^{-1}$.

coefficient of linear expansion an alternative name for mean linear expansivity (↑).

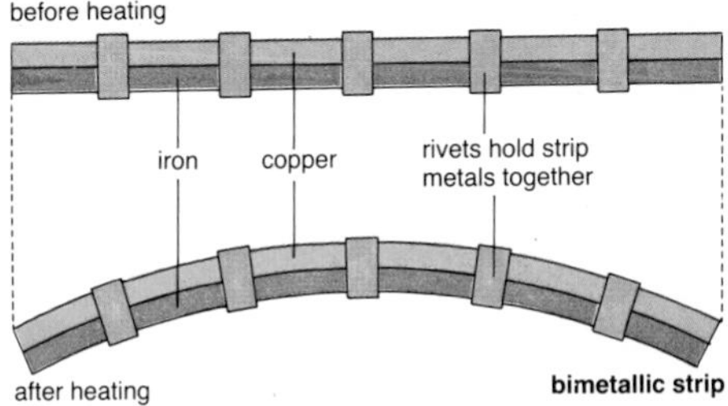

bimetallic strip

bimetallic strip a strip is made from equal lengths of 2 metals of different mean linear expansivity (↑), usually brass and steel with λ approximately 18×10^{-6} and $12 \times 10^{-6}K^{-1}$, welded together at room temperature (p.139); on heating or cooling, the difference in expansion (↑) causes the strip to bend; it is used as a switching device in a thermostat (p.132), in automatic flasher units, in the compensated balance wheel of a watch and as an automatic zero adjuster in the sensitive millivoltmeter recording thermoelectric e.m.f. in a thermoelectric thermometer (p.120).

thermostat (*n*) a device for maintaining an externally heated physical system at constant temperature (p.116); it operates using a bimetallic strip (p.131) to switch off the heat supply when the required temperature is reached and to switch it on again when the system falls below this temperature, e.g. electric iron, central heating system; a gas thermostat uses thermal expansion (p.131) effects to control gas supply to burners. **thermostatic** (*adj*).

thermostat

adjustment knob

insulating support block

contacts

bimetallic strip

to heater circuit

electrically controlled thermostat

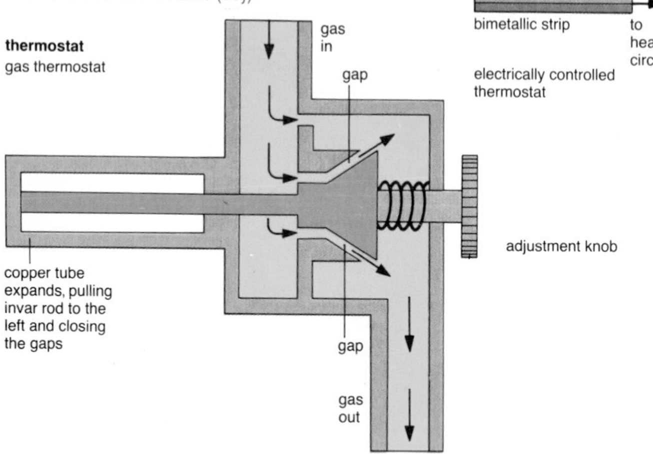

thermostat
gas thermostat

temperature compensation the metallic parts controlling the time-keeping mechanisms of clocks and watches are affected by thermal expansion (p.131) effects due to temperature changes; the different expansion effects of 2 metals, usually brass and steel with λ approximately 18×10^{-6} and $12 \times 10^{-6} K^{-1}$, can be used to compensate for this and maintain correct time keeping, e.g. the compensated balance wheel of a watch, compensated clock pendulum. **compensate** (*v*), **compensating** (*adj*), **compensated** (*adj*).

superficial expansion thermal expansion (p.131) for solids considered in 2 dimensions only, because the other dimension is relatively small, e.g. expansion of a metal plate on heating.

surface expansion an alternative name for superficial expansion (↑).

area expansion an alternative name for superficial expansion (↑).

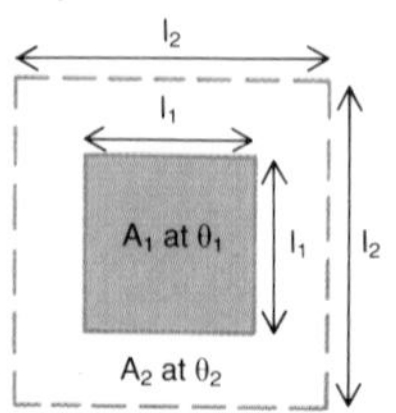

$A_2 = l_2^2 = l_1^2(1 + \lambda\theta)^2$
$A_2 \simeq A_1(1 + 2\lambda\theta)$
where $\theta = (\theta_2 - \theta_1)$ (K)

superficial expansion

mean superficial expansivity the measure of superficial expansion (↑); the change in area a (m^2) per unit of original area A_o (m^2) at 0°C (273K) per unit of temperature interval (K) (p.116); its value is approximately 2λ, where λ is the mean linear expansivity (p.131); it is measured in units of K^{-1}: $2\lambda = a/A_o\theta$, where θ = temperature change (K); for areas A_1 and A_2 (m^2) at temperatures θ_1 and θ_2 (°C), the approximate relation is: $2\lambda = (A_2 - A_1)/A_1\theta$, where $\theta = (\theta_2 - \theta_1)$ (K), and $A_2 = A_1(1 + 2\lambda\theta)$; the average value of 2λ for metals is $36 \times 10^{-6}K^{-1}$, for glass it is $16 \times 10^{-6}K^{-1}$.

coefficient of superficial expansion an alternative name for mean superficial expansivity (↑).

cubical expansion

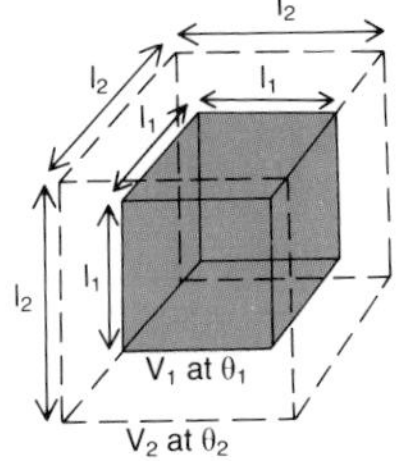

$V_2 = l_2^3 = l_1^3(1 + \lambda\theta)^3$
$v_2 \simeq V_1(1 + 3\lambda\theta)$
where $\theta = (\theta_2 - \theta_1)$ (K)

cubical expansion the thermal expansion (p.131) for solids, liquids and gases considered in 3 dimensions, e.g. the expansion of a metal cube on heating, the apparent expansion (↓) of a sample of liquid or gas; for liquids and gases, linear (p.131) and superficial expansion (↑) is not practically important.

volume expansion an alternative name for cubical expansion (↑).

mean cubical expansivity the measure of cubical expansion (↑); the change in volume v (m^3) per unit of original volume V_o (m^3) at 0°C (273K) per unit of temperature interval (K) (p.116); its value is approximately 3λ, where λ is the mean linear expansivity (p.131) measured in units of K^{-1}; $3\lambda = v/V_o\theta$, where θ = temperature change (K); for volumes V_1 and V_2 (m^3) at temperatures θ_1 and θ_2 (°C), the approximate relation is: $3\lambda = (V_2 - V_1)/V_1\theta$, where $\theta = (\theta_2 - \theta_1)$ (K), and $V_2 = V_1(1 + 3\lambda\theta)$; the average value of 3λ for metals is $54 \times 10^{-6}K^{-1}$, for glass it is $24 \times 10^{-6}K^{-1}$.

coefficient of cubical expansion an alternative name for mean cubical expansivity (↑).

apparent expansion the amount by which a liquid or gas appears to expand when heated in its containing vessel; it does not take account of the expansion of the vessel.

mean apparent expansivity for a fixed mass of liquid, the apparent change in volume v_a (m^3) per unit of original volume V_o (m^3) at 0°C (273K) per degree K; $\gamma_a = v_a/V_o\theta$, where θ = temperature change (K).

real expansion the real amount by which a liquid or gas expands when heated in its containing vessel; it is greater than apparent expansion (↑) and takes account of the expansion of the vessel.

absolute expansion an alternative name for real expansion (p.133).

mean absolute expansivity the measure of absolute expansion (↑) for a fixed mass of liquid, it is the real change in volume v (m^3) per unit of original volume V_o (m^3) at 0°C (273K) per unit of temperature interval (K) (p.116); denoted by γ and measured in units of K^{-1}; $\gamma = v/V_o\theta$, where θ = temperature change (K); for volumes V_1 and V_2 (m^3) at temperatures θ_1 and θ_2 (0°C), the approximate relation is: $\gamma = (V_2 - V_1)/V_1\theta$, where $\theta = (\theta_2 - \theta_1)$ (K), and $V_2 = V_1(1 + \gamma\theta)$; values of γ for liquids are higher than values of the mean cubical expansivity (p.133) for solids and vary with temperature, e.g. mercury $18.2 \times 10^{-5}K^{-1}$, aniline $85 \times 10^{-5}K^{-1}$, paraffin oil $90 \times 10^{-5}K^{-1}$ from 0°C to 100°C; for mean apparent expansivity γ_a (p.133) for a liquid: $\gamma = \gamma_a + 3\lambda$, where 3λ = mean cubical expansivity for the containing vessel, e.g. for parrafin oil $\gamma_a = \gamma - 3\lambda = (90 - 2.4) \times 10^{-5} = 87.6 \times 10^{-5} K^{-1}$; the volume coefficient (↓) for a fixed mass of gas at constant pressure is similarly defined.

coefficient of absolute expansion an alternative name for mean absolute expansivity (↑).

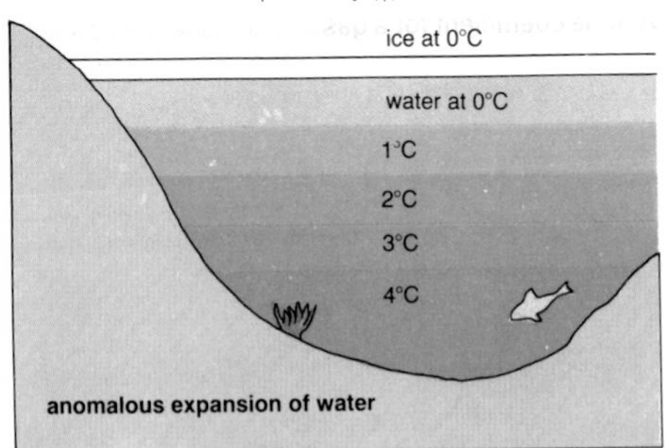

anomalous expansion of water

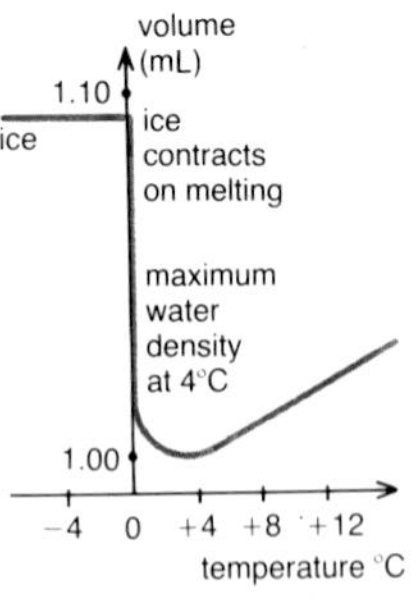

anomalous expansion of water a fixed mass of water does not expand uniformly on heating from 0°C to 100°C, but has a minimum volume, and maximum density (p.7), at 4°C; water at temperatures between 0°C and 4°C floats on top of water at 4°C, so that exposed water surfaces outdoors freeze from the top downwards, leaving lower layers unfrozen and aquatic life viable.

Hope's apparatus demonstrates the anomalous expansion of water (↑).

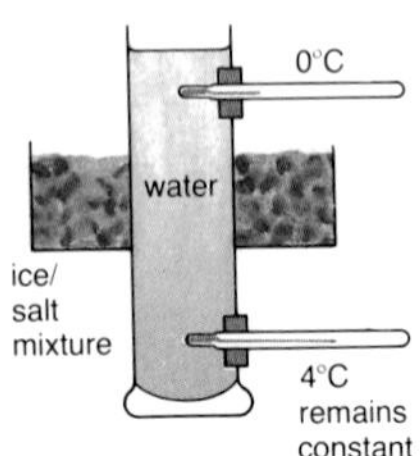

Hope's apparatus

Charles' Law

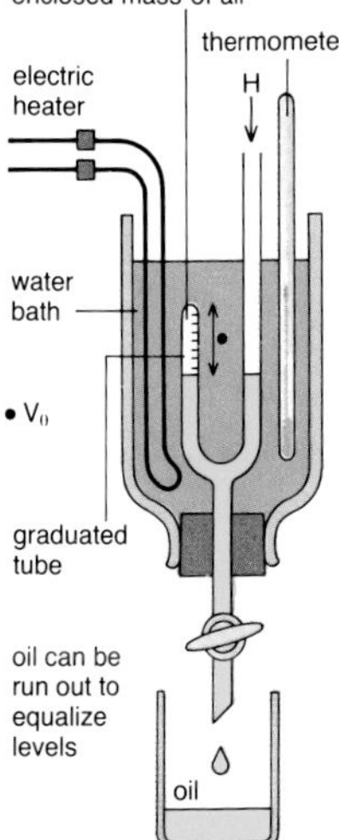

gas expansion the real expansion (p.133) of a gas or vapour (p.16) on increase in temperature (p.116) and/or decrease in pressure (p.39); unlike solids and liquids (p.15), gases and vapours are readily compressible, and the variation in volume V (m^3) of a fixed mass of gas depends on the variation in both gas pressure p (Nm^{-2}) (p.39) and temperature T (Kelvin) (p.145); the variables p, V and T are interdependent and, if any one is kept constant, experiment shows a relation exists between the other 2, giving the 3 gas laws (p.138).

mean expansivity of a gas refers to the measure of gas expansion (↑); for a fixed mass of gas at constant pressure p (p.39), under conditions of ideal behaviour (p.138), it is the real change in volume v (m^3) per unit of original volume V_o (m^3) at 0°C (273K) per unit of temperature interval (K) (p.116); denoted by α_p and measured in units of K^{-1}; $\alpha_p = v/V_o\theta$, where θ = temperature change (K); for volumes V_θ and V_o (m^3) at temperatures θ_1 and 0 (°C), where θ_1°C = (θ +273)K and 0°C = 273K, the relation is: $\alpha_p = (V_\theta - V_o)/V_o\theta$, where $\theta = (\theta_1 - 0)$ (K), and $V_\theta = V_o(1 + \alpha_p\theta)$; the value of α_p is 1/273K^{-1} at constant pressure for all gases.

volume coefficient for a gas an alternative name for the mean expansivity α_p (↑) of a gas at constant pressure.

Charles' Law relates the variation in volume V (m^3) with temperature θ (°C) (p.116), for a fixed mass of gas at constant pressure p (Nm^{-2}) (p.39) under conditions of ideal behaviour (p.138): volume V varies by 1/273 of its original value V_o (m^3) at 0°C (273K) for every degree Centigrade (p.119) change in temperature θ; for volume V_θ (m^3) at temperature θ (°C), or (θ + 273)K, the relation is: $(V_\theta - V_o)/V_o\theta = 1/273 = \alpha_p$ (K^{-1}), where α_p is mean expansivity (↑) of the gas; so $V_\theta = V_o(1 + \alpha_p\theta)$, and the graph of V_θ against θ is linear.

Charles' Law relationship

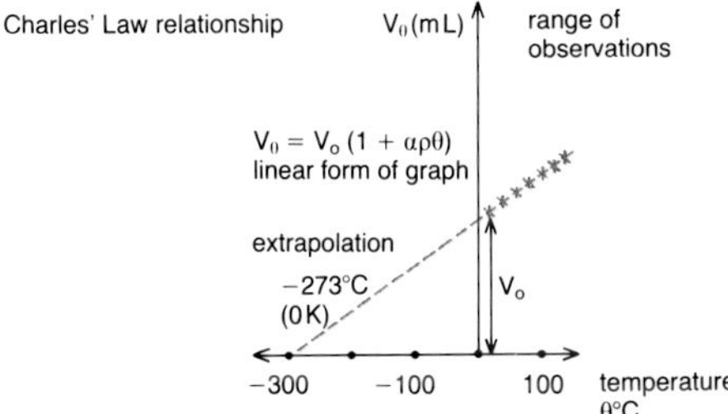

pressure coefficient for gas for a fixed mass of gas at constant volume V (m^3), under conditions of ideal behaviour (p.138), it is the change in pressure p (Nm^{-2}) (p.39) per unit of original pressure p_o (Nm^{-2}) at 0°C (273K) per unit of temperature interval (K) (p.116); denoted by α_v with units of K^{-1}; $\alpha_v = p/p_o\theta$, where θ = temperature change (K); for pressures p_θ and p_o (Nm^{-2}) at temperatures θ_1 and 0 (°C), where θ_1°C = (θ + 273)K and 0°C = 273K, the relation is: $\alpha_v = (p_\theta - p_o)/p_o\theta$, where $\theta = (\theta_1 - 0)$ (K), and $p_\theta = p_o(1 + \alpha_v\theta)$; the value of α_v is $1/273K^{-1}$ at constant volume for all gases; $\alpha_v = \alpha_p = 1/273K^{-1}$, where α_p = volume coefficient (p.135) for gas at constant pressure.

pressure law this relates the variation in pressure p (Nm^{-2}) (p.39) with temperature θ (°C) (p.116), for a fixed mass of gas at constant volume V (m^3) under conditions of ideal behaviour (p.138); pressure p varies by 1/273 of its original value p_o (Nm^{-2}) at 0°C (273K) for every degree Centigrade (p.119) change in temperature θ; for pressure p_θ (Nm^{-2}) at temperature θ (°C); or (θ + 273)K, the relation is: $(p_\theta - p_o)/p_o\theta = 1/273 = \alpha_v$ (K^{-1}), where α_v is the pressure coefficient (↑) for the gas; so $p_\theta = p_o(1 + \alpha_v\theta)$, and the graph of p_θ against θ is linear; the relation can also be expressed as a proportionality when temperatures are stated on the Kelvin scale (p.145) with respect to zero at −273°C: $p_T/p_o = T/273$, where p_T = pressure at TK, and TK = (θ + 273); in general $p_T \propto T$ and $p_1/p_2 = T_1/T_2$ for pressures at T_1K and T_2K.

pressure law apparatus

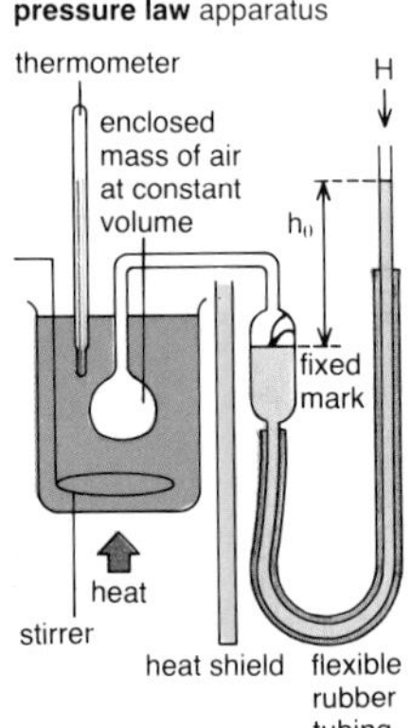

readings of h_0 and θ taken with mercury at fixed mark:
p_0 = (H + h_0) (mmHg)
H = atmospheric pressure (mmHg)

pressure law relationship

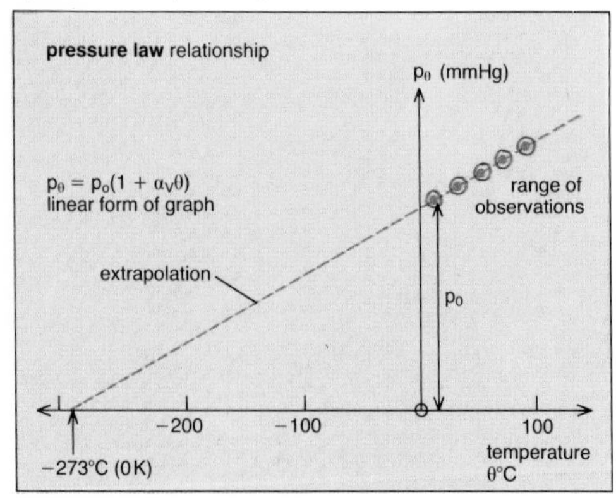

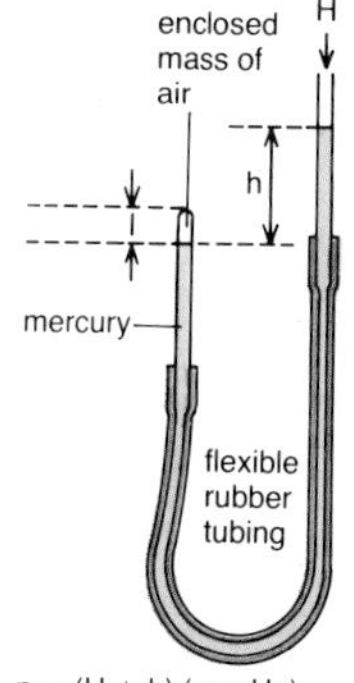

p = (H + h) (mm Hg)

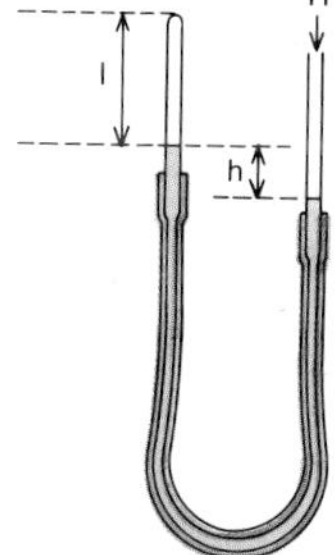

p = (H – h) (mm Hg)

Boyle's Law apparatus

absolute zero of temperature extrapolation of Charles' Law (p.135) and the pressure law (↑) graphs gives intercepts at −273°C on the temperature axes, corresponding to a fixed mass of gas with zero volume and pressure, having zero molecular vibrational energy (p.28); all gases liquefy (p.16) and solidify (p.15) above −273°C, but this point is made the theoretical origin 0 (zero) K of an absolute scale of temperature (↓) on which all Celsius scale temperatures θ can be given values; θ°C = (θ + 273) K.

absolute scale of temperature the scale of temperature (p.116) having its zero at the absolute zero of temperature −273°C (0 K) (↑) and fixed points at the freezing and boiling points (p.16) of pure water at standard atmospheric pressure 760 mm (p.40); it is also called the ideal gas scale (p.117) because the practical means of realizing the scale is the standard constant volume gas thermometer (p.117) with its gas under conditions for ideal behaviour (p.138); the scale is now replaced by the Kelvin thermodynamic scale (p.145), with which its temperatures coincide; Celsius scale (p.119) temperature θ = (θ + 273) K, e.g. 0°C = 273 K; 100°C = 373 K.

Boyle's Law relates the variation in pressure p ($\mathrm{N\,m^{-2}}$) (p.39) with volume V (m^3), for a fixed mass of gas at constant temperature (p.116) under conditions of ideal behaviour (p.138): pressure p is inversely proportional to volume V; $p \propto 1/V$, so pV = constant and the graph of p against 1/V is linear.

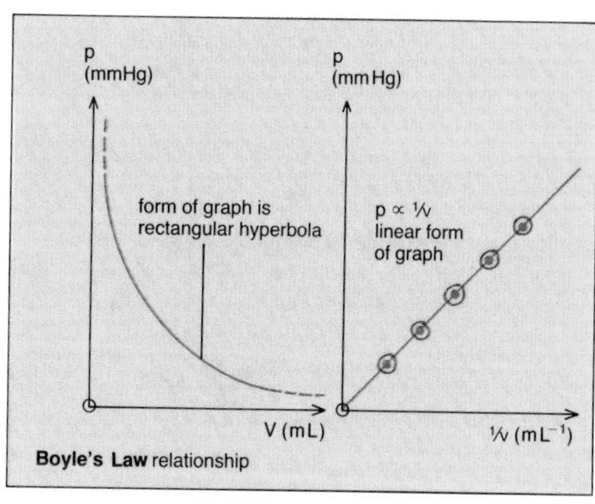

Boyle's Law relationship

gas laws refer to laws describing gas behaviour under ideal behaviour (↓) conditions of high temperature (p.116) and low pressure (p.39), when intermolecular attractive forces of cohesion (p.14) are negligible; they include Boyle's Law (p.137), Charles' Law (p.135) and the pressure law (p.136) as the basis for equations of state (↓); and also Avogadro's Law (↓), Dalton's Law of partial pressure (p.17) and Graham's Law of diffusion; the gas laws do not accurately describe gas behaviour near the liquid state (p.15), or in the critical state (p.140) and other equations of state must be applied under these conditions.

ideal gas behaviour gas behaviour conforming with the gas laws (↑), e.g. for oxygen, nitrogen, hydrogen and helium under normal laboratory conditions, when they are far from the liquid state and intermolecular attractive forces of cohesion (p.14) are negligible; their boiling points (p.16) at standard atmospheric pressure 760 mm Hg (p.40) are −183°C (90 K), −196°C (77 K), −252°C (21 K) and −269°C (4 K) respectively.

non-ideal gas behaviour gas behaviour not conforming with the gas laws (↑), e.g. for sulphur dioxide, ammonia, carbon dioxide and ethylene, under normal laboratory conditions, when they are too near the liquid state (p.15) for intermolecular attractive forces of cohesion to be negligible; their boiling points (p.16) at standard atmospheric pressure 760 mm Hg (p.40) are −10.8°C (262.2 K), −33.5°C (239.5 K), −78.2°C (194.8 K) and −102.7°C (170.3 K) respectively.

real gases former name for gases showing non-ideal behaviour (↑) under normal laboratory conditions (↓).

Avogadro's Law equal volumes of different gases (p.15) contain the same number of molecules when under the same conditions of temperature (p.116) and pressure (p.39).

Avogadro's hypothesis an alternative name for Avogadro's Law (↑).

standard temperature and pressure (s.t.p.) defines reference conditions of standard temperature 0°C (273 K) (p.116) and standard atmospheric pressure 760 mm Hg (p.40); used in quoting data about physical properties of matter varying with temperature, e.g. gas densities (p.7), for hydrogen 0.09 kg m^{-3} at s.t.p.

normal temperature and pressure (n.t.p.) an alternative name for standard temperature and pressure (↑); not now used.

ideal gas behaviour

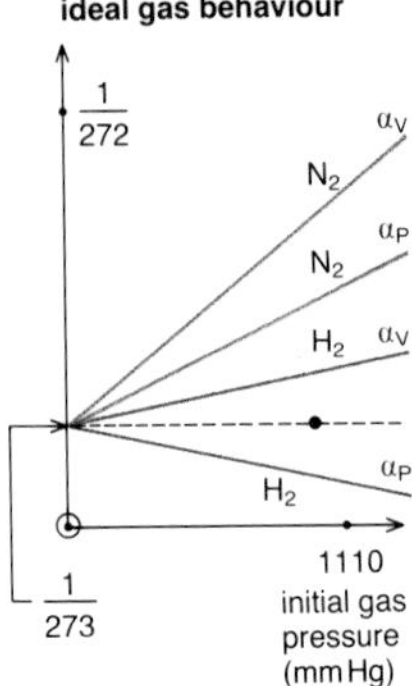

• ideal gas $\alpha_P = \alpha_V = 1/273\ (K^{-1})$

variation in pressure and volume coefficients, α_V and α_P (K^{-1}), for nitrogen and hydrogen: at very low pressure all tend to 1/273 (K^{-1})

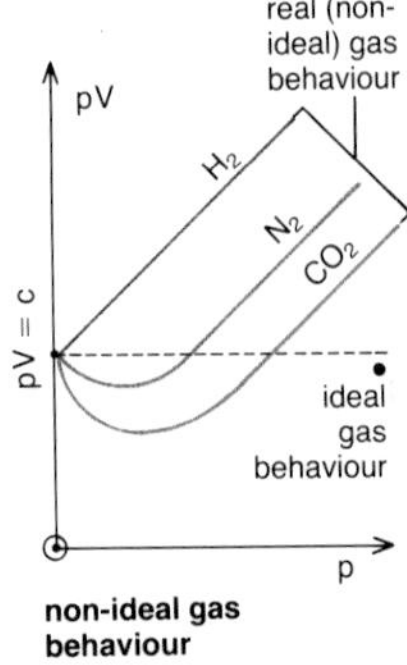

non-ideal gas behaviour

normal laboratory conditions refer to experimental conditions easily reproducible in a laboratory without specialized apparatus or equipment; usually refer to what is practicable in North-Central European latitudes, e.g. conditions near standard temperature and pressure (↑).

room temperature an approximate average value 15°C in U.K. climatic conditions.

equation of state the relation between pressure p (Nm^{-2}) (p.39), volume V (m^3) and temperature T (K) (p.116) of a fixed mass of gas; it should account for both ideal gas behaviour (↑), e.g. general gas equation (↓), and non-ideal gas behaviour (↑), e.g. Van der Waals' equation (↓).

general gas equation the equation of state (↑) accounting for ideal gas behaviour (↑): pV/T = constant r for 1 kilogram of gas, where r = gas constant for 1 kg of a specified gas; for m kg of a specified gas: pV/T = mr; pV/T = constant R for 1 mole (p.11) of any gas, where R is the molar gas constant (↓); R = r/M, for a gas of molecular weight M.

molar gas constant the constant R for 1 mole of gas in the general gas equation (↑): pV/T = R; under conditions of standard temperature and pressure (↑) $p = 1.013 \times 10^5$ (Nm^{-2}) (760 mm Hg), $V = 22.4 \times 10^{-3}$ (m^3) and T = 273K, so R = 8.31 joule/mole/K ($J mol^{-1} K^{-1}$); 1 mole (p.11) of any gas occupies 22.4×10^{-3} (m^3) at s.t.p.; formerly known as **universal gas constant**.

van der Waals' equation an equation of state (↑) accounting for ideal (↑) and non-ideal gas behaviour (↑): $(p + a/V^2)(V - b) = rT$ for 1 kilogram of a specified gas, where a and b are constants; b = a constant depending on the volume of the gas molecules at their closest possible packing, approximately 4 × volume of molecules; a = a constant accounting for the effect of intermolecular attractive forces of cohesion (p.14); this equation has cubic form and can be related to Andrews' isothermals (p.140) below the critical temperature.

isothermal process a process carried out at constant temperature (p.116), e.g. variation in pressure with volume of a fixed mass of gas according to Boyle's Law (p.137), changes of state (p.14); ideally the process should be very slow with perfect heat exchanges between the physical system and its surroundings.

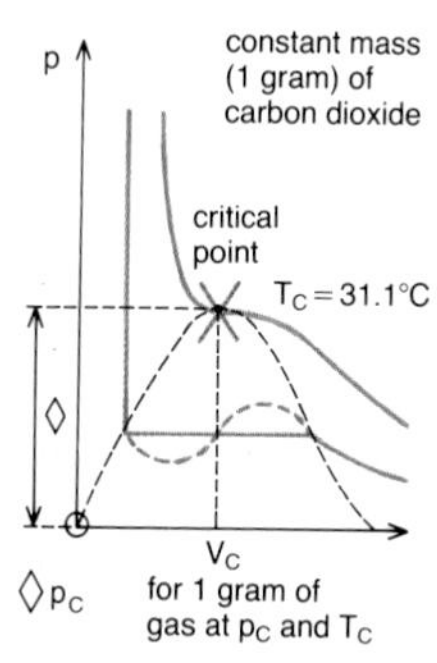

van der Waals' equation
$(p + a/V^2)(V - b) = rt$
expands to cubic equation in V; critical and subcritical isothermals show cubic form

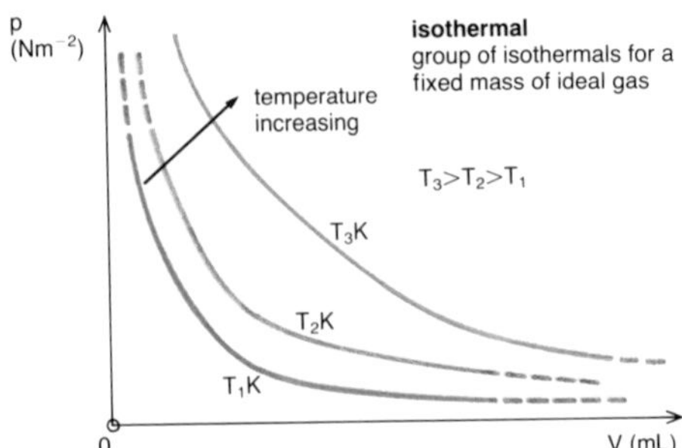

isothermal
group of isothermals for a fixed mass of ideal gas

$T_3 > T_2 > T_1$

isothermal (*n*) a graph of the variation of pressure p (Nm^{-2}) (p.39) with volume V (m^3) of a fixed mass of substance during an isothermal process (p.139). **isotherm** (*n*), **isothermal** (*adj*).

Andrews' isothermals a series of experiments made on carbon dioxide gas over the approximate pressure range 1 to 100 atmospheres (p.18) and temperature range 0 to 50°C, to investigate non-ideal gas behaviour (p.138) near the liquid state; they predict a characteristic critical temperature (↓) for every gas, above which it cannot be liquefied (p.16) by compression.

critical temperature the characteristic temperature (p.116) for a specified gas, above which it cannot be liquefied (p.16) by compression; denoted by T_c; for carbon dioxide $T_c = 31.1$°C.

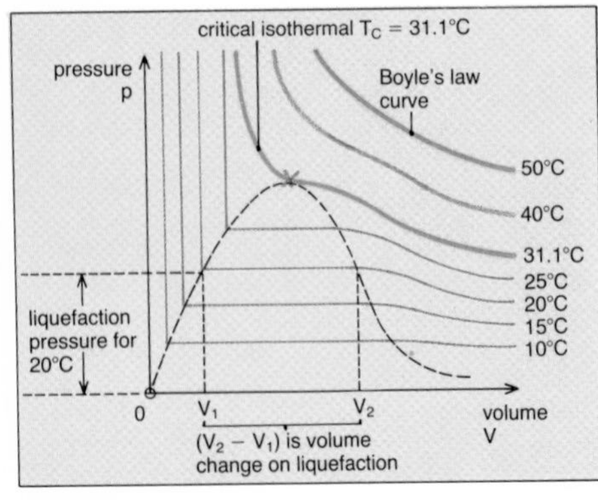

Andrew's isothermals
for a fixed mass of carbon dioxide (approximate temperatures except T_C)

Joule-Kelvin effect when a gas at room temperature (p.139) is made to expand through a narrow opening its temperature falls; for hydrogen and helium the Joule-Kelvin cooling effect is replaced by a slight heating effect on expansion; the effect depends on the inversion temperature (↓) of the gas.

inversion temperature the characteristic temperature for a specified gas above which the Joule-Kelvin cooling effect (↑) changes to a heating effect; denoted by T_I: for hydrogen $T_I = -83°C$ (190K) and for helium $T_I = -240°C$ (33K), so on expansion at room temperature (p.139) a slight heating effect occurs; for all other gases $T_I > 0°C$ (273K) so a cooling effect occurs.

refrigeration (*n*) process by which internal heat energy (p.28) is continuously removed from a physical system or substance already at a temperature (p.116) below that of its surroundings; the working substance or refrigerant of the refrigerator is a vapour (p.16), whose critical temperature T_c (↑) is above room temperature under normal laboratory conditions (p.139), e.g. freon CCl_2F_2; it is liquefied by compression and then made to evaporate (p.15) under reduced pressure, withdrawing latent heat of vaporization (p.16) partly from molecular vibrational energy (p.13) of its own molecules and then from the refrigerated compartment and its contents. **refrigerator** (*n*), **refrigerant** (*n*), **refrigerate** (*v*).

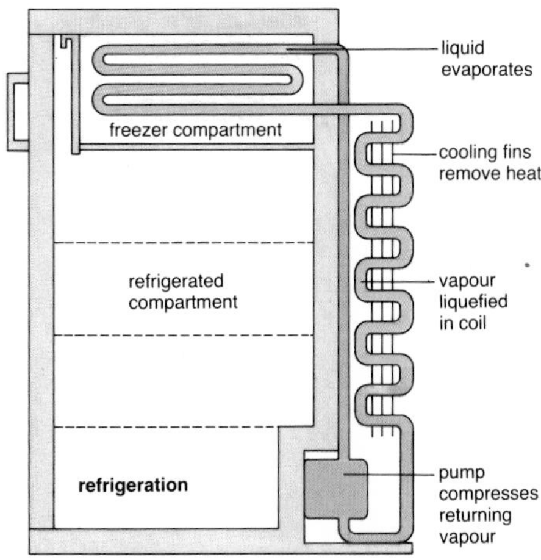

refrigeration

heat pump (*n*) a device for extracting heat from large quantities of material, such as air or water, at a low temperature and supplying it at a higher temperature; the principle is similar to that of the refrigerator; work must be done to operate the process.

adiabatic process a process carried out without heat being exchanged or lost between a physical system and its surroundings, e.g. rapid compression and expansion of gaseous fuel during the working cycle of the internal combustion engine (p.146); during the process, pressure p (Nm^{-2}) (p.39), volume V (m^3) and temperature T (K) (p.116) of a fixed mass of gas under conditions of ideal behaviour (p.138), are related: pV^{γ} = constant and $TV^{\gamma-1}$ = constant, where $\gamma = C_p/C_v$ and C_p and C_v are the principal specific heat capacities (↓) of the gas; the process must be very rapid and the physical system lagged to prevent heat exchanges.

adiabatic (*n*) a graph of the variation of pressure with volume for a fixed mass of gas during an adiabatic process (↑).

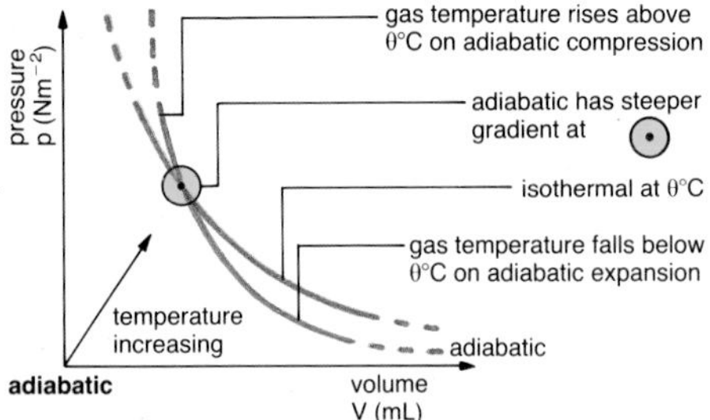

adiabatic

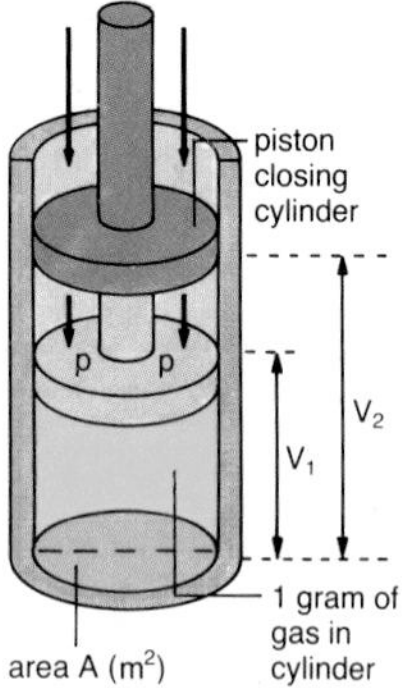

specific heat capacity of a gas

For 1 kg of a specified gas, heated through 1 K at constant volume V_1: C_V is its specific heat capacity. For 1 kg of the gas, heated through 1 K at constant pressure p, the gas expanding to V_2: C_p is its specific heat capacity. For a very small change, $(V_2 - V_1) = \delta V$ and the work done on expansion against p is $p.\delta V$ (joule).

specific heat capacity of a gas the quantity of heat energy (p.28) required to raise the temperature (p.116) of 1 kilogram (kg) of a specified gas (p.15) by 1 degree Kelvin (1 K) (p.145), under specified conditions of volume and pressure (p.39); measured in units of joule/kg/K ($Jkg^{-1}K^{-1}$); it can have an infinite (p.67) number of values lying between 2 extreme values for constant volume conditions (C_V) and constant pressure conditions (C_p); C_v and C_p are the principal specific heat capacities for the gas; C_p always exceeds C_v since the gas expands during heating under constant pressure conditions, keeping its pressure constant and requiring extra heat energy to do the work of expansion against the external pressure.

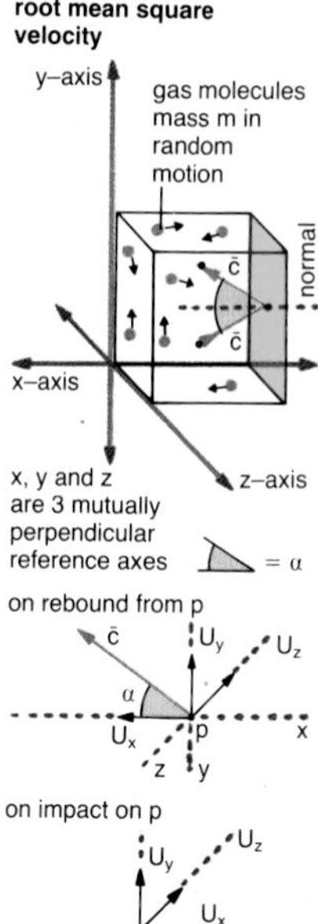

kinetic theory of matter a theory that atoms and molecules have molecular vibrational energy (p.13) for matter in the solid and liquid states (p.15) and translational kinetic energy (p.13) for matter in the gaseous state (p.15); practical supporting evidence is provided by Brownian movement (p.14).

kinetic theory of gases a theory developed in the 19th century from the kinetic theory of matter (↑) applied to gases; atoms and molecules in a gas (p.15) are considered to be in continuous random movement (p.14), having translational kinetic energy $\frac{1}{2}mv^2$ (joule) (p.13) and momentum mv ($kg\,m\,s^{-1}$), where m = particle mass (kg) and v = particle velocity ($m\,s^{-1}$) (p.30); the internal energy of the gas (p.15) depends only on the gas temperature (p.116); gas pressure (p.39) is caused by momentum (p.37) change on bombardment of the containing surfaces by gas molecules; intermolecular collisions are assumed to be elastic (p.38) and of very short duration, compared with the time interval between collisions; the conditions are assumed to be those of ideal gas behaviour (p.138), so intermolecular attractive forces and volume of the molecules are both assumed to be negligible.

root mean square (r.m.s.) velocity the square root of the mean square velocity (p.30) of all molecules in a gas (p.15) sample denoted by $\bar{c}$, e.g. for 3 molecules with velocities u_x, u_y and u_z in 3 mutually perpendicular directions: $\bar{c} = \sqrt{(u_x^2 + u_y^2 + u_z^2)/3}$; it is used in the kinetic theory of gases (↑) as a convenient way of representing the mean velocity of all molecules in a gas sample at a particular temperature, as shown by the Maxwell distribution curves, e.g for oxygen molecules at standard temperature and pressure (p.138) $\bar{c}$ = 450 $m\,s^{-1}$.

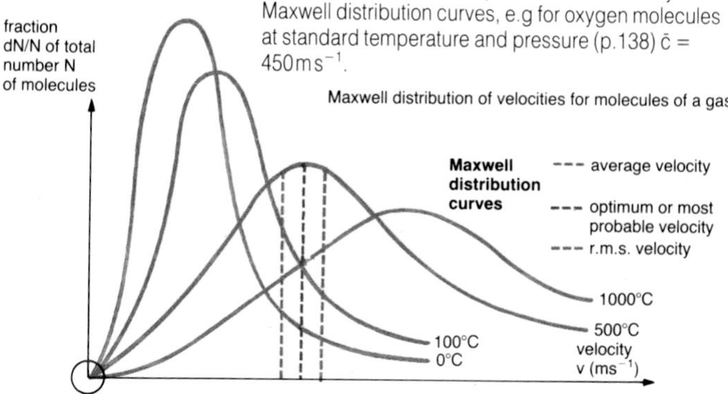

Maxwell distribution of velocities for molecules of a gas

gas pressure equation the relation derived between gas pressure p (Nm^{-2}) (p.39) and r.m.s. velocity $\bar{c}$ (ms^{-1}) (p.143), using the kinetic theory of gases (p.143): $pV = \frac{1}{3} nm\bar{c}^2$, where V = gas volume ($m^3$), n = number of molecules in the gas sample, m = mass per molecule (kg); an alternative form of the relation is: $p = \frac{1}{3} \rho\bar{c}^2$, where ρ = gas density (kgm^{-3}) (p.7) at the pressure and temperature (p.116) of the gas sample, from which $\bar{c} = \sqrt{3p/\rho}$; $pV = \frac{1}{3} nm\bar{c}^2 = \frac{2}{3} n.\frac{1}{2} m\bar{c}^2$, so $pV \propto \frac{1}{2} m\bar{c}^2 \propto kT$, where k = constant and T = Kelvin temperature (p.145).

diffusion (*n*) movement of solute molecules from a region of high concentration towards a region of low concentration in a liquid solution, giving a mixture of uniform composition, e.g. copper sulphate in water; for 2 different gases in contact, movement of molecules occurs in both directions until the mixture has uniform concentration of both gases. **diffuse** (*v*), **diffusing** (*adj*), **diffused** (*adj*).

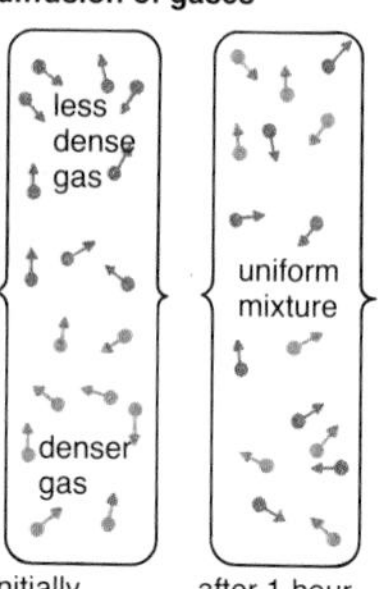

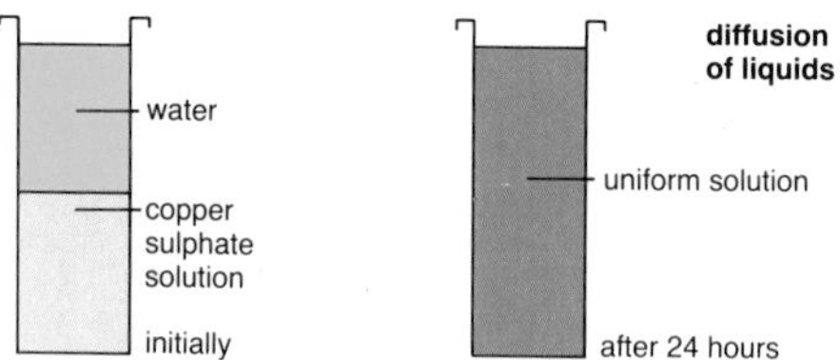

reversible process a sequence of changes in a physical system which can be carried out in reverse, so that the physical system returns to its original state without energy loss, e.g. a simple pendulum oscillating without damping (p.103) due to air friction; compression or expansion of a gas sample contained in a frictionless (p.21) cylinder and piston.

two-way process alternative name for reversible process (↑).

irreversible process a sequence of changes in a physical system which can be reversed, but not without energy loss; the process is not truly reversible (↑), e.g. a simple pendulum oscillating with damping due to air friction; compression or expansion of a gas sample contained in a cylinder and piston not free from friction (p.21).

one-way process alternative name for irreversible process (↑).

working cycle the cycle of changes experienced by the working substance in a physical system, e.g. the internal combustion engine (p.146).

heat engine a physical system producing mechanical work (p.26) output from heat energy (p.28) input.

thermodynamics (*n*) the study of the equivalence between heat (p.28) and mechanical work (p.26) as forms of energy (p.26); a given amount of mechanical work can be totally converted to heat but a given amount of heat energy can never be totally converted to mechanical work; these changes, and the restrictions on them, are expressed in the laws of thermodynamics. **thermodynamic** (*adj*).

Zeroth law of thermodynamics if bodies A and B are each separately in thermal equilibrium with body C, then A and B are in thermal equilibrium with each other.

1st law of thermodynamics when heat energy Q is taken into a system, and when external work W (p.26) is done, there is an increase U (joule) in the internal energy (p.13) of the system; for infinitesimally small changes, $\delta Q = \delta W + \delta U$.

2nd law of thermodynamics no working physical system, e.g. a heat engine (↑), can continuously transfer heat from a lower to a higher temperature (p.116) without a continuous mechanical work (p.26) input from an external source, e.g. a refrigerator (p.141) must be continuously driven by gas or electricity in order to function; the 2nd law can be stated in terms of entropy (↓) changes viz. the total entropy of a closed system cannot decrease.

3rd law of thermodynamics the Kelvin thermodynamic zero (0K) temperature (↓) can never be attained.

entropy (*n*) the property of a system that changes, when the system undergoes a reversible change (↑), by an amount $\delta S = \delta Q/T$ where δQ is the energy absorbed by the system and T is the thermodynamic temperature (↓); increasing entropy in a physical system gives increasing randomness or disorder.

Kelvin thermodynamic temperature scale the scale of temperature (p.116) based on the operation and working conditions of an ideal heat engine (↑) using ideal gas as the working substance.

thermodynamic temperature a value of temperature on the Kelvin thermodynamic scale (↑).

Kelvin temperature an alternative name for thermodynamic temperature (↑).

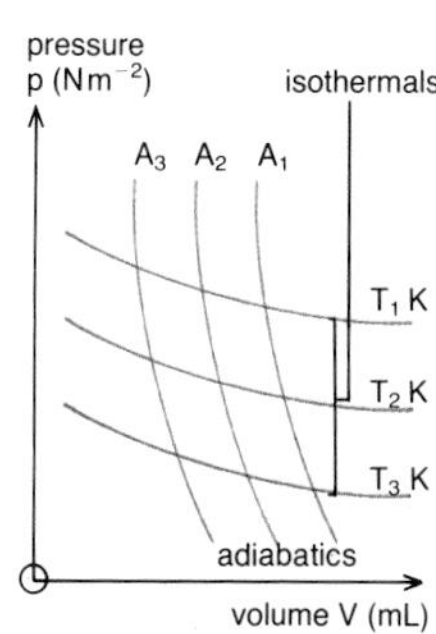

in going from A_1 to A_2 along an isothermal, heat δQ (J) absorbed depends on temperature T K:
$$\frac{\delta Q_1}{T_1} = \frac{\delta Q_2}{T_2} = \frac{\delta Q_3}{T_3} = \text{constant } \left(\frac{\delta Q}{T}\right)$$

entropy
adiabatics are distinguished by entropy

triple point refers to co-ordinates of the single point on the graph of pressure p (p.39) against temperature t (p.116) for a specified substance, at which all 3 phases of matter (p.14) can exist in dynamic equilibrium (p.18), e.g. for water p = 4.66 mm Hg, t = 0.01°C; the point of intersection of 3 lines on the graph separates the 3 phases of matter and specifies the conditions under which the substance can exist in its solid, liquid or vapour phase.

triple point of water the fixed point (p.118) on the Kelvin thermodynamic temperature scale (p.145) used to define the Kelvin (K) as the unit of temperature interval (p.116); 1 K is defined as 1/273.16 of the thermodynamic temperature (p.145) of the triple point (↑) of water, 0.01°C, where the zero of the temperature scale is −273.16°C = 0 K; good practical approximations are: 0°C = 273 K, 100°C = 373 K.

internal combustion engine a heat engine (p.145) burning fuel in the cylinder where energy conversion occurs, e.g. petrol engine (↓), diesel engine (↓), jet engine (↓), rocket engine (↓).

petrol engine an internal combustion engine (↑) burning petrol as fuel.

4-stroke cycle the continuous working cycle of operations carried out during the working of a petrol engine (↑); its strokes are: 1. induction (↓); 2. compression (↓); 3. ignition (↓); 4. exhaust (↓).

triple point of water

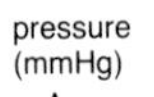

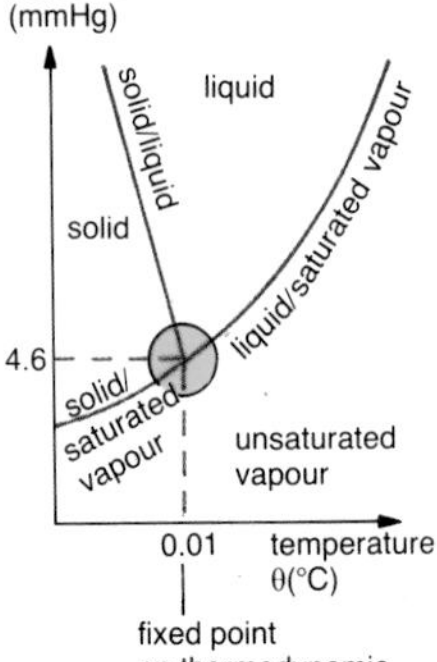

4 stroke cycle

of internal combustion engine

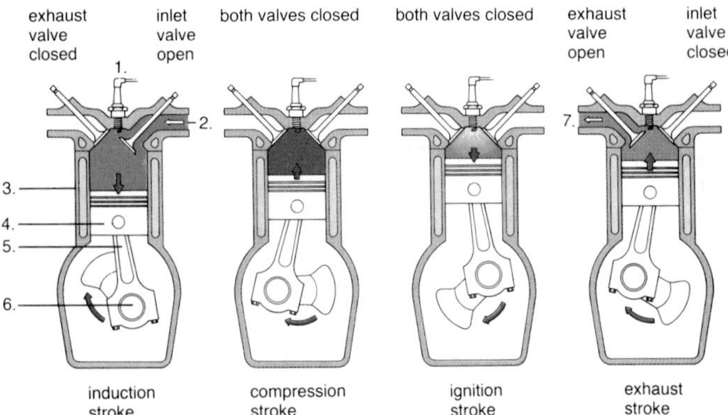

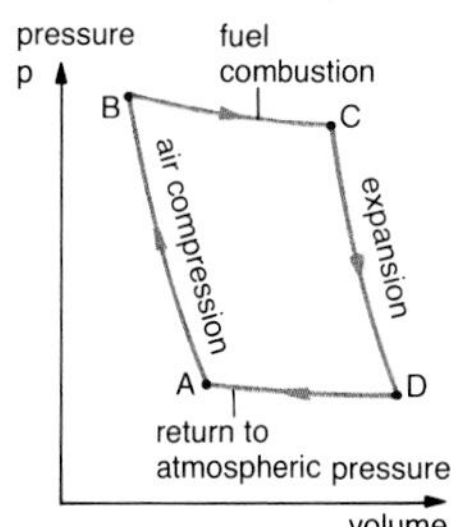

4-stage working cycle of gas-turbine engine

induction stroke the intake of fuel in a petrol engine (↑) as the piston moves down, lowering pressure (p.39) in the cylinder.

compression stroke the compression of fuel in a petrol engine (↑) as the piston moves up the cylinder.

ignition stroke explosive combustion of fuel in a petrol engine (↑), followed by expansion, pushing the piston down the cylinder.

exhaust stroke the removal of burnt fuel gases from a petrol engine (↑) after fuel combustion, the piston moving up the cylinder.

diesel engine an internal combustion engine (↑) burning oil as fuel; it is heavier than the petrol engine (↑) but has greater efficiency (p.36).

gas turbine engine

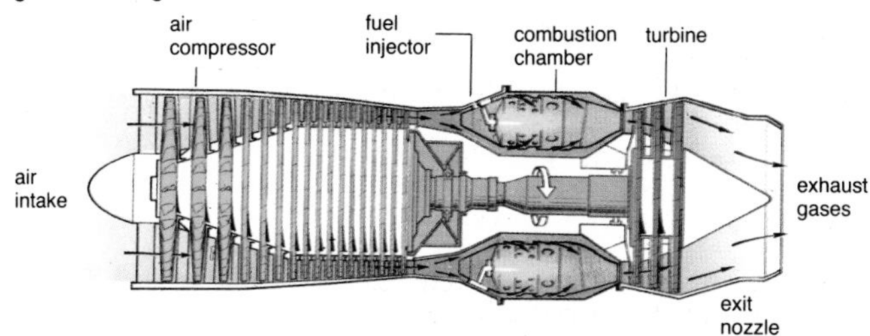

rocket engine

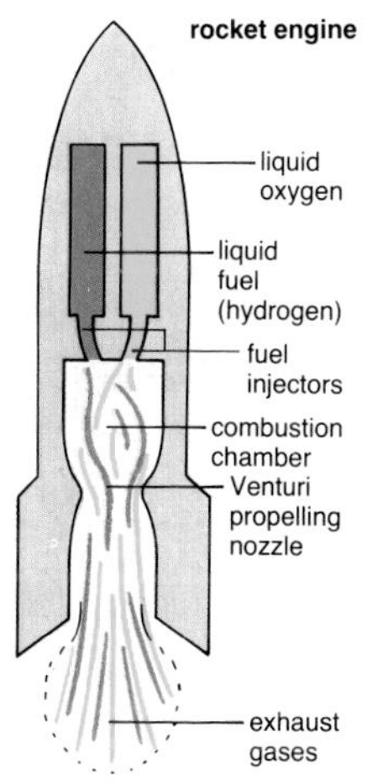

gas turbine engine a heat engine (p.145) using air as its working substance to provide thrust (p.23), e.g. turbo-jet engine; it has a 4-stage working cycle: 1. atmospheric air is drawn in and compressed; 2. fuel is introduced and burned; 3. combustion product gases expand through the turbine and jet pipe in a propulsive (↓) gas stream; 4. the gas stream reaches atmospheric pressure (p.40).

jet propulsion the forward movement of an aircraft or rocket produced by the reaction force (p.36) as exhaust gases are ejected from the gas turbine (↑); momentum is conserved (p.37) along the line of flight, forward momentum of the aircraft being equal and opposite to the momentum of the exhaust. **propeller** (*n*), **propel** (*v*), **propellant** (*adj*), **propulsive** (*adj*).

rocket engine an internal combustion engine (↑) using its own transported fuel as the means of jet propulsion (↑); it can operate outside the Earth's atmosphere as no lift (p.22) forces are needed.

electric cell a source of direct electric current (p.150) consisting of 2 cell poles (↓) acquiring different electric potentials (p.170) when in contact with a suitable electrolyte (p.10), e.g. Leclanché cell (p.151); electric potential difference (p.154) between the poles can be applied to a closed conducting circuit (p.154) outside the cell.

primary cell an electric cell (↑) deriving its electrical energy (p.28) from the chemical energy (p.28) liberated when the cell poles are in contact with the electrolyte; chemical changes within the cell are irreversible, the cell capacity is small and electric current (p.152) from the cell is 1 ampere or less, e.g. Leclanché cell (p.151), dry battery (p.151).

secondary cell an electric cell (↑) storing electrical energy (p.28) as chemical energy (p.28) to be released when required; the cell requires initial charging (p.153) from a d.c. supply (p.150); chemical changes within the cell are reversible within certain practical limits and a discharged cell can be recharged; the cell capacity is much larger than for a primary cell (↑) and electric current (p.152) can be several amperes, e.g. nickel-iron alkaline battery, nickel-cadmium cell, lead-acid accumulator (p.152).

fuel cell a source of electrical energy (p.28) derived from the chemical energy (p.28) of oxidation of a fuel, e.g. hydrogen by oxygen; its e.m.f. (↓) is about 1 volt.

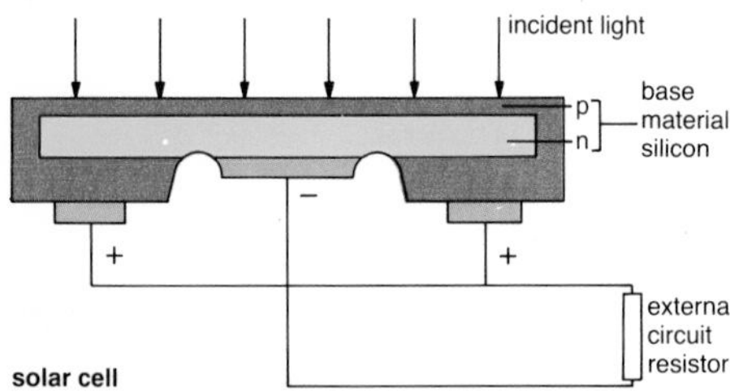

solar cell

solar cell a semiconductor p-n junction (p.224) device converting solar energy (p.29) into electrical energy (p.28).

solar battery an array of solar cells (↑) used in space satellites (p.44).

cell poles 2 distinct conducting components of an electric cell (↑), separated by electrolyte and made of different conducting materials, e.g. carbon, zinc; they acquire different electric potentials with respect to the electrolyte; the electric potential difference between the poles acts across the electrolyte inside the cell when the cell is on open circuit (p.154); the poles are identified as positive and negative according to conventional current direction (p.150) in the external circuit.

cell terminals points outside an electric cell (↑) or d.c. source at which electrical contact can be made with the cell poles (↑); the positive terminal is coloured red and the negative terminal black.

electromotive force (e.m.f.) the electric potential difference (p.154) between the poles of an electric cell (↑) on open circuit (p.154); it is determined only by the electric potential difference between each pole and the electrolyte and is independent of the cell dimensions and quantity of electrolyte; it is measured accurately by potentiometer (p.161) and approximately by a high resistance voltmeter (p.198) connected between the cell terminals (↑) on open circuit (p.154); it gives a measure of the work done (joule) by a cell when an electric charge (p.165) of 1 coulomb is transferred once round the complete circuit (p.154); it is defined as the ratio of the electrical power (watt) (p.159) developed by a cell per ampere of electric current (p.152) flowing, watt (W)/ampere (A); e.m.f. is denoted by E and measured in units of the volt; 1 volt (V) = 1 joule/coulomb (JC^{-1}) = 1 watt/ampere (WA^{-1}); an e.m.f. is also generated by photovoltaic cell (p.230), thermocouple (p.163) and alternator (p.214).

simple voltaic cell a primary cell (↑) with a positive copper pole and negative zinc pole in dilute sulphuric acid electrolyte.

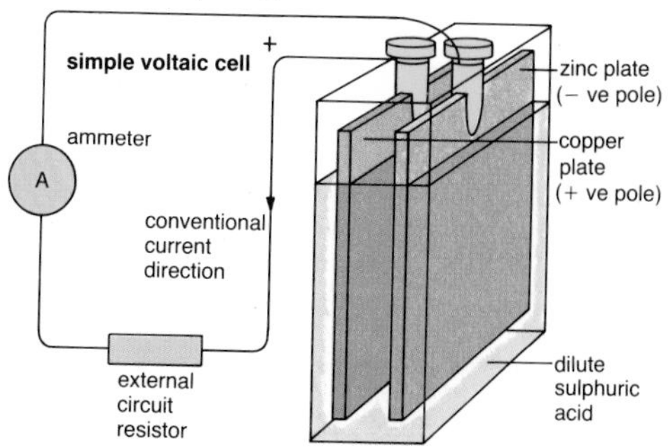

current as moving charge carriers if a conductor contains n charge carriers per m^2, each of charge q (C) and moving with drift velocity (↓) v (m/s) and if the area of cross-section of the conductor is A (m^2) then the current flowing I (ampere) is given by $I = nAqv$.

drift velocity the rate at which charge carriers (p.9) diffuse through a material when an electric field (p.165) is applied; the value for electrons in a wire carrying current is low, (of the order $10^{-4}\,m\,s^{-1}$), compared with the velocity with which the conducting wire transmits the electric field, close to the velocity of light in free space, $3 \times 10^8\,m\,s^{-1}$ (p.70).

conventional current direction the direction of positive electric charge (p.165) transfer during electric current (p.152) flow in an external conducting circuit; it is equivalent to negative charge transfer at an equal rate in the opposite direction, e.g. electron charge carriers in a cathode-ray tube (p.230) beam; conventional current flows between positive and negative terminals external to a d.c. source (↓), e.g. from copper to zinc for a simple voltaic cell (p.149).

cell polarization hydrogen gas bubbles, liberated at the positive pole of a simple voltaic cell (p.149) or a Leclanché cell (↓), electrically insulate the positive pole causing the e.m.f. (p.149) to fall and the electric current (p.152) flow to cease; a change in the concentration of ions (p.9) near the cell poles (p.148) causes the fall in e.m.f.; chemical depolarization is needed in Leclanché cells.

depolarizer (*n*) chemical substance used in a primary cell (p.148) to prevent polarization (↑), e.g. Leclanché cell (↓).

direct current (d.c.) source an electrical energy (p.28) source providing unidirectional current, e.g. primary, secondary, solar cells (p.148); d.c. generator (p.212).

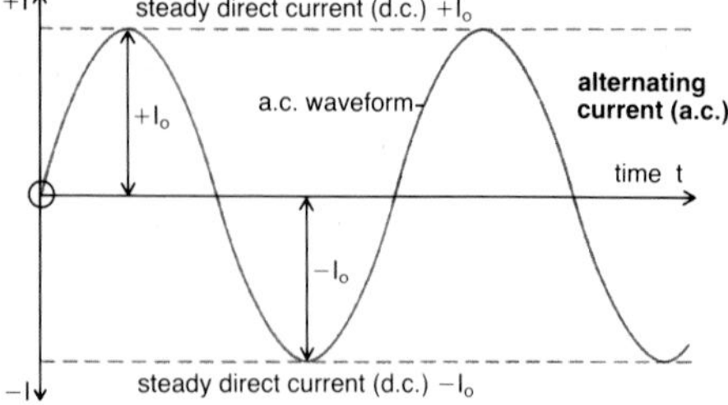

alternating current (a.c.) source an electrical energy (p.28) source providing an alternating current (p.213) of specific frequency (p.54), e.g. dynamo (p.213), a.c. generator (p.212).

Leclanché cell a primary cell (p.148) with a carbon positive cell pole (p.148) and a zinc negative pole, using ammonium chloride solution as electrolyte and manganese dioxide and powdered carbon as depolarizer (↑); its internal resistance (p.154) is about 1 ohm and the depolarizing action is slow, so the cell is suitable for intermittent use giving an electric current (p.152) of less than 1 ampere; its e.m.f. (p.149) is about 1.5 volt; it is in common use as the dry cell (↓).

Daniell cell a primary cell (p.148) of e.m.f. 1.08 volt, with copper (+) and zinc (−) poles, zinc sulphate electrolyte and saturated copper sulphate solution as depolarizer.

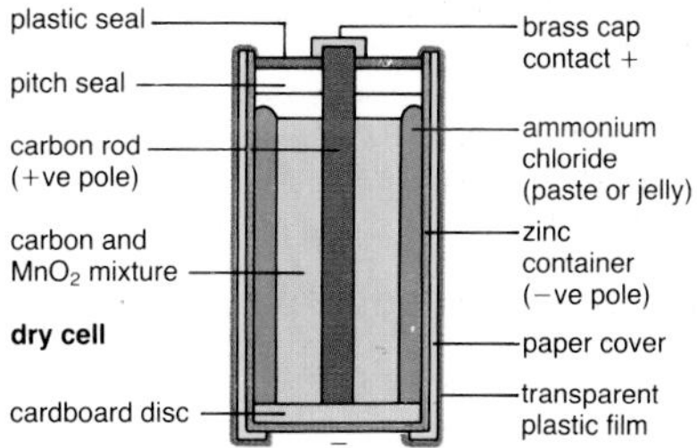

dry cell

dry cell a Leclanché cell (↑) with its liquid components in paste form, making the cell smaller and practically more convenient; it is used in a dry battery (↓).

battery (*n*) a number of electric cells (p.148) connected in series so that their total e.m.f. (p.149) is the sum of their individual e.m.f.s; parallel connection gives increased electric current (p.152) at the same e.m.f.

dry battery a number of dry cells (↑) in battery (↑) array; used in torches and radios, for intermittent current.

Weston cadmium cell a primary cell (p.148) with mercury as the positive cell pole (p.148) and a negative pole of mercury-cadmium amalgam, using saturated cadmium sulphate solution as electrolyte and mercurous sulphate paste as depolarizer (↑); its internal resistance (p.154) is several hundred ohm, so the electric current (p.152) from the cell is only a few micro-ampere; it has very steady e.m.f. (p.149) = 1.0186 volt at 20°C and is used internationally as a primary laboratory standard of e.m.f.

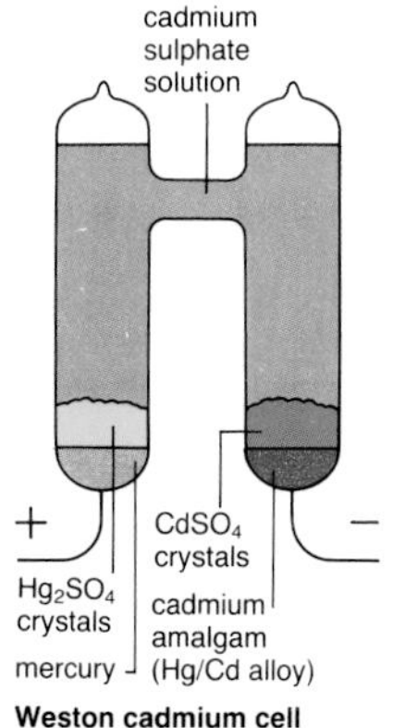

Weston cadmium cell

electric current refers to any charge transfer (p.9), through an electrical conductor (p.155) or conducting medium, occurring at a measurable rate; the rate of electric charge transfer past a point in a conducting circuit; it is expressed in units of the ampere (A), milliampere (mA), microampere (μA); 1 ampere (A) = 1 coulomb/second ($C s^{-1}$).

ampere (A) the SI unit of electric current (↑); when electric charge (p.165) is transferred through an electrical conductor at a rate of 1 coulomb/second, the current is 1 ampere (A).

lead-acid accumulator a secondary cell (p.148) with its positive cell pole (p.148) of lead dioxide and negative pole of lead, using dilute sulphuric acid as electrolyte of specific gravity 1.25 (p.7); its internal resistance (p.154) is less than 1 ohm, so the electric current (p.152) from the cell can be several amperes; its e.m.f. (p.149) is 2.2 volt when fully charged (↓); on discharge (↓) the e.m.f. falls to 1.8–1.9 volt and recharging is required; several cells in series are used as a battery (p.151), e.g. a 12 volt car battery; the current capacity (↓) can be up to 100 ampere hour (↓).

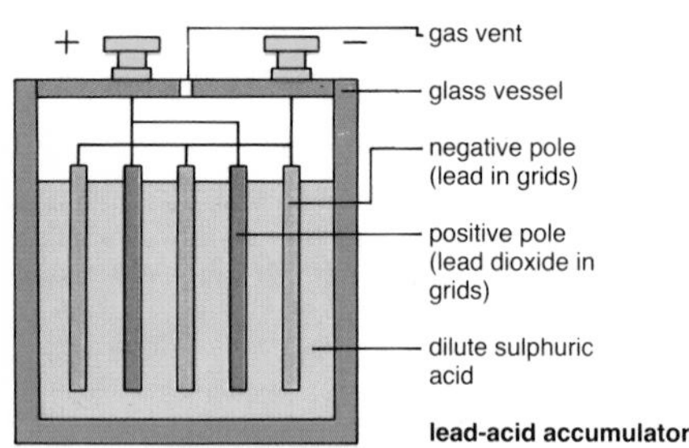

lead-acid accumulator

nickel-cadmium (Ni-Cad) alkaline cell a secondary cell (p.148) that has a nickel hydroxide/nickel oxide mixture as the anode, a cadmium cathode and potassium hydroxide as the electrolyte; e.m.f. is 1.25 volt; sealed Ni-Cad cells are manufactured for portable equipment; these cells are mechanically and electrically robust.

ampere-hour (Ah) electrical energy (p.28) is associated with stored electric charge (p.165) in a secondary cell (p.148); if a cell gives 1 ampere (A) steady d.c. electric current (↑) for 1 hour (h) the charge delivered from the cell is 1 ampere hour; 1 Ah = 3600 coulomb (C).

current capacity electrical energy (p.28) is associated with stored chemical energy (p.28) in a secondary call (p.148) or battery (p.151) on charging (↓); on discharging (↓), energy is released as electric current (p.152); units are ampere-hour (↑), e.g. a battery of current capacity 80 A h can supply l A for 80 h, 2 A for 40 h, 5 A for 16 h or 8 A for 10 h; for commercial value it is quoted for a discharge (↓) time of 10 hours.

secondary cell efficiency the ratio of the energy output (joule) on discharging (↓) ÷ energy input (joule) on charging (↓); it has no units and is expressed as %; cell efficiency (%) = ampere-hour (↑) × average e.m.f. (p.149) on discharge (↓) ÷ ampere-hour × average e.m.f. on charge (↓); its value is about 80%.

working cell power output for an electric cell of efficiency η (↑) and internal resistance r (ohm) (p.154), supplying electric current I (A) (↑) to an external circuit of resistance R (ohm) (p.154), the power generated (watt) by the source of e.m.f. E (volt) = I^2R; since $I = E/(R + r)$ the power output P from $E = E^2R/(R + r)^2$; differentiation to obtain a maximum or minimum value gives $dP/dR = 0$ for $R = r$; so maximum power output P (watt) occurs when the external circuit resistance R (ohm) = internal cell resistance r (ohm).

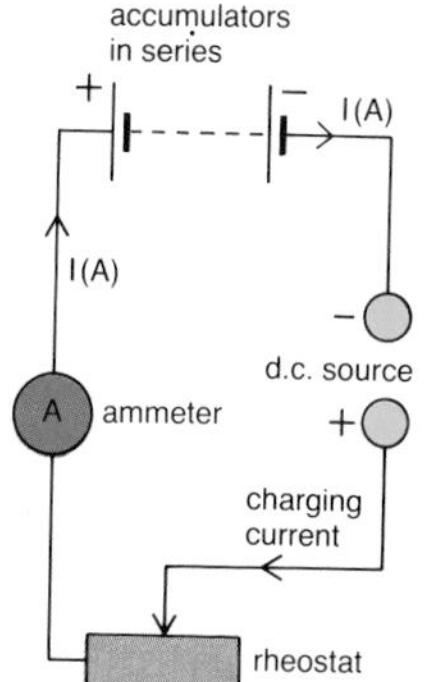

battery charging electrical energy (p.28) input to a secondary cell (p.148) or battery (p.151) as a d.c. (p.150) charging current flows; it is stored as chemical energy (p.28) of the cell constituents. **charge** (*n*), **charging** (*v*), **charged** (*adj*).

battery discharging electrical energy (p.28) output from a secondary cell (p.148) or battery (p.151) in association with d.c. electric current (p.150) flow; it is released from the stored chemical energy (p.28) of the cell constituents. **discharge** (*n*), **discharging** (*n*), **discharge** (*v*), **discharged** (*adj*).

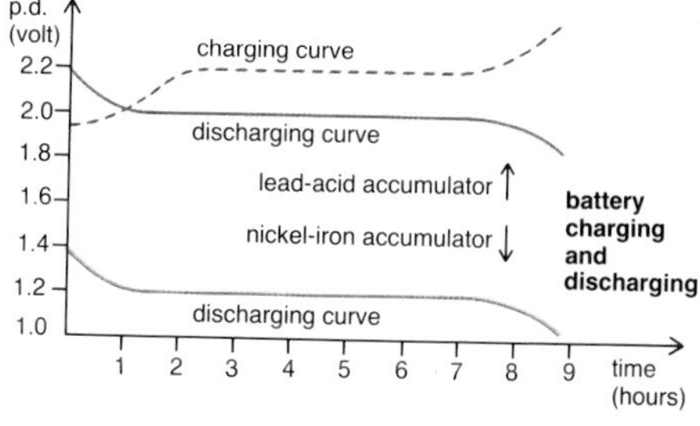

battery charging and discharging

closed circuit a conducting circuit with all contacts and connections made so that electric current can flow on circuit make; the final closing connection can be a switch, plug, key, tapping key, pressure or press button switch.

open circuit a conducting circuit with all contacts and connections made, except the final connection to complete the closed circuit (↑); current cannot flow until circuit make; the terminal p.d. (↓) of the source of e.m.f. is now a maximum and equal to the source e.m.f.

complete circuit a closed conducting circuit (↑).

internal resistance refers to the electrical resistance (↓) of the source of e.m.f. in the complete circuit (↑); it is denoted by r and measured in ohms; for an electric cell (p.148) it depends on the cell dimensions and pole arrangements, and on the quantity and conducting properties of the electrolyte; r can vary for low currents.

external resistance the electrical resistance (↓) of the conducting components external to the source of e.m.f. in a complete circuit (↑); denoted by R and measured in ohms; also called load resistance (↓).

load resistance an alternative name for external resistance (↑); resistance through which a source of e.m.f. drives a current.

total circuit resistance the electrical resistance (↓) of a complete circuit (↑), including internal (↑) and external resistance (↑); denoted by (R + r) and measured in ohms.

terminal potential difference the electrical potential difference (↓) between the cell terminals (p.149) in a conducting circuit carrying an electric current (p.152); it is denoted by V and measured in volts (V); it varies with the electric current I (ampere) up to a maximum value E (volt), the electromotive force (p.149) of the cell, on open circuit (↑); for the complete circuit (↑): E (volt) = V + v, where v = p.d. (↓) driving current I through an internal resistance r (ohm) (↑); applying Ohm's Law (↓) to the complete circuit of total circuit resistance (R + r) (↑): $E = I(R + r) = V + Ir$, so $V = E - Ir$.

electric potential difference[1] (p.d.) gives a measure of the work done (joule) by the source of e.m.f. (p.149) in driving an electric charge (p.165) of 1 coulomb between 2 points in a current-carrying circuit; 1 volt (V) = 1 joule/coulomb (JC^{-1}); applying Ohm's Law (↓) to a resistor of electrical resistance (↓) R (ohm) carrying current I (ampere): $V = IR$ (volt).

closed circuit

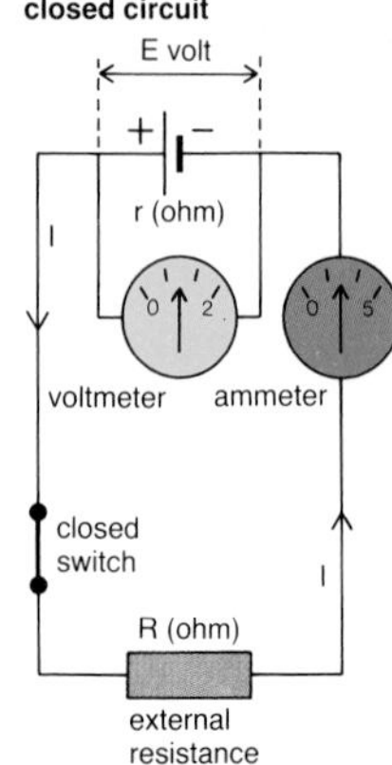

open circuit

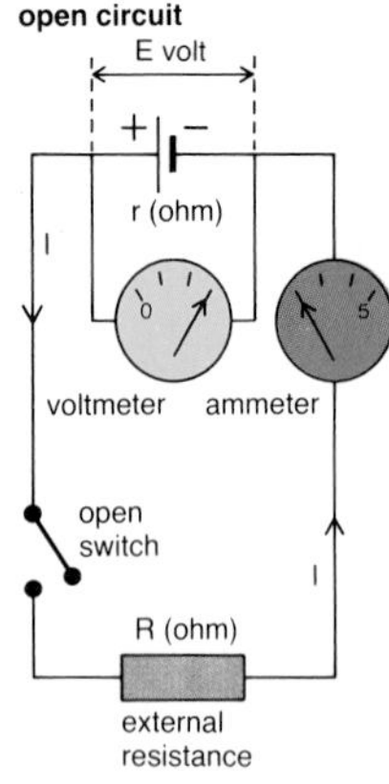

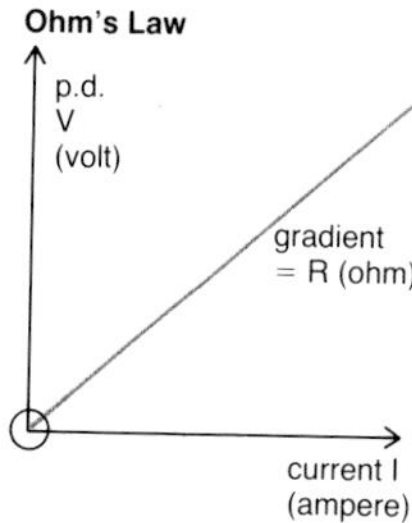

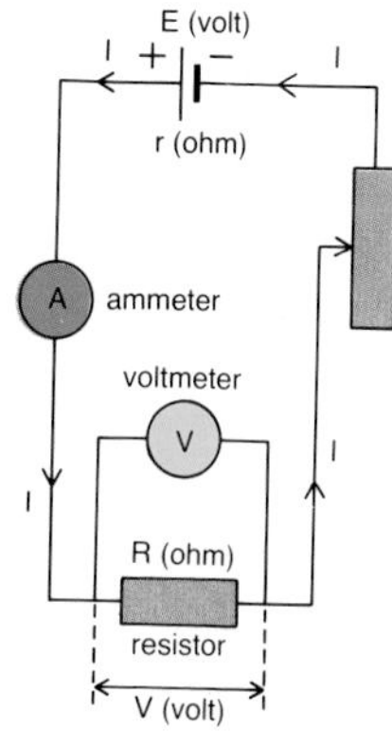

voltage an alternative name for electric potential difference (↑).

volt (V) the unit of electric p.d. (↑); when work done in driving an electric charge of 1 coulomb (C) between 2 points in a circuit is 1 joule (J) the p.d. between the 2 points is 1 volt (V); $1\,V = 1\,JC^{-1}$; also the unit of e.m.f. (p.149) for d.c. (p.150) and a.c. (p.151) sources, defined as 1 volt (V) = 1 watt/ampere (WA^{-1}).

Ohm's Law the electric current (p.152) I (ampere), flowing in a conductor (↓) at constant temperature, is directly proportional to the p.d. (↑) V (volt) applied to the conductor; $I \propto V$, so V/I = constant (R), where R = electrical resistance (↓) of the conductor.

electrical conductor a circuit component made from electrically conducting material, e.g. a resistor, rheostat (p.156); an electrically conducting part of a circuit, e.g. electrolyte in an electrolytic cell (p.10) or electric cell (p.148), gas in a discharge tube, a solid state device (p.15). **conductor** (*n*), **conduction** (*n*), **conduct** (*v*), **conducting** (*adj*).

electrical insulator a component made from electrically insulating material; used for electrical safety and to prevent loss of electric current from a closed conducting circuit (↑), e.g. oil, pure de-ionized water, dry atmospheric air, ceramic block insulators used where high voltage overhead transmission lines are connected to transformers, rubber or plastic sheathing on a.c. mains (p.160) connecting wires, bakelite covering on plugs and switches. **insulator** (*n*), **insulation** (*n*), **insulate** (*v*), **insulating** (*adj*), **insulated** (*adj*).

ohmic conductor an electrical conductor (↑) conforming with Ohm's Law (↑), e.g. a resistor (p.156) at constant temperature.

non-ohmic conductor an electrical conductor (↑) not conforming with Ohm's Law (↑), e.g. junction diode (p.225), electrolyte showing back e.m.f.; V/I graph is non-linear, I is not directly proportional to V for all values of V.

electrical resistance the ratio of the electric p.d. (↑) V (volt) ÷ electric current I (ampere); it is denoted by R and measured in ohms. **resistive** (*adj*).

ohm (Ω) the unit of electrical resistance (↑); the resistance of an electrical conductor carrying an electric current of 1 ampere (A) when a p.d. (↑) of 1 volt (V) is applied; 1 ohm = 1 volt/ampere (VA^{-1}).

resistor (*n*) an electrically conducting (p.155) circuit component of specified electrical resistance (p.155), e.g. 5 ohm, 5 megohm; made from high resistivity (p.158) alloy wire, e.g. nichrome, wound on an electrically insulating (p.155) former; accurate in value to ±1%; high resistance resistors, e.g. kilohm (10^3 ohm), megohm (10^6 ohm) are made from a solidified carbon mixture and are accurate to ±20% unless otherwise specified by high resistance colour code (↓) markings.

standard resistor a resistor (↑) made from high resistivity alloy wire, e.g. manganin of very low temperature coefficient of resistance (p.158) non-inductively wound and accurate to ±0.01%.

resistance box a range of wire-wound standard resistors (↑) mounted in a box; selection is made by removing a plug key from the copper bar running along the top of the box, or by turning a dial pressure contact; the usual range is 1 to 10000 ohm.

rheostat (*n*) a variable carbon or wire-wound resistor (↑), used to control and vary the electric current (p.152) flowing in a conducting circuit; 2 fixed connections give a specified maximum resistance; 1 fixed connection and 1 variable sliding contact give resistances between 0 and maximum.

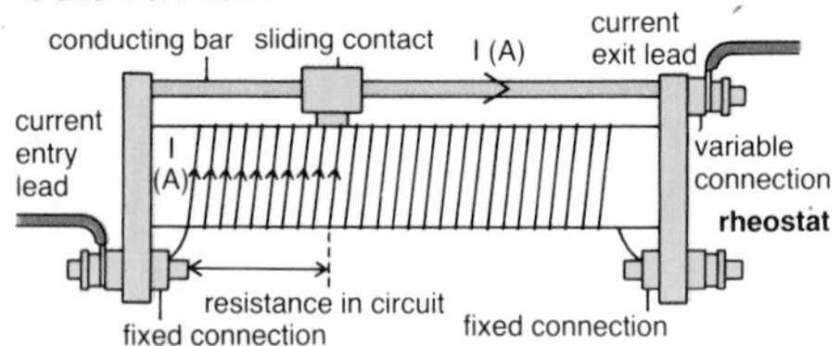

rheostat

variable resistor alternative name for rheostat.

light dependent resistor (l.d.r.) a photoconductive cell consisting of a piece of semiconductor (p.222) such as cadmium sulphide covered by a window to admit light; when light falls on the cell its conductivity increases as its resistance decreases; typically, the resistance of the light dependent resistor varies from 10 MΩ in darkness to 1 kΩ in daylight.

non-inductive winding a method of double-winding the coils of standard resistors (↑), so that the varying magnetic fields caused by varying electric currents (p.152) produce equal and opposite self-induced (p.206) e.m.f.s (p.149), giving a nett zero e.m.f.

resistor
wire wound resistor

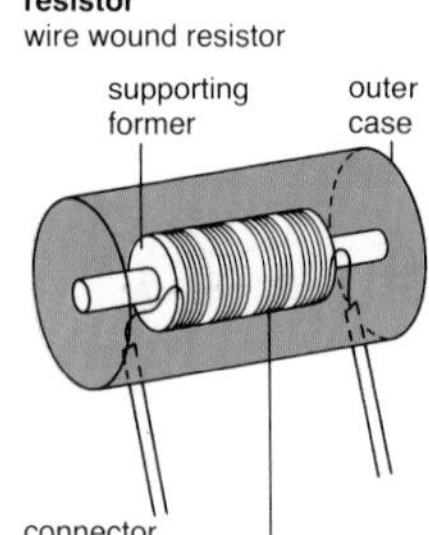

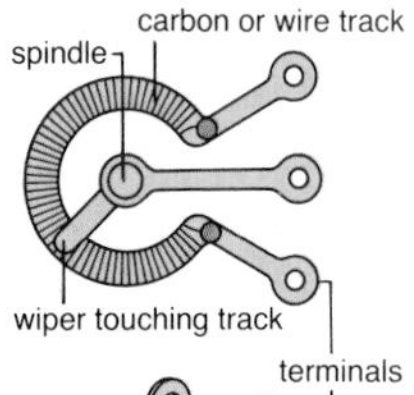

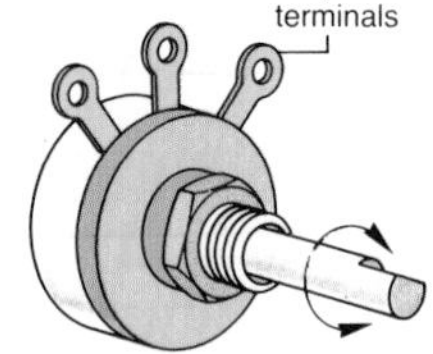

variable resistor
used in radio receivers etc.

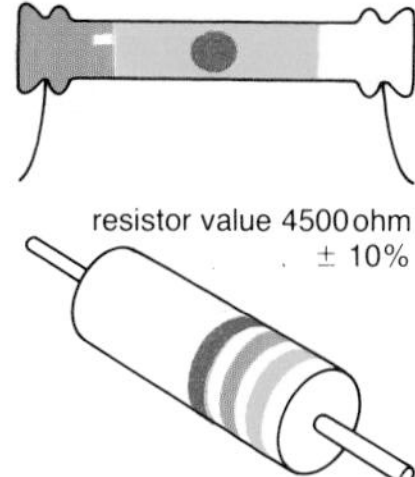

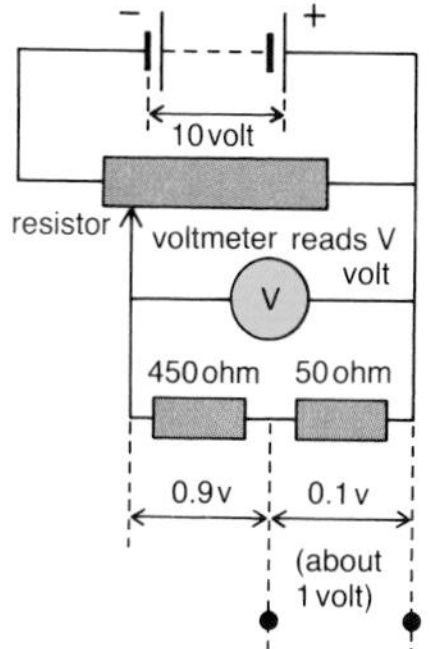

high resistance colour code a method of indicating approximate values of carbon mixture resistors (↑) by colour markings; Standard Colour Code: Black 0; Brown 1; Red 2; Orange 3; Yellow 4; Green 5; Blue 6; Violet 7; Grey 8; White 9; modern marking: first ring from the end of the resistor is the first digit, second ring is the second digit, third ring is the number of zeros; former marking: body colour is first digit, end colour is second digit, body spot is number of zeros; the tolerance is expressed as % accuracy in value indicated by: gold ring ± 5%, silver ring ± 10%, no marking ± 20%; red ring at the opposite end ± 2%, brown ring ± 1%, pink ring indicates high stability.

potential divider circuit a conducting circuit including a rheostat (↑) used to vary the electric p.d. (p.154) applied to a conducting component or another circuit, e.g. a wire-wound potentiometer (↓) used in radio and electronic circuitry; it can be used to obtain a known fraction of a higher p.d. above 2 volt, e.g. voltmeter calibration.

wire-wound potentiometer a rheostat used as a potential divider (↑) in radio circuitry.

switch (*n*) a mechanical or electronic device for rapid circuit make or break. **switch** (*v*), **switching** (*adj*).

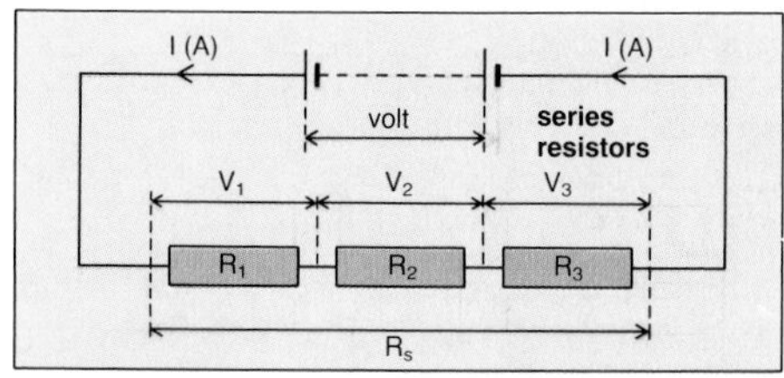

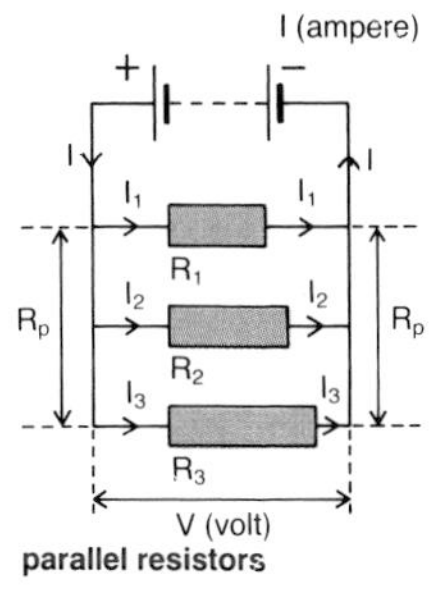

parallel resistors

series resistors a method of connecting 2 or more resistors (↑) so that each carries the same electric current; the effective resistance R_s (ohm) of the single circuit component is equivalent to the series arrangement: $R_s = R_1 + R_2 + R_3$.

parallel resistors a method of connecting 2 or more resistors (↑) so that the same electric p.d. is applied to each; the effective resistance R_p (ohm) of the single circuit component equivalent to the parallel arrangement is given by the equation $1/R_p = 1/R_1 + 1/R_2 + 1/R_3$; the value of R_p is always less than the value of any of the individual components.

circuit symbol a simplified diagrammatic representation of a circuit component used in circuit diagrams (↓).

circuit diagram a diagrammatic representation of the essential components and connections of a conducting circuit, using circuit symbols (↑); it is used as a precise guide for assembling a circuit.

Kirchoff's Laws are laws used in conducting circuit networks: 1. the total electric current (p.152) approaching the junction of any number of conducting components equals the total current leaving that junction; 2. in any closed path in a network of conducting components, the total e.m.f. (p.149) equals the algebraic sum of the products of currents and resistances (p.155).

Kirchoff's Laws

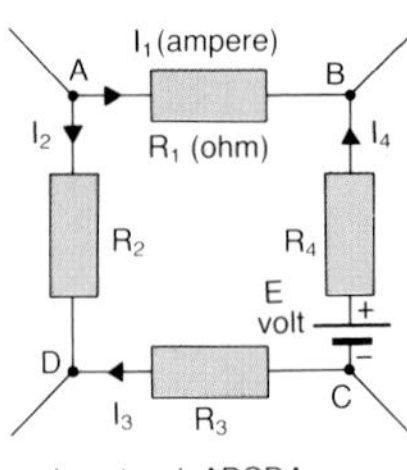

in network ABCDA:
$E = I_4R_4 - I_1R_1 + I_2R_2 - I_3R_3$

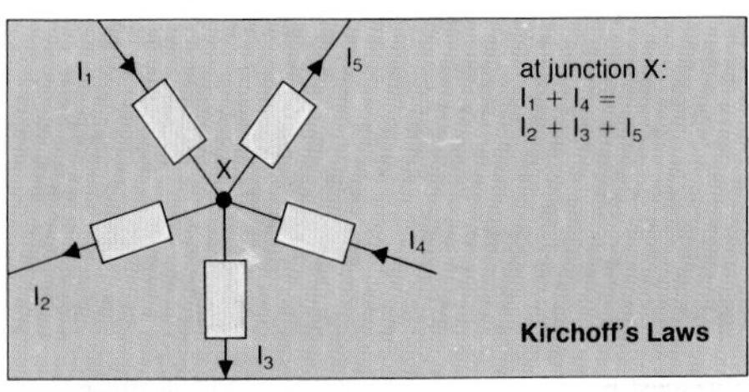

Kirchoff's Laws

wire resistance the electrical resistance R (ohm) (p.155) of a length l (m) of uniform conducting wire of cross-sectional area A (m^2): $R \propto l/A$, so $R = \rho l/A$, where ρ = proportionality constant depending on the conducting properties of the wire material, its electrical resistivity (↓).

wire resistance

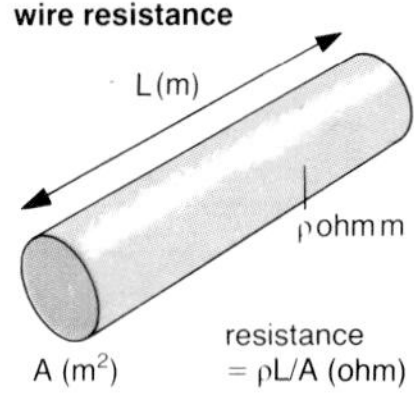

resistivity (*n*) a constant representing the electrical conducting properties of a specific material at constant temperature (p.116); denoted by ρ; wire resistance R (↑) = $\rho l/A$; units are ohm metre, e.g. for copper at 20°C, $\rho = 1.69 \times 10^{-8}$ ohm metre = 1.69×10^{-6} ohm cm = 1.69 microhm cm.

temperature coefficient of resistance electrical resistance (p.155) of conducting components varies with temperature (p.116), due to variation in resistivity (↑) and conductor dimensions; for wire-wound resistors (p.156) the resistance increases linearly with temperature within specified temperature limits; the coefficient α is defined as the ratio of the resistance change per degree K/resistance at 0°C (273K); $\alpha = (R_\theta - R_o)/R_o\theta$, where R_θ (ohm) = resistance at θ°C (θ + 273K), and $R_\theta = R_o(1 + \alpha\theta)$; units of α are K^{-1}.

temperature coefficient of resistance
(linear in range 0–100°C)

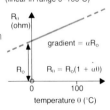

R_θ (ohm)
gradient = αR_o
R_o
$R_\theta = R_o(1 + \alpha\theta)$
0 100
temperature θ (°C)

electric power the rate at which electrical energy (p.28) is generated by a source of e.m.f., e.g. a battery (p.151), or dissipated in driving an electric current (p.152) through a load resistance in an external circuit; units are watt (W), where 1 watt (W) = 1 joule/second (Js^{-1}); the power output from an e.m.f. source E (volt), driving current I (ampere), is EI watt; the power dissipated in a resistor R (ohm) (p.156), carrying current I at p.d. V (volt) (p.154), is VI watt; since Ohm's Law (p.155) gives $V = IR$ and $I = V/R$, power $VI = I^2R = V^2/R$ watt.

watt the unit of electric power (↑); 1 watt (W) = 1 joule/second (Js^{-1}); also 1 watt (W) = 1 volt ampere (VA).

electrical heating effect the rise in temperature (p.116) of a conducting component carrying an electric current (p.152); it is due to the conversion of electrical energy into heat (p.28) as electric power (↑) is dissipated in the conducting component; the effect is maximized in high resistivity (↑) heating elements and minimized in conducting leads.

Joule heating effect an alternative name for electrical heating effect (↑).

resistive energy losses refer to energy losses due to the Joule heating effect (↑).

kilowatt-hour a commercial unit of electrical energy (p.28); the energy consumed in 1 hour when electric power (↑) is dissipated at the rate of 1 kilowatt (kW); 1 kilowatt-hour (kWh) = 10^3 (watt) × 60^2 (s) = 3.6×10^6 joule (J) = 3.6 megajoule (MJ) (↓); units consumed are recorded by an electricity meter (p.160).

megajoule a commercial unit of electrical energy, it can be used as an alternative to the kilowatt-hour (↑); 1 megajoule (MJ) = 10^6 (J).

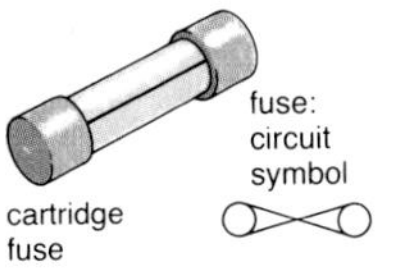

fuse (*n*) a device for preventing overloading of an electrical appliance with excess electric current (p.152); it prevents excessive electrical heating effects (↑) and damage to the appliance; a fuse wire or cartridge is located in the live lead side of a fused plug or consumer unit (p.160); fuse wire thickness is increased for increased current load; the fuse melts rapidly (blows) on high current overload and more slowly on sustained low excess current, breaking the conducting circuit and preventing further current supply to the appliance; a fuse blow-out disconnects the live lead side of the a.c. main (p.160) from the appliance, leaving it safe for handling. **fuse** (*v*), **fused** (*adj*).

fused plug a device for connecting an electrical appliance into a socket point from the a.c. mains (↓); it is fitted with a fuse (p.159) of suitable current rating in the live lead side; 3-pin connections are wired to live (L), neutral (N) and earth (E) leads; socket outlet and lighting switches (p.157) are also located in the live lead side.

a.c. main an electrical installation system carrying electrical power supplied by the electricity grid system (p.217) into buildings for industrial, commercial and domestic use; in the U.K. the supply is an a.c. voltage (p.214) of peak value 250 volt at a frequency of 50 hertz (p.54); underground supply cables from a transformer substation enter premises through the Electricity Board's sealed fuse box, passing through an electricity meter (↓) to a consumer unit (↓), from which leads, fused on the live side, carry electricity to component parts of the installation; the neutral (N) lead is earthed (↓) at the substation, live (L) lead is positive and negative with alternating voltage cycles.

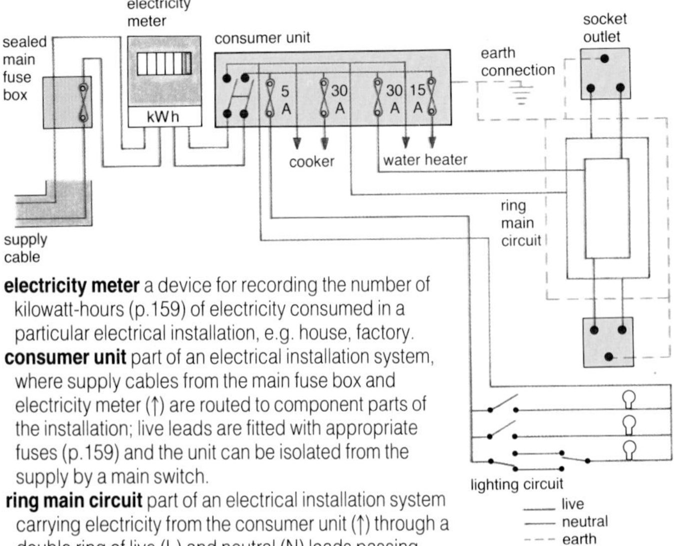

a.c. main
domestic electrical installation system

electricity meter a device for recording the number of kilowatt-hours (p.159) of electricity consumed in a particular electrical installation, e.g. house, factory.

consumer unit part of an electrical installation system, where supply cables from the main fuse box and electricity meter (↑) are routed to component parts of the installation; live leads are fitted with appropriate fuses (p.159) and the unit can be isolated from the supply by a main switch.

ring main circuit part of an electrical installation system carrying electricity from the consumer unit (↑) through a double ring of live (L) and neutral (N) leads passing round the building being supplied; access to a.c. main (↑) is through power point socket outlets by means of

3-pin fused plugs (↑) connecting with electrical appliances; the ring has a 30 A fuse (↑) in the consumer unit, limiting the total current supplied to all sockets to 30 A; all power points are earthed (↓) from the top socket pin, either through a metal water pipe passing out of the building to earth, or through an earth connection in the supply cable.

earthing (*n*) making a conducting connection between an electrical appliance, or part of an electrical installation, and the earth, giving the connection earth or zero potential (p.171) and providing a low resistance (p.155) current path to earth; should a short-circuit (↓) occur between a live conductor and the earthed part of the equipment a large current flows to earth blowing the fuse, thereby cutting off the current; safety precaution in a.c. mains (↑) installations and electrical appliances. **earth** (*v*), **earthed** (*adj*).

short-circuit (*n*) an alternative low resistance (p.155) path offered to an electric current in a closed conducting circuit (p.154), usually arising as a result of a circuit fault in the wiring; it is dangerous as a source of localized high heating effect (p.159), or when current-carrying high voltage lines make contact without a load resistance (p.154) in series, or as a cause of electric shock when the human body makes contact with a faulty unearthed electrical appliance; earthing (↑) as a safety precaution offers a lower resistance path to earth than does the human body if the appliance is fused (p.159).

Wheatstone bridge a network of resistors arranged so that when no current flows in the galvanometer the p.d.s across the four arms are balanced and P/Q = R/X, where P, Q and R are known and X can be calculated.

potentiometer principle when a d.c. source (p.150) of constant e.m.f. (p.149) drives a constant electric current I (p.152) through a uniform length of conducting material, the potential difference V (p.154) between any 2 points along the length is directly proportional to the distance L between the 2 points; V (volt) $\propto$ L (m) and V = kL, where k = V/L (volt m^{-1}) is a constant for that particular conductor carrying current I; the p.d. kL (volt) driving current I (A) between the 2 points would also tend to drive current through any conducting component connected in parallel between the 2 points, e.g. voltmeter (p.198); the laboratory potentiometer is a practical application of the circuit.

Wheatstone bridge

potentiometer principle

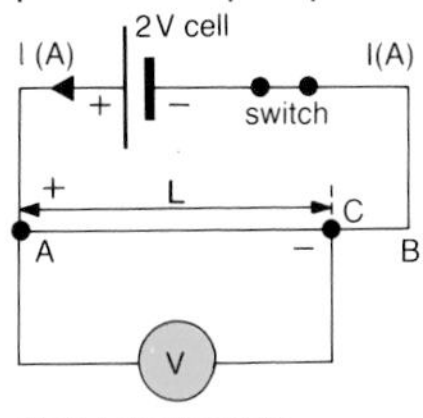

AB is potentiometer
$V_{AC} \propto L$ so V = kL

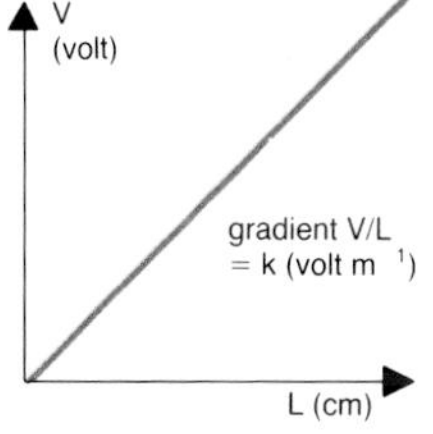

laboratory potentiometer the potentiometer principle (p.161) is employed in the form of a 1 or 2 metre length of resistance wire of uniform cross-sectional area A (m^2) and uniform resistivity ρ (ohm m) (p.158), mounted over a metre scale on a baseboard and soldered or clamped at both ends to copper strip of negligible resistance; a constant current I (ampere) is supplied by a 2 volt d.c. source, e.g. lead-acid accumulator (p.152), with a rheostat in series so that I, and hence constant k, can be varied; resistance R (ohm) (p.155) of a length L (m) of wire is ρL/A, so Ohm's Law (p.155) gives the p.d. V (volt) across L (m) as $V = IR = I.\rho L/A = I\rho/A \times L = kL$, where the potentiometer constant k (volt m^{-1}) = $I\rho/A$; the value of k (Vm^{-1}) is found by potentiometer calibration (↓); the calibrated potentiometer is used to make accurate measurements of e.m.f., p.d. and quantities related to p.d., e.g. electric current (p.152) and resistance (p.155); measurement is limited to the range 0–2 volt; for p.d. measurement above 2 volt a potential divider circuit is required.

potentiometer calibration refers to calibration of a laboratory (↑) or industrial potentiometer using a standard cell of known e.m.f. E_o (volt), e.g Weston cadmium cell (p.151); the cell and galvanometer in series are connected in parallel with the potentiometer wire, electrical contact on the wire being made by a sliding contact or tapping key; a p.d. kl (volt) tends to drive electric current via cell E_o upwards through the galvanometer; a terminal p.d. (p.154) less than E_o (volt) tends to drive current through the wire and downwards through the galvanometer; the contact point on the wire is adjusted to length l_o for zero galvanometer deflection; at potentiometric balance point l_o, $E_o = kl_o$ and $E_o/l_o = k$ (Vcm^{-1}), where k is the calibration constant of the potentiometer; the value of k can be varied by including a series rheostat (p.156) in circuit.

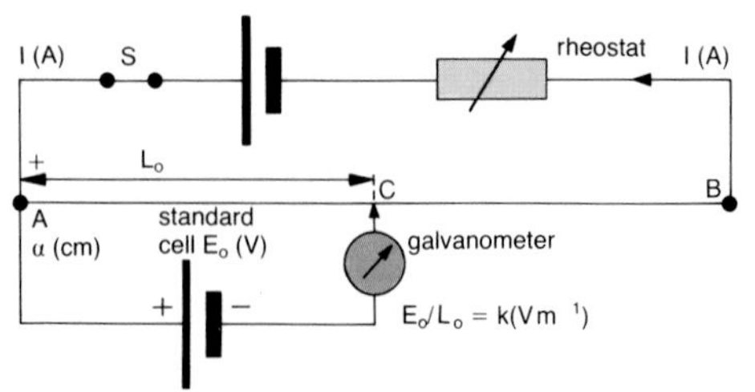

AB is potentiometer wire with soldered clamped ends A and B; α (cm) is end-correction at A.

laboratory potentiometer calibration

thermoelectric effect when a closed conducting circuit (p.154), made from 2 different metals, has a temperature difference between the metal junctions, a small e.m.f., in the range of microvolt ($1\ \mu V = 10^{-6}$ V) to millivolt ($1\ mV = 10^{-3}$ V), is generated in the circuit; the variation of thermoelectric e.m.f. with the temperature difference between the metal junctions is non-linear and the thermocouple characteristic curve (p.164) is approximately parabolic.

Seebeck effect an alternative name for thermoelectric effect (↑).

thermoelectric e.m.f. refers to the electromotive force (p.149) generated by the thermoelectric effect (↑); it is a process by which thermal energy (p.28) supplied as heat to one junction is converted to electrical energy (p.28) in a thermocouple (↓).

thermocouple
copper-iron thermocouple

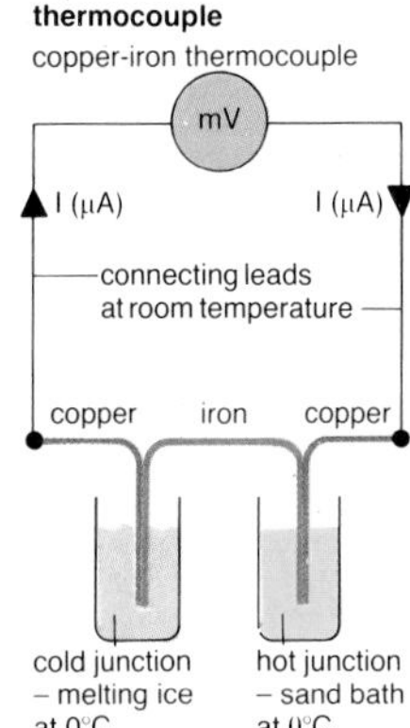

thermocouple (*n*) a closed conducting circuit, made from 2 metals selected from the thermoelectric series (↓), demonstrating the thermoelectric effect (↑) when one junction is heated; a thermocouple characteristic (p.164) is specific for the 2 selected metals and, after calibration or by reference to standard thermocouple tables, the thermocouple can be used as a thermoelectric thermometer (p.120); the metal couple is chosen for the value of its maximum thermoelectric e.m.f. (↑), and to have melting points (p.15) well above the temperature range of measurement; when a number of thermocouples are connected in series, their total e.m.f. is the sum of the individual e.m.f.s, giving a sensitive method of detecting heat, e.g. a thermopile (p.127).

thermoelectric series the direction and magnitude of the thermoelectric e.m.f. (↑) for a number of suitable thermocouple (↑) metals is given by arranging them as a series: antimony, nichrome (80% nickel; 20% chromium), iron, zinc, copper, gold, silver, lead, aluminium, platinum-rhodium alloys, platinum, nickel, constantan (Eureka) (60% copper; 40% nickel), bismuth; across the cold junction of a thermocouple made from 2 metals in this series, the e.m.f. acts in a direction driving current from a metal earlier in the series towards one later in the series, e.g. from iron to copper across the cold junction in a copper-iron thermocouple; the e.m.f. is also greater in magnitude the further apart in the series the 2 metals are.

thermocouple characteristic the characteristic curve of a specific thermocouple (p.163) with its cold junction at a fixed specified temperature, e.g. 0°C (273K), and its hot junction temperature variable throughout the range of measurement of thermoelectric e.m.f. E (microvolt); E (μV) varies with temperature difference θ°C (θ + 273K) between hot and cold junctions and the relation is parabolic of the form $E = \alpha\theta + \beta\theta^2$, where α and β are constants characteristic of the thermocouple; for a specific thermocouple the maximum e.m.f. E (μV) is at the neutral temperature θ_N (↓), characteristic of the thermocouple; e.m.f. E reverses direction at the inversion temperature θ_I (↓); thermocouple calibration throughout the temperature range of measurement enables a thermocouple to be used in temperature measurement.

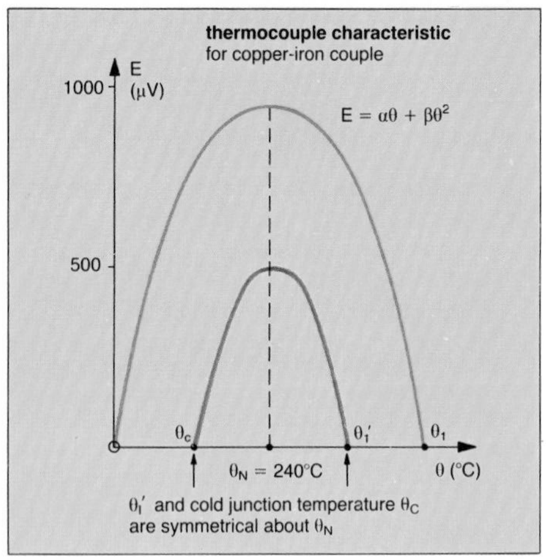

neutral temperature the characteristic temperature for a specific thermocouple (p.163) at which the thermoelectric e.m.f. (p.163) is a maximum, e.g. 240°C for copper-iron thermocouple; denoted by θ_N.

inversion temperature the temperature at which the thermoelectric e.m.f. (p.163) for a thermocouple (p.163) reverses direction; it is denoted by θ_I; θ_I and the cold junction temperature are symmetrical with respect to the neutral temperature (↑) on the thermocouple characteristic (↑).

electric field between 2 equal similar charges

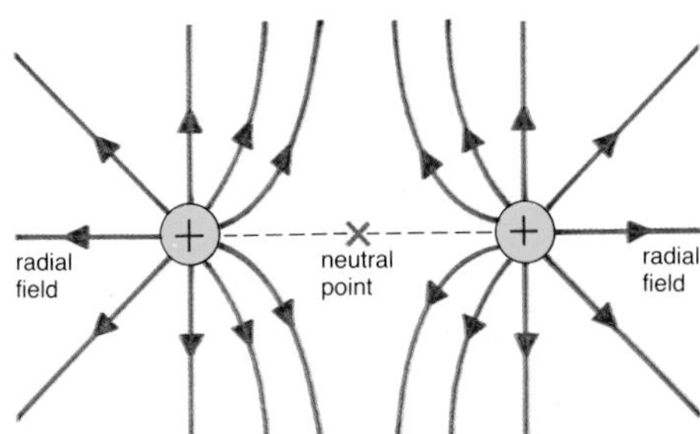

electric field between 2 equal opposite charges

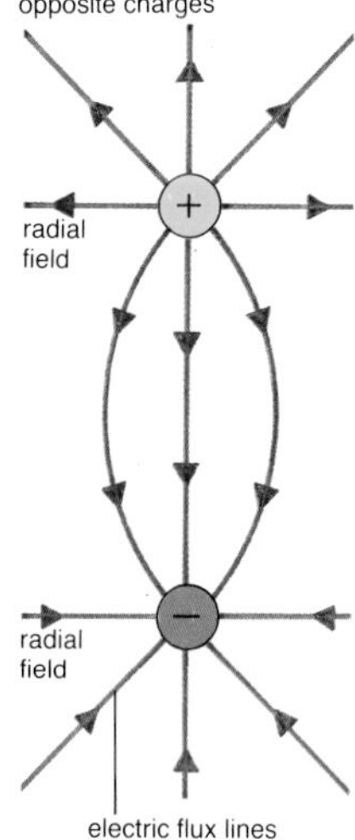

electric field defines the field (p.26) in the region of space around an electric charge (↓), throughout which electric forces (p.166) act; an electric field is a conservative field (p.27).

electric charge a source of electric field forces (p.166); 2 kinds of electric charge are identifiable as electrically opposite in nature, and designated as positive (+) and negative (−); electric charge can only exist in whole number multiples of the electronic charge (p.7); equal amounts of positive and negative electric charge, maintained separate in a region of space, can give a state of electrical neutrality, e.g. neutral atom (p.8); the unit of charge is the coulomb (↓); electric charge transfer through a conducting medium is measured as electric current (p.152).

Millikan's experiment an experiment in which the electric charge (↑) on an oil drop was repeatedly measured with the result that the charge was always a whole number multiple of a basic quantity of charge, now regarded as the fundamental unit of charge, the electronic charge (p.7), 1.6×10^{-19}C.

coulomb (C) the unit of electric charge (↑); the charge/second (Cs^{-1}) transferred through an electrical conductor when electric current 1 ampere (A) (p.152) is flowing.

point electric charge the electric charge (↑) concentrated into a very small region around a point or small spherical charged body; electric flux lines (p.166) are radial.

like charges refers to electric charges (↑) of the same sign, either both positive (+) or both negative (−); repulsive electric forces (p.166) act between them; several like charges all have the same sign.

unlike charges refers to electric charges (p.165) of opposite signs, one positive (+) and the other negative (−); attractive electric forces (↓) act between unlike or dissimilar charges.

electric forces electric field (p.165) forces acting on bodies carrying electric charge (p.165), or on uncharged bodies in which electrical neutrality has been disturbed by inductive charge separation (p.169); the forces can be repulsive or attractive, depending on whether the body's electric charge is of like (p.165) or unlike (↑) sign to the source charge creating the electric field; the strength and direction of electric field forces are used to define electric field intensity (p.169) and direction (p.169); electric forces between point electric charges vary according to an inverse square law (↓); electric deflecting forces act on a cathode-ray electron beam passing through the deflecting plate system of a cathode-ray oscilloscope.

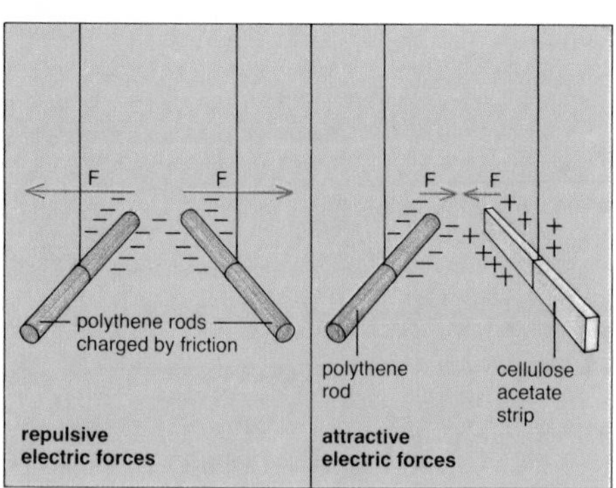

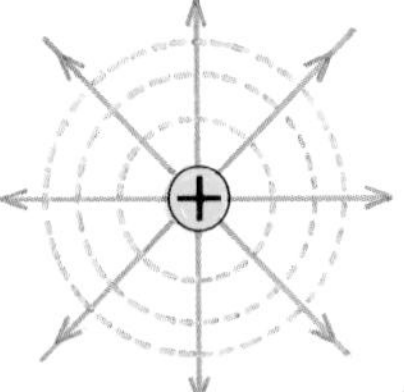

electric flux lines and equipotentials around a point charge

electrostatic forces an alternative name for electric forces (↑), referring specifically to stationary, not moving, electric charges.

electric flux lines lines used to represent diagrammatically the directions in which electric forces (↑) would act on a point positive electric charge (p.165) in an electric field (p.165), and in which the charge would move if free to do so; conventionally a flux line starts on a positive (+) and ends on a negative (−) charge.

electric lines of force an alternative name for electric flux lines (↑).

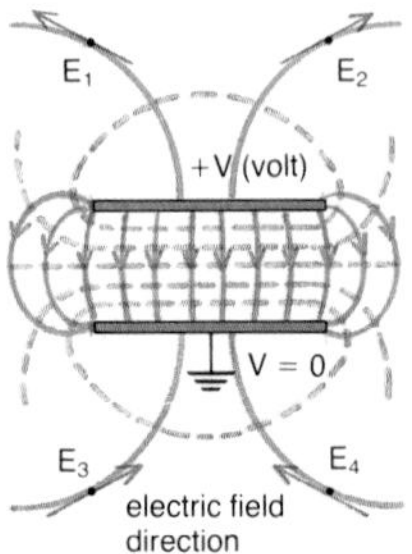

electric flux lines and equipotentials between parallel plates

radial electric field refers to the electric field (p.165) in the region of 3-dimensional space around a point electric charge (p.165) or spherical capacitor; electric flux lines (↑) extend radially outwards from the charge in all directions.

uniform electric field an electric field (p.165) in a region of space in which electric flux lines (↑) are parallel and the electric field strength (p.169) is uniform and its direction constant, e.g. between the plates of a parallel plate capacitor (p.172).

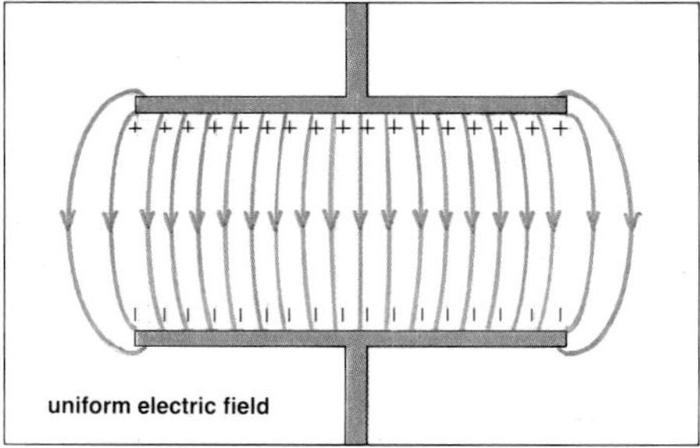

Coulomb's Law the law expressing the variation in attractive or repulsive electric force (↑) F (newton) (p.26) between 2 point electric charges (p.165), q_1 and q_2 (coulomb) (p.165), with distance r (metre) between them, F is directly proportional to the product q_1q_2 of the charges and inversely proportional to the square of distance r between them; $F \propto q_1q_2/r^2$ so the force equation $F = kq_1q_2/r^2$, where k = constant, depending on the medium between the charges; for charges in free space or a vacuum, $k = 1/4\pi\varepsilon_o$, where ε_o = electric permittivity (p.168) of free space; for other dielectric media (p.168), $k = 1/4\pi\varepsilon$, where ε = electric permittivity of the medium; $\varepsilon > \varepsilon_o$ so F is reduced by the presence of a medium.

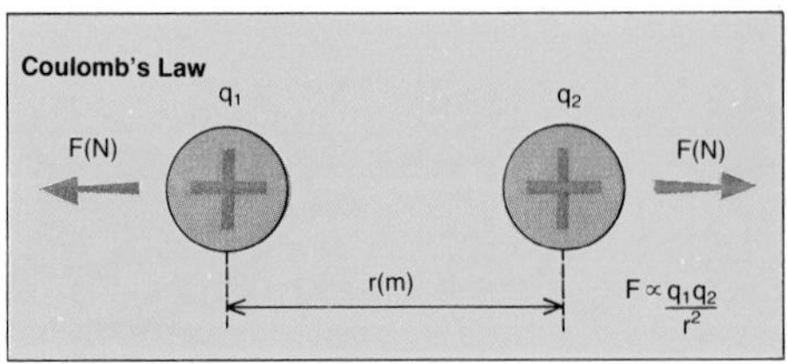

electric permittivity the factor ε_o or ε in Coulomb's Law (p.167) accounting for the effect on force F (N), between electric charges q_1 and q_2 (C), of the medium transmitting the electric field effects; from the force equation, $F = q_1q_2/4\pi\varepsilon_o r^2$ in free space, so $\varepsilon_o = q_1q_2/4\pi Fr^2$, and ε_o and ε have units $C^2N^{-1}m^{-2}$; units are also farad/metre (Fm^{-1}), where 1 farad (F) (p.172) = 1 coulomb/volt (CV^{-1}): for free space, $\varepsilon_o = 8.854 \times 10^{-12}\ Fm^{-1}$, and $1/4\pi\varepsilon_o = 9 \times 10^9\ Nm^2C^{-2}$ approximately; for other dielectric media (↓) the force equation is $F = q_1q_2/4\pi\varepsilon r^2$, where ε = absolute electric permittivity (Fm^{-1}) of the medium; force F is reduced by the factor ε_r compared with its value in free space where ε_r = relative permittivity (↓).

relative permittivity the ratio of permittivity in the medium to permittivity in free space, i.e. $\varepsilon_r = \varepsilon/\varepsilon_o$ and has no units; for free space or a vacuum $\varepsilon_r = 1$, for air at 20°C and atmospheric pressure (p.40) $\varepsilon_r = 1.0005$, so the practical difference made to force F by the presence of air is negligible; for glass ε_r = 5–10, for mica = 6, for pure de-ionized water at 20°C = 81.

dielectric constant a former name for relative electric permittivity (↑) ε_r.

dielectric medium an electrically insulating material (p.155) readily transmitting electric field (p.165) effects, e.g. in a parallel plate capacitor (p.172); its relative electric permittivity (↑) is high compared with air; a solid medium will retain its surface electric charge (p.165) on frictional electrification (↓), unless discharged, and it can insulate a charged conducting body from earth during inductive electrification (↓); the medium becomes conducting when the insulation breakdown intensity (Vm^{-1}) is exceeded.

electrification (*n*) the process of giving an electric charge (p.165) to an uncharged body or surface; usually a frictional (↓) or inductive (↓) process. **electrify** (*v*), **electrical** (*adj*).

frictional electrification electrification (↑) brought about by rubbing the surface of certain solid dielectric media (↑), so that electron (p.7) transfer occurs betweeen the surface and the rubbing agent, giving the surface either a positive (+) or negative (−) electric charge (p.165), e.g. polythene rubbed with cloth receives electrons and acquires a negative (−) surface charge; cellulose acetate rubbed with cloth loses electrons and acquires a positive (+) surface charge.

frictional electrification

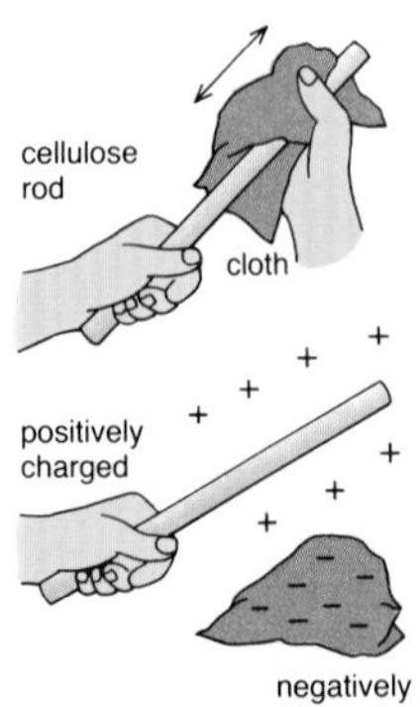

inductive electrification

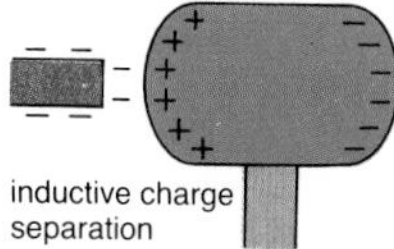
inductive charge separation

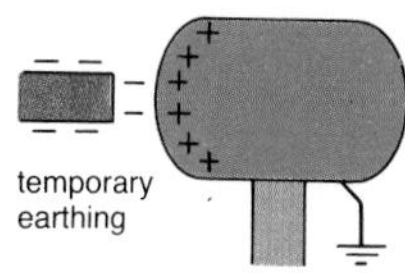
temporary earthing

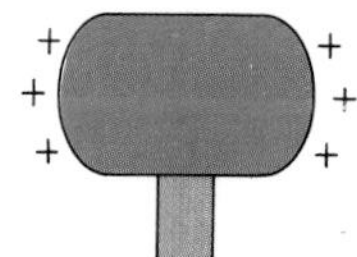
charge redistribution

inductive charge separation when an electrically charged body is held near an electrically insulated metal body, conduction electrons (p.9) move through the metal either towards a positive (+) inducing charge or away from a negative (−) inducing charge; if the metal is now temporarily earthed (p.161) by touching, only the separated charge near the inducing charge is retained on the metal body.

inductive electrification refers to electrification (↑) of an electrically insulated metal body by inductive charge separation (↑); after earthing, the inducing charge is removed and the induced charge spreads over the metal body. **induction** (*n*), **induce** (*v*), **inducing** (*adj*), **induced** (*adj*), **inductive** (*adj*).

electric field direction conventionally defined at a specific point in an electric field (p.165) as the direction in which an isolated point positive (+) electric charge (p.165) located at that point would move if it were free to do so.

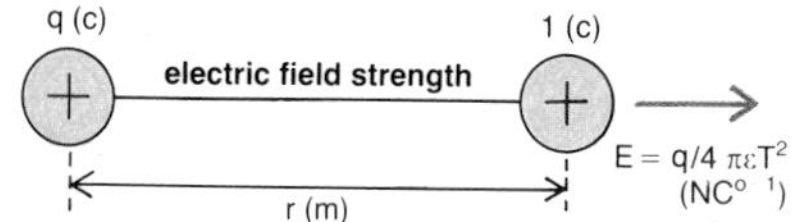

electric field strength
resultant electric field strength E_R at X, due to 4 equal charges of different signs

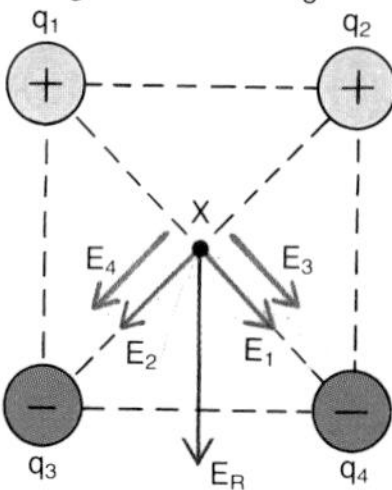

electric field strength a vector quantity (p.30), conventionally defined at a specific point in an electric field (p.165) as the electric field force (p.166) exerted on an isolated point positive (+) electric charge (p.165) of value 1 coulomb (C) (p.165) located at that point; denoted by E and measured in units of newton/coulomb (NC^{-1}); for charge q (C), force F = qE (N); the field vector is tangential to an electric flux line (p.166) at the specific point of definition, giving the electric field direction (↑) at that point; the electric field strength at distance r (m) from a point + charge q (C) is given by Coulomb's Law (p.167) from the force equation: $F = q \times 1/4\pi\varepsilon_0 r^2$ (N), where ε_0 = electric permittivity (↑) of free space, so $E = q/4\pi\varepsilon_0 r^2$ (NC^{-1}) at r (m) from charge q (C); E is also defined from the potential gradient dV/dr (Vm^{-1}) (p.170) as $E = -dV/dr$; a negative (−) sign shows that the potential V decreases in the direction of the field; units are volt/metre (Vm^{-1}); for a uniform field (p.167), $-dV/dr = E$ is constant.

electric intensity a former name for electric field strength (↑).

electric potential a scalar quantity (p.30), defined for a specific point in an electrical field (p.165), or for a specific point on or inside the surface of a charged body, as the work done W (joule) in transferring electric charge 1 coulomb (C) (p.165) from a point outside the field up to the point concerned, work (p.26) being done against the repulsive electric forces exerted by the source charge of the field on a charge of similar sign; the value of work done W (J) is independent of the charge path as the electric field is conservative (p.27); denoted by V; units are volt (p.155), where 1 volt (V) = 1 joule/coulomb (JC^{-1}); the electric potential V at a point distant r (m) from point charge q (C) is calculated from the electric potential difference (↓) between that point and points at infinity (p.67) outside the field, so V (volt) $= q/4\pi\varepsilon_o r$, where ε_o = electric permittivity (p.168) of free space; the electric potential V (volt) is also defined as the physical quantity in an electric field whose variation with distance r (m) at a specific point gives the electric field strength E (volt/m) (p.169) at that point; $E = -\,dV/dr$ where dV/dr (Vm^{-1}) is the electric potential gradient (↓) at the point.

electric potential gradient a vector quantity (p.30) defined as the variation of electric potential V (volt) (↑) with distance r (m) at a specific point in an electric field; $-\,dV/dr = E$ defines the electric field strength E (p.169) at that point; dV/dr is constant for a uniform field (p.167).

electric potential difference² (p.d.) a scalar quantity (p.30) defined, for 2 specific points in an electric field (p.165), as the work done W (joule) in transferring an electric charge of 1 coulomb (C) (p.165) between the 2 points, work (p.26) being done against the repulsive electric forces (p.166) exerted by the source charge of the field on a charge of similar sign; the value of work done W (J) is independent of the charge path between the 2 points as the electric field is conservative (p.27); it is denoted by V and measured in units of volt (V) (p.155), where 1 volt (V) = 1 joule/coulomb (JC^{-1}); for 2 points A and B distant a and b (m) from point charge q (C): $V_{AB} = \int_A^B dW = \int_a^b \text{force}.dr$

$= \int_a^b q/4\pi\varepsilon_o r^2.dr$

$= q/4\pi\varepsilon_o(1/a - 1/b)$ volt, where ε_o = electric permittivity (p.168) of free space; to define electric potential (↑), $b = \infty$, so $V_A = q/4\pi\varepsilon_o a$ for point A.

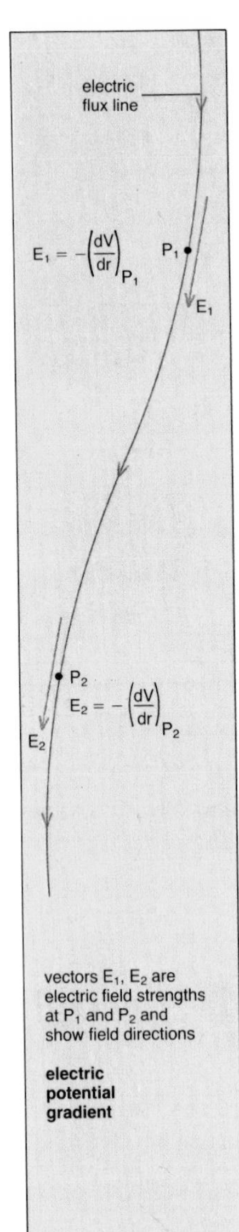

vectors E_1, E_2 are electric field strengths at P_1 and P_2 and show field directions

electric potential gradient

equipotentials

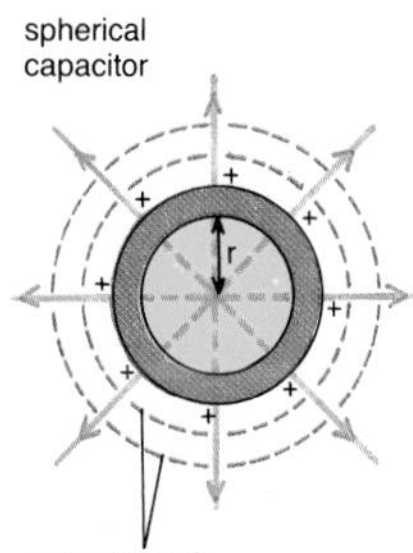

zero potential a reference potential of constant value with respect to which the values of other electric potentials (↑) can be stated; a point at an infinite distance from a point charge is conventionally taken to be at zero potential; the Earth's surface remains at constant potential and provides a suitable electrical contact for earthing (p.161).

earth potential the electric potential (↑) of the Earth's surface remains constant and is assigned the value of zero potential (↑).

equipotential (*adj*) having a constant electric potential (↑), e.g. space inside a closed hollow charged conductor; a surface in space over which all points have the same potential, e.g. concentric surfaces around a point electric charge (p.165), plane surfaces between charged parallel plates; equipotentials are always perpendicular to electric flux lines (p.166) so that no work is done against electric forces in moving a small charge in any direction on an equipotential surface.

capacitor (*n*) a physical system having the property of capacitance (↓), e.g. parallel plate capacitor (↓).

condenser (*n*) a former name for a capacitor (↑).

capacitance (*n*) the property of a conducting physical system, insulated from earth, of retaining an electric charge (p.165) placed upon it at an electric potential (↑) appropriate to the size and shape of the system, e.g. spherical capacitor, parallel plate capacitor (p.172); electrical energy (p.28) associated with the stored electric charge is derived from the work done in charging the system against repulsive electric field forces (p.165) from the source charge of the same sign already on the system; sometimes part of the system is earthed (p.161) to modify its capacitative properties, these are also modified by the presence of a dielectric medium (p.168) of electric permittivity ε_0 or ε (p.168); capacitance is defined as the ratio of charge q (coulomb)/potential V (volt) for a charged system; denoted by C in units of the farad (F), where 1 farad = 1 coulomb/volt (CV^{-1}); practically, the microfarad (μF) which equals 10^{-6}F and the picofarad (pF) which equals 10^{-12}F are the most useful units; the value of C (F) = q/V (CV^{-1}) depends on the size and shape of the charged system and on the presence of dielectric medium; its value is identified by the capacitor colour code (p.175). **capacitative** (*adj*).

farad (F) unit of capacitance; 1 (F) = 1 coulomb/volt (CV^{-1}); 1 microfarad (μF) = 10^{-6}F.

electric shielding the surrounding of a body or volume of space with a conducting enclosure, to maintain it at constant electric potential within an equipotential (p.171) volume free of electric field (p.165) effects; the shielding surface need not be solid; a wire cage can be used as a safety shield for a person working in a region of high electric field, e.g. near high voltage generating equipment (p.178), where the danger exists of a discharge to earth through the body.

parallel plate capacitor a capacitor consisting of 2 parallel metal plate electrodes (p.10), of common area A (m^2), separated by a dielectric medium (p.168) of electric permittivity ε or ε_0 (p.168) maintaining them d (m) apart; electric field strength E (p.169) = σ/ε for uniform electric field (p.165) between the plates, where σ = electric charge density (Cm^{-2}) (p.177); for plates d (m) apart: work done in transferring 1 coulomb (C) between the plates = (σ/ε) × d (joule), giving a potential difference of V (volt) (p.170) between the plates; the capacitance C (F) = charge σA (C)/p.d. V (volt) = σA ÷ (σ/ε) × d, so C = εA/d (F); the dielectric medium can be paper, mica, ceramic, air, aluminium oxide, silicon dioxide.

charged parallel plate capacitor

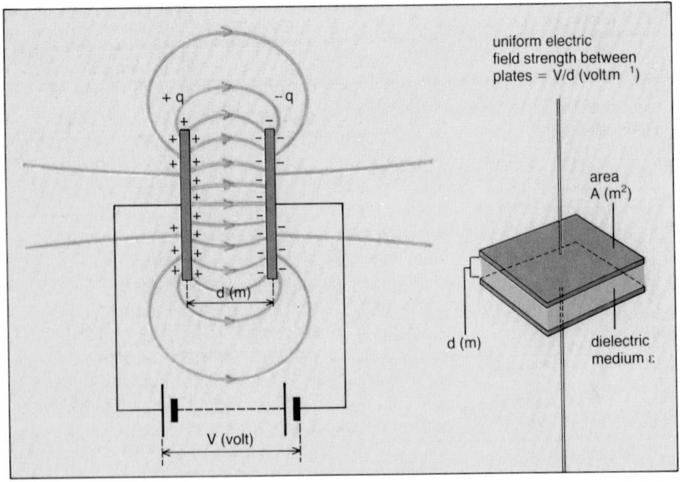

paper capacitor defines a capacitor in which the electrodes (p.10) are 2 long strips of thin metal foil, interleaved with 2 similar strips of waxed or oiled paper, as a dielectric medium (p.168) of electric permittivity ε (p.168) about 5; the layered strip is rolled up tightly into a cylinder, external connections being made from the electrodes to outside terminals on the cylindrical or rectangular metal container; the range of capacitance (p.171) values is from about 10^{-3} to 10μF, ± 10%; the frequency (p.54) range for paper dielectric is 100 Hz to 1 MHz, the system forms an inexpensive and widely used parallel plate capacitor (↑) of larger surface area but poor stability.

paper capacitor

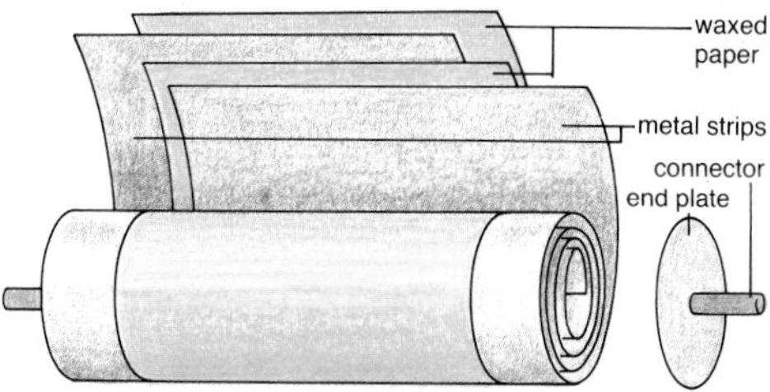

polyester capacitor a capacitor made on the same principle as the paper capacitor (↑) but using polyester film as the dielectric (p.168); in the metallized version, films of metal are deposited on the plastic to act as plates; capacitance values range from 0.01μF to 10μF; they can be used over a wide frequency range.

mica capacitor a capacitor in which the electrodes (p.10) are interleaved sheets of metal foil, separated by thin sheets of the mineral mica as dielectric medium (p.168) of electric permittivity ε (p.168) about 6; alternate metal sheets are joined to a common connecting wire, each electrode being a stack of sheets; external connections are made from the electrodes to outside terminals on the sealed plastic case; construction of electrodes by deposition of metal on mica is also used; the range of capacitance (p.171) values is from about 5pF to 0.01μF, ± 1%; its frequency (p.54) range is up to about 300MHz; it has sufficient stability and accuracy in its specified capacitance value for use as a laboratory standard; the system forms a parallel plate capacitor (↑).

mica capacitor

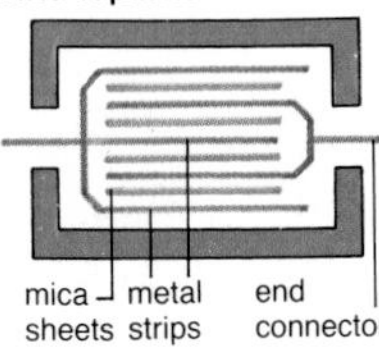

ceramic capacitor the capacitor uses a ceramic dielectric medium (p.168) in tubular shape, with silver metal deposited on its inside and outside surfaces as electrodes (p.10); external connections are made to metal wires soldered round the silvered ends of the ceramic tube; the tube is lacquered, or placed in another containing tube; electric permittivity ε (p.168) depends on the ceramic mixture and can be 100 or more, though stability is reduced at high ε; different mixtures can be used for different parts of the frequency (p.54) range; the system forms a parallel plate capacitor (p.172).

electrolytic capacitor the capacitor electrodes (p.10) are 2 long strips of thin aluminium or tantalum foil, interleaved with 2 similar strips of paper soaked in ammonium borate electrolyte (p.10); the layered strip is rolled up tightly into a cylinder, external connections being made from the electrodes to outside terminals on the container; on passing an electric current (p.152) through the system, a layer of oxide forms on the anode (p.10), forming the dielectric medium (p.168) of the capacitor; a layer thickness of about 10^{-4} mm gives a capacitor of very small size; a small d.c. current (μA) must flow in the initial direction during capacitor action to maintain deposition of the oxide layer and the positive (+) terminal is marked externally; capacitance values are large but not always stable, the frequency (p.54) range being up to 50 kHz; an electrolyte in contact with metal foil acts as the cathode (p.10); the system forms a parallel plate capacitor (p.172).

variable air capacitor the capacitor electrodes (p.10) are 2 stacks of interleaved parallel metal plates, separated by air as dielectric medium (p.168); one stack can be rotated relative to the other, varying area A (m^2) of overlap and hence capacitance $C = \varepsilon A/d$ (F) for the parallel plate capacitor (p.171); it can be used at all frequencies (p.54), particularly for tuning the acceptor circuit of a radio receiver to resonant frequency.

microelectronic capacitor the capacitor electrodes (p.10) are a semiconductor (p.222) material layer and an aluminium layer, separated by a thin layer of silicon dioxide as a dielectric medium (p.168); capacitance (p.171) values are very small and the unit is often replaced by an arrangement of microelectronic transistors (p.226) in large scale integrated circuits (p.227).

ceramic capacitor

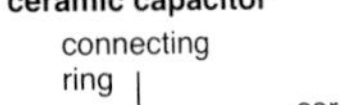

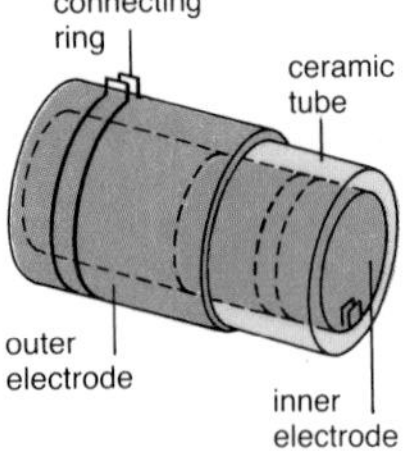

electrolytic capacitor

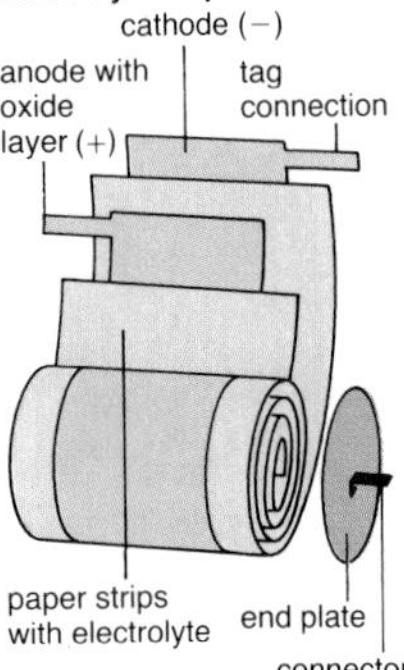

variable air capacitor

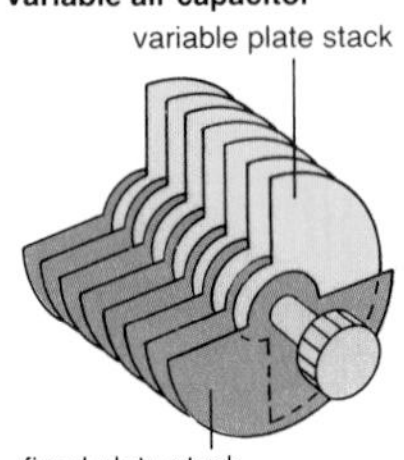

capacitor colour code
mica capacitor

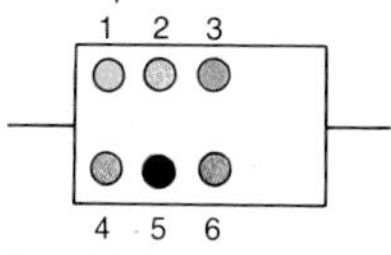

C = 460 pF ± 20%
foil type + 350 V d.c.

ceramic capacitor

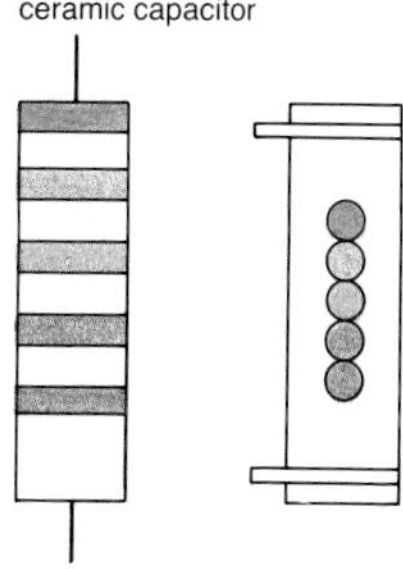

C = 56 000 pF ±2%
temperature coefficient
− 350 K⁻¹

capacitor colour code a method of indicating approximate values of mica and ceramic capacitors (p.171) by colour markings similar to that used for carbon mixture resistors (p.156); the Standard Colour Code is Black 0: Brown 1; Red 2; Orange 3; Yellow 4; Green 5; Blue 6; Violet 7; Grey 8; White 9; for mica capacitors with body spot markings 1–6: the value of capacitance (pF) (p.171) is given by the 1st and 2nd spots on the top row for the digits and 3rd spot on the bottom row for the number of zeros; tolerance is expressed as % accuracy in value, indicated by the 2nd spot on the bottom row: gold ±5%, silver ±10%, black or no marking ±20%; 1st spot on the bottom row indicates the maximum d.c. working voltage (V): red 350 V, green 750 V, white 2000 V; 3rd spot on the top row is green for metal foil electrodes and red for metal deposition on mica electrodes; for ceramic capacitors with a body band or spot markings the value of capacitance (pF) is given by the 1st and 2nd bands for the digits and the 3rd band for the number of zeros; tolerance is given by the 4th spot colour: brown ±1%, red ±2%, orange ±2.5%, green ±5%. white ±10%, black or no marking ± 20%; an end band indicates the variation in stability of value with temperature, the temperature coefficient (K^{-1}) ranging from −750 violet to +100 white, black or no marking being ± zero.

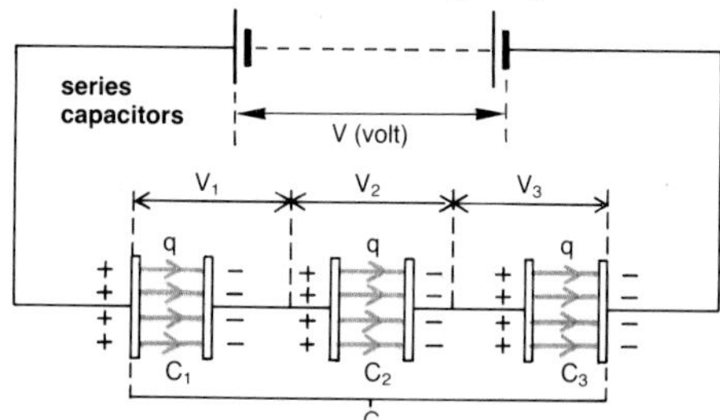

series capacitors a method of connecting 2 or more capacitors (p.171) so that each capacitor C_1, C_2 and C_3 (μF) receives the same charge q (coulomb); the total p.d. V across the series arrangement of C_1, C_2 and C_3 is the sum of p.d.s V_1, V_2 and V_3 across the individual capacitors; the effective capacitance C_s (μF) of the single circuit component equivalent to the series arrangement: $1/C_s = 1/C_1 + 1/C_2 + 1/C_3$.

parallel capacitors a method of connecting 2 or more capacitors (p.171) so that each has the same electric potential difference V (volt) (p.170) between its plates; the effective capacitance C_p (μF) of C_1, C_2 and C_3 in parallel is $C_p = C_1 + C_2 + C_3$.

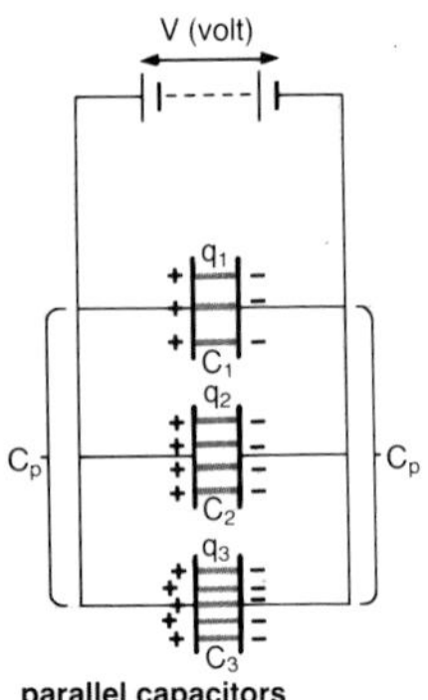

parallel capacitors

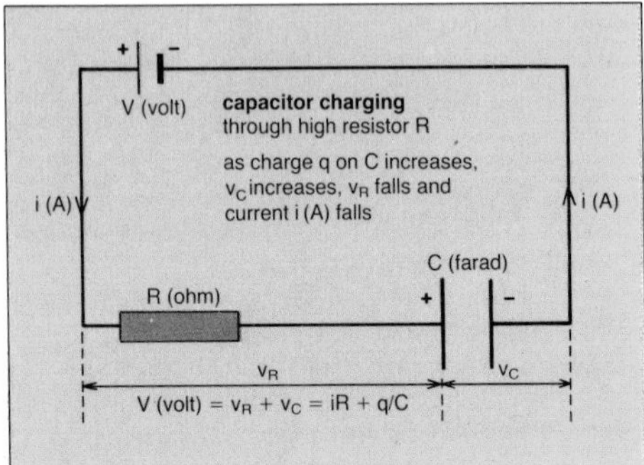

capacitor energy for a parallel plate capacitor of capacitance (p.171) C (F), having electric charge q (coulomb) (p.165) and electric potential difference V (volt) between its plates, work is done against the repulsive forces of plate charge q in adding further increments of charge δq to give a total charge Q (C) at a final p.d. V (volt); the work done in increasing plate charge from zero to Q is stored as energy of the charged capacitor, for release as electrical energy (p.28) on capacitor discharge; the total work done $= \int_0^Q V.dq = \int_0^Q q/C.dq$, since $v = q/C$, and the capacitor energy $= \frac{1}{2}Q^2/C = \frac{1}{2}CV^2 = \frac{1}{2}QV$ (joule).

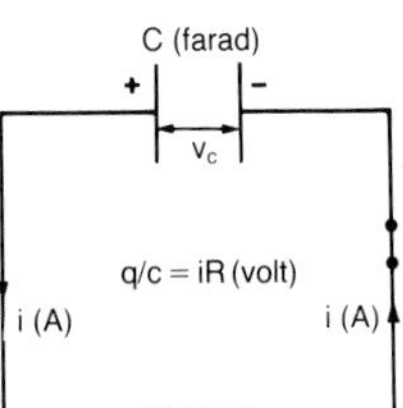

capacitor discharging
through high resistor R

as charge q on C increases, v_c decreases and current i (A) falls

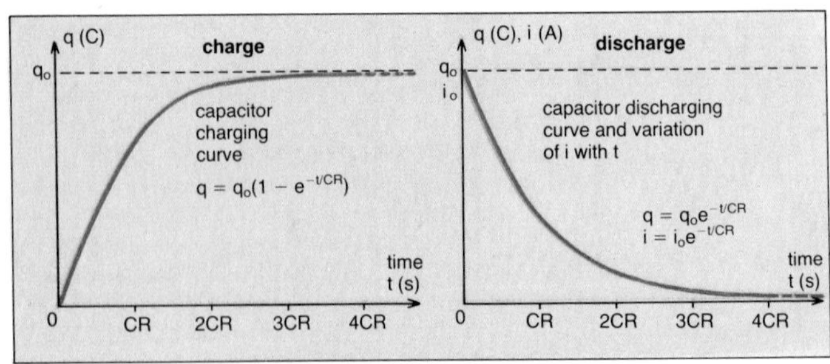

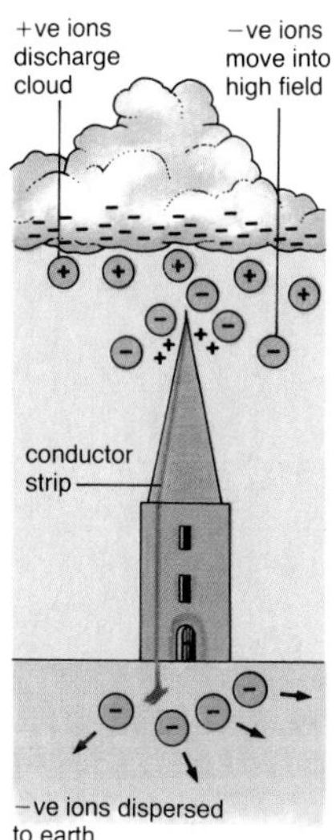

corona discharge

electric charge density the ratio of the electric charge q (C)/area A (m^2) for a charged surface; denoted by σ, units are coulomb/m^2 ($C\,m^{-2}$).

electric charge distribution the electric charge density (↑) is uniform for a charged sphere; σ is inversely proportional to the radius of curvature r of the surface and the charge distribution depends on the surface curvature 1/r; for a charged body of non-uniform curvature, e.g. pear-shaped conductor, σ is a maximum where r = 0, and the charge concentrates in regions of maximum curvature around points, giving sufficiently high electric field strength $E = \sigma/\varepsilon$ ($V\,m^{-1}$) to cause air insulation around the charged conductor to fail and corona discharge (↓) to begin.

corona discharge the electric charge distribution (↑) over the surface of a charged body gives maximum electric charge density σ (↑) near points or sharp edges; the electric field (p.165) strength $E = \sigma/\varepsilon$ ($V\,m^{-1}$) can be very high and the few ions (p.9) present in normally non-ionized air are attracted towards the point; as accelerating ions are drawn into the high field region they collide with neutral air molecules, producing more ions in the process of ionization by collision; electric charge (p.165) is exchanged between the conductor surface and incoming ions carrying charge of opposite sign, and surface charge is gradually lost; for surface charge above about 3×10^{-5} ($C\,m^{-2}$) the charge loss is extremely rapid and corona breakdown occurs.

action of points an alternative name for corona discharge (↑).

sparking potential the critical value of electric field strength (p.165) for which spark discharge occurs in a dry gas at normal temperature and pressure (p.39); its value is characteristic of the gas concerned, e.g. for air 3×10^6 ($V\,m^{-1}$).

lightning conductor a sharply pointed conducting rod, projecting above a tall building and leading to earth; storm clouds above the earth, highly charged by frictional electrification (p.168) create a sufficiently high electric field strength (p.165) between clouds and earth for spark discharge to occur as lightning; inductive charge separation (p.169) from the earth gives a sufficiently high electric field strength near the rod tip to produce corona discharge (↑) and prevent spark discharge; the charge passes to earth through the conductor, by-passing the building.

high voltage generator a machine using electrical energy input to generate high voltage (p.155) output, e.g. d.c. induction coil (p.209), a.c. step-up transformer, Van de Graaff electrostatic generator (↓); the equipment must be designed to avoid corona discharge (p.177) by surrounding all its points and sharp edges with uniformly curved surfaces.

Van de Graaff generator

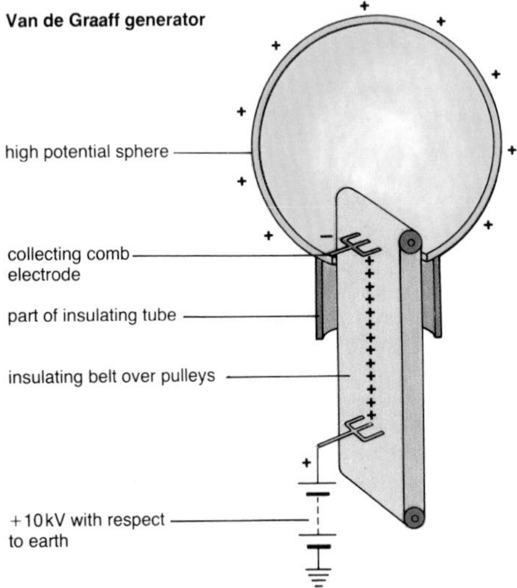

Van de Graaff generator a high voltage electrostatic generator (↑) giving an output up to 5 million volt (V); the electric charge (p.165) is stored at high electric potential (p.170) on a hollow metal sphere of large diameter, e.g. 5m, supported on an insulating tube of height about 15m and immersed in a high pressure insulating gas of high sparking potential, e.g. freon CCl_2F_2; an insulating belt, passing over 2 pulleys, is driven up the tube by a high horsepower (p.27) motor; near the bottom, a row of pointed comb electrodes, at + 10000 volt (V) with respect to earth potential (p.171), pass very close to the belt; an electric field of high electric intensity (p.169) near the points produces inductive charge separation (p.169) in the belt, withdrawing electrons to the points by corona discharge action (p.177) and leaving the belt positively

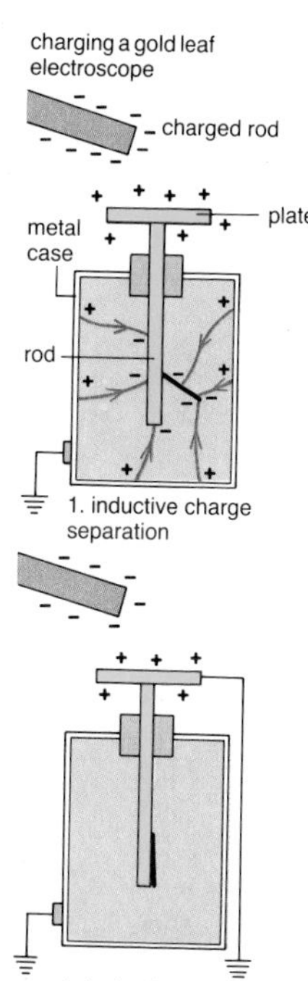

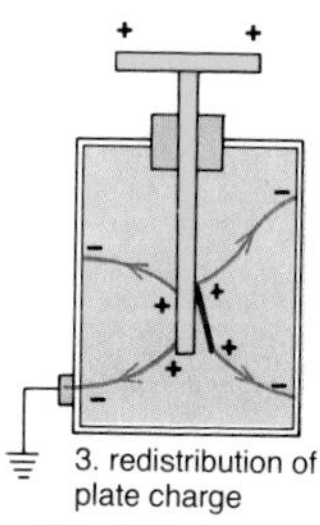

gold leaf electroscope

(+) charged; a second set of pointed comb electrodes, fixed to the inside of the sphere, pass very close to the top of the charged belt; inductive charge separation in the metal sphere gives high electric field strength (p.165) near the points, and electrons pass to the belt by corona discharge action, discharging the belt and leaving the sphere positively (+) charged; the energy of charge, gradually built up on the sphere, comes from the work done by the motor in driving the charged belt upwards against the repulsive electric forces (p.166) exerted by the similarly charged sphere; the machine is used to operate high voltage X-ray machines and high energy particle accelerators in nuclear fission (p.238) experiments.

electrometer a device for measuring potential difference, usually consisting of a very high impedance (p.219) amplifier and commonly using semiconductor (p.222) devices.

directly coupled (d.c.) amplifier an amplifier whose stages are directly coupled and are not joined by capacitors as in a normal amplifier; it acts as a very high resistance voltmeter (about $10^{13}\Omega$) and can be adapted to measure current and charge; in the latter case it is used as an electrometer (↑).

gold leaf electroscope an instrument for the detection of electric charge (p.165) and for approximate measurement of charge magnitude; a circular metal plate and rod are insulated by a perspex collar from the metal case, usually made from a thin metal strip lining a wooden box with glass windows; a thin gold leaf is attached to the rod; an external terminal connects to the metal lining and the electroscope is usually earthed (p.161) through the box; inductive charge separation (p.169) occurs in the plate, rod and leaf system when a charged rod is held near the plate; temporary earthing, followed by removal of the charging rod, leaves the electroscope inductively charged with a sign opposite to that of the source charge, causing the leaf to deflect; when a charge of the same sign is brought near the cap, the charge repelled to the leaf and rod increases the leaf deflection; for an external charge of opposite sign, the charge withdrawn from the leaf and rod decreases the leaf deflection; leaf deflection indicates the electric potential difference (p.170) between the plate, rod and leaf system and the metal case and varies as this p.d. varies.

magnetic field
due to bar magnet

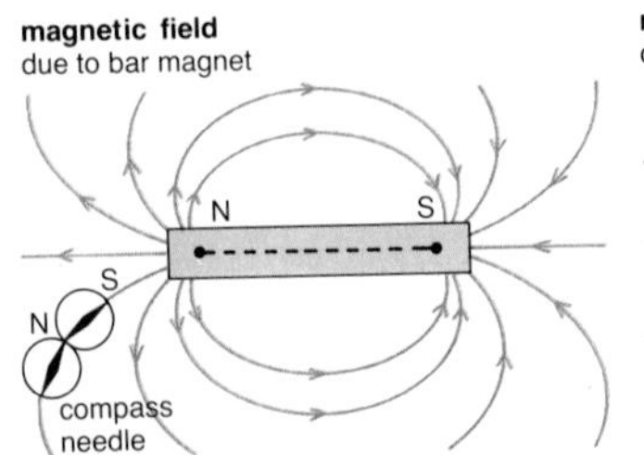

magnetic field
due to current-carrying solenoid

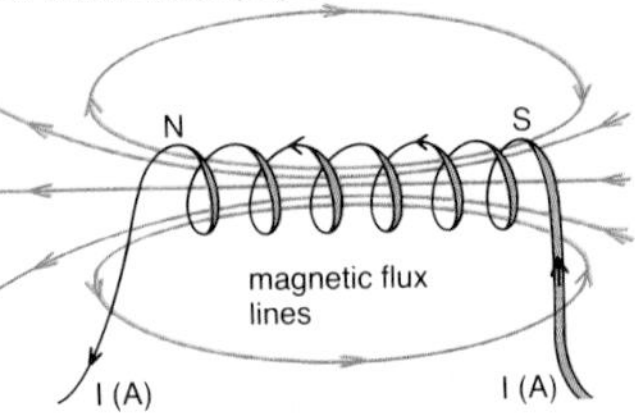

magnetic field refers to the physical field in the region of space around a source of magnetism (↓), throughout which magnetic forces (↓) can act.

magnetic forces magnetic field (↑) forces (p.26) acting on other sources of magnetism (↓) or upon unmagnetized ferromagnetic (p.184) materials, e.g. iron, steel; the forces can be repulsive or attractive depending on whether the magnetic polarity (↓) of the source exerting them is like or unlike that of the source experiencing them; for unmagnetized ferromagnetics, the forces are always attractive due to induced magnetism; the strength and direction of magnetic field forces are used to define magnetic field intensity and direction (p.183); magnetic forces between point magnetic poles (↓) vary according to an inverse square law.

magnetization (*n*) the process of making an unmagnetized sample of ferromagnetic material into a magnet (↓) using the magnetizing action of a source of strong magnetic field (↑), e.g. current-carrying solenoid; magnetism (↓) can be temporary, e.g. for soft iron, or permanent, e.g. for hard steel, depending on the magnetic remanence (p.186) of the sample. **magnetize** (*v*), **magnetized** (*adj*), **magnetizing** (*adj*).

magnetic field
due to straight wires carrying current

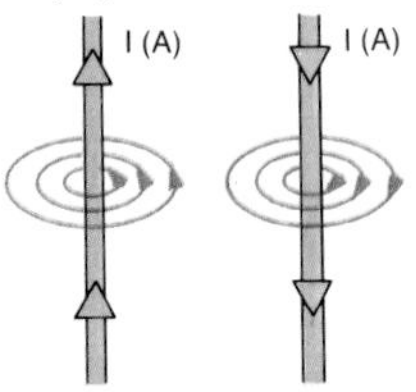

magnetic field
due to plane coil carrying current

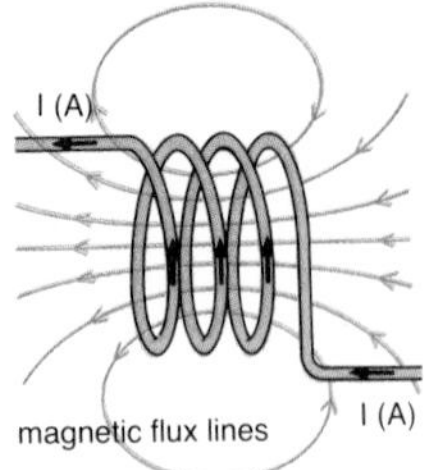

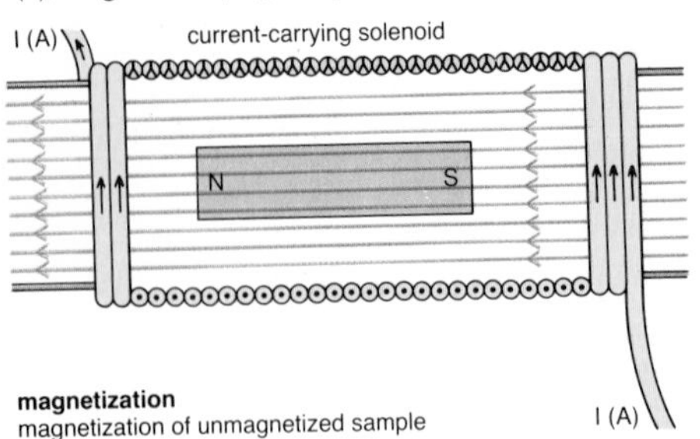

magnetization
magnetization of unmagnetized sample

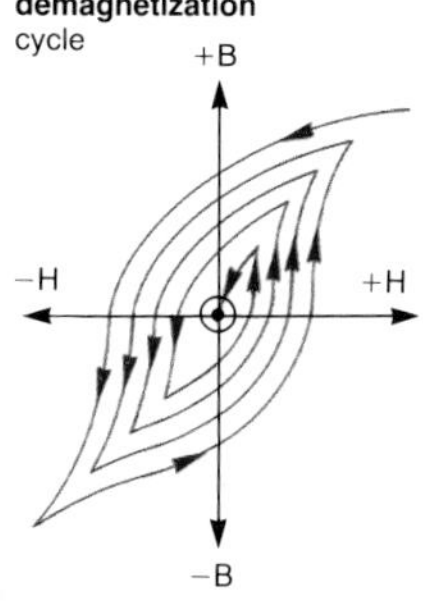

demagnetization cycle

demagnetization (*n*) the process of making a magnetized (↑) sample of ferromagnetic material into an unmagnetized sample, using the demagnetizing action of a source of strong reverse magnetic field (↑), e.g. current-carrying solenoid; demagnetization can be easy, e.g. for soft iron, or more difficult, e.g. for hard steel, depending on the coercive field (p.186) for the sample; the sample is placed in a solenoid carrying alternating current (p.151) and slowly withdrawn to a distance beyond the influence of the field; alternatively the sample is placed in the solenoid and the a.c. current is reduced slowly to zero, taking the sample round a hysteresis loop (p.187) of slowly decreasing area. **demagnetize** (*v*), **demagnetized** (*adj*), **demagnetizing** (*adj*).

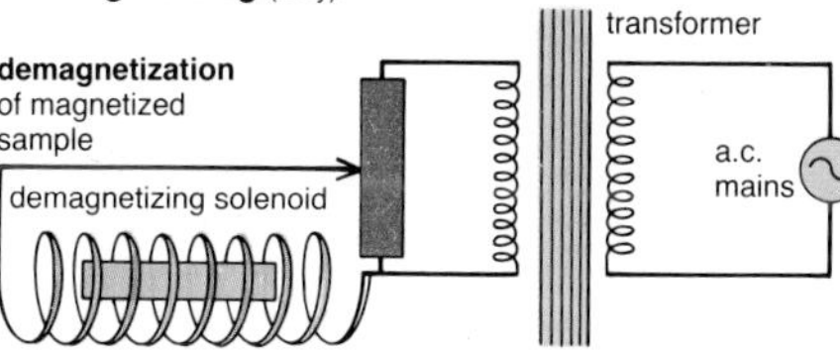

non-magnetic stirrup
Earth's field direction
N
N
twist-less thread
S
magnetic axis
compass needle

alignment of suspended magnet and compass needle in Earth's field

magnetism (*n*) the study of magnetic field phenomena; the property of a source of magnetism by which it can exert magnetic forces (↑) throughout its magnetic field (↑).

magnet (*n*) any source of magnetism (↑) demonstrating characteristic magnetic bipolarity (↓); usually a ferromagnetic (p.184) specimen in the shape of a bar, horse-shoe, or the pole-pieces of an electromagnet (p.195). **magnetic** (*adj*).

magnetic poles points within a magnet (↑) from which magnetic forces (↑) apparently originate; centres of attraction or repulsion for bodies experiencing magnetic field forces; 2 types of pole are identifiable as magnetically opposite in nature, always occurring in opposite pairs, and are designated as North-seeking (N) poles or South-seeking (S) poles, from the direction in which the freely suspended magnet or a pivoted compass needle (p.182) would take in the Earth's magnetic field (p.189); the 2 poles are usually symmetrically located at short distances from the magnet's ends. **polarity** (*n*).

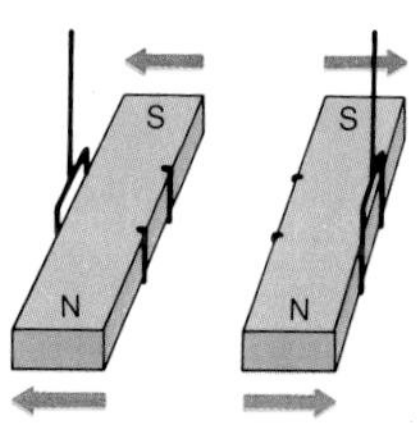

repulsive forces between **like poles**

like poles magnetic poles (↑) of the same polarity, either both N poles or both S poles; repulsive magnetic forces (↑) act between them.

unlike poles magnetic poles (p.181) of opposite polarity, one N pole and one S pole; attractive magnetic forces (p.180) act between them.

compass needle a magnetized needle pivoted (p.33) at the centre of a circular scale graduated in degrees °; in the absence of other magnetic influences, the needle takes the direction of the Earth's magnetic field (p.189).

induced magnetism refers to the magnetization (p.180) of an unmagnetized sample of ferromagnetic material (p.184) by placing it near to, or in contact with, a strong source of magnetism (p.181), e.g. a permanent bar magnet (p.181); the magnetic field (p.181) of the magnetizing source induces in the sample opposite magnetic polarity to the nearest pole of the magnetizing source, so that the magnetized sample experiences an attractive force between unlike poles; induced magnetism is lost from soft iron on removal from the magnetizing source, but is retained by hard steel, due to higher remanence (p.186); the Earth's magnetic field (p.189) is capable of producing induced magnetism in structures made from ferromagnetic materials, e.g. steel, and also in ferromagnetic rocks, giving information about the geomagnetic (p.188) history of the Earth. **induction** (*n*), **induce** (*v*), **inducing** (*adj*).

unlike poles
attractive forces between unlike poles

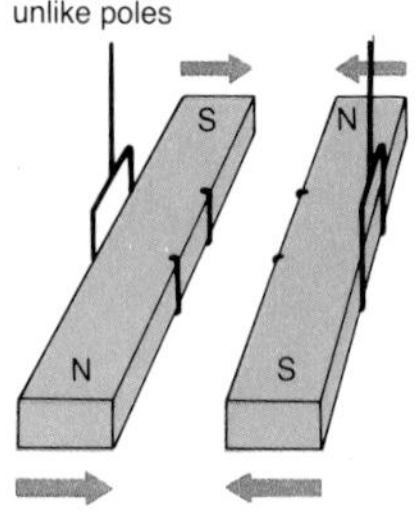

compass needle

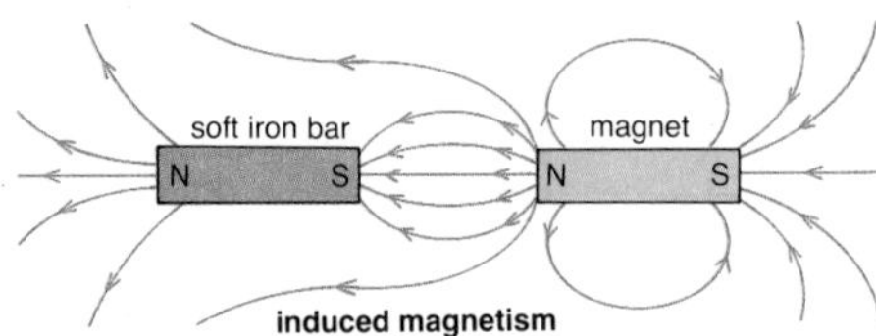

induced magnetism

magnetic flux lines refers to lines used to represent diagrammatically the directions in which magnetic forces (p.180) would act on an isolated point North pole in a magnetic field (p.180); conventionally, a flux line starts on a N pole and ends on a S pole.

magnetic lines of force an alternative name for magnetic flux lines (↑).

radial magnetic field the magnetic field (p.180) in the region of 3-dimensional space around an isolated point pole; magnetic flux lines (↑) extend radially outwards from the pole in all directions; for cylindrical pole pieces around the cylindrical central core of a moving coil galvanometer, the field in the separating air gap is radial with respect to the armature axis of rotation.

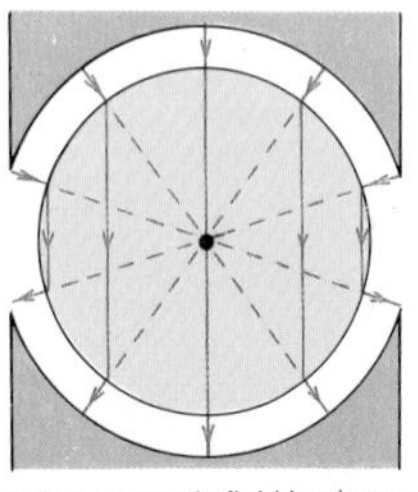

radial magnetic field in air gap between cylindrical pole faces and armature

radial magnetic field

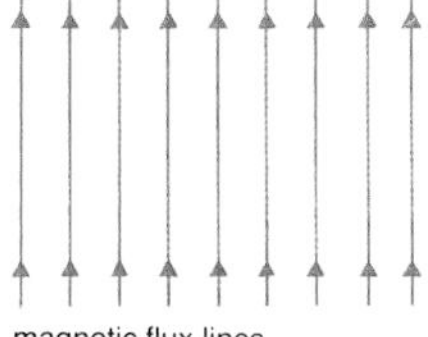

uniform magnetic field the magnetic field (p.180) in a region of space in which magnetic flux lines (↑) are parallel, and the magnetic field strength (↓) is uniform and its direction (↓) constant, e.g. between the pole-faces of a large electromagnet (p.195), at localized regions on the Earth's surface (p.189).

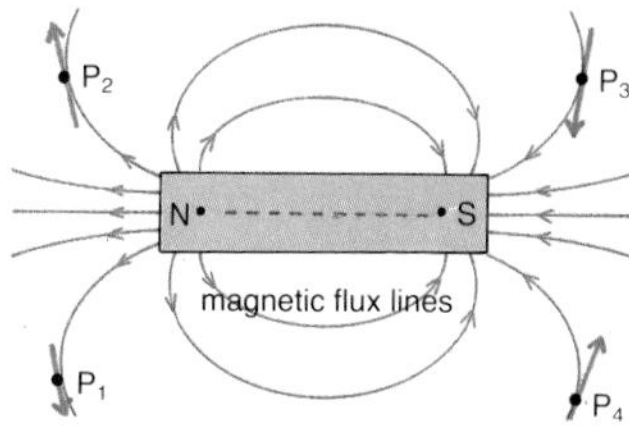

magnetic field direction
at points P_1, P_2, P_3 and P_4

magnetic field direction it is conventionally defined at a specific point in a magnetic field (p.180) as the direction of the force experienced by a N pole (p.181) located at that point, or the direction in which an isolated point N pole would move if it were free to do so; the direction is tangential to the flux line at the point.

Hall effect when a magnetic field (p.180) of magnetic flux density B ($Wb\,m^{-2}$) (p.185) is applied perpendicular to the direction of flow of electric current I (ampere) (p.152) in a conductor, e.g. metal wire or semiconducting layer, the charge carriers (p.9) experience a force perpendicular to B and I in a direction given by the Left Hand Rule (p.192), and move with this force to opposite sides of the conductor, establishing an electric potential difference (p.170) perpendicular to B and I; this p.d. increases to maximum value V_H, the Hall voltage, of value a few millivolts in a semiconductor (p.222), when the drift of charge carriers ceases; $V_H = BI/net$, where n = number of charge carriers/m^3, e = electric charge (C) on each carrier, and t = conductor thickness (m) in direction of field B.

Hall probe a semiconductor device using the Hall effect (↑) to measure the magnetic flux density B ($Wb\,m^{-2}$) of a magnetic field applied perpendicular to the direction of flow of electric current I (A) through the device; the Hall voltage $V_H = BI/net$, so $B = V_H net/I$; since e and t are constant, and n is constant under fixed conditions for the semiconductor, B can be calculated if V_H and I are measured.

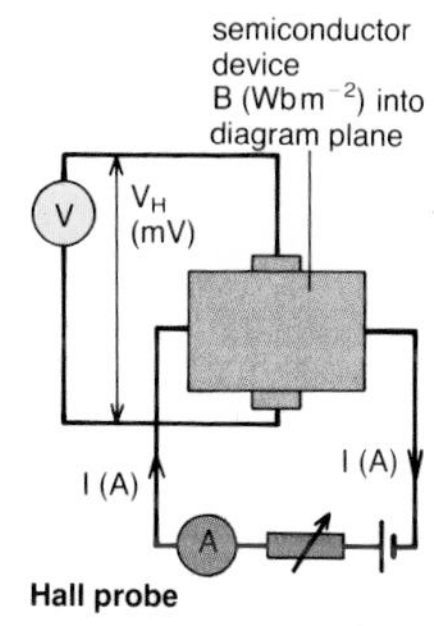

Hall probe

ferromagnetism (*n*) a magnetic property of certain metals, e.g. iron, cobalt, nickel and their alloys, which acquire very strong magnetism in an applied magnetic field; magnetization curve (p.186) of B/H is non-linear and dependent on the applied field H (↓), showing magnetic hysteresis (p.186) and remanence (p.186); magnetic permeability (↓) $\mu > 1$ and varies with applied field H and temperature; the magnetic flux density B is very much greater than H and the magnetic domain structure (↓) can be observed in ferromagnetic samples. **ferromagnetic** (*adj*).

magnetic domain structure refers to internal regions in ferromagnetic (↑) materials in which individual molecular magnets are already aligned parallel to each other, the domain having saturation magnetization (p.186) in the absence of an applied magnetic field; the application of a magnetizing field B (↓) causes growth by boundary displacement of domains already aligned with the field, then rotation of non-aligned domains to saturation; domain volumes are 10^{-6} to 10^{-2} cm^3, with 10^{17} to 10^{21} molecules, depending on the crystal lattice structure of the metal or alloy; domains and boundary movements are observable under a microscope illuminated with polarized light (p.99).

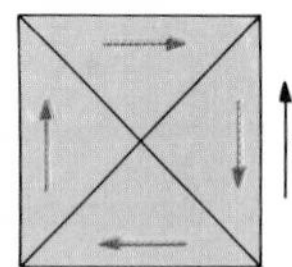

1.random domain orientation in unmagnetized ferromagnetic sample

magnetic domain structure

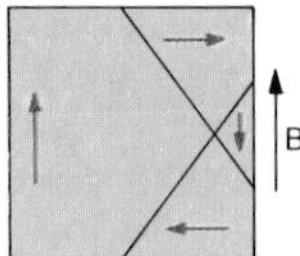

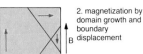

2. magnetization by domain growth and boundary displacement

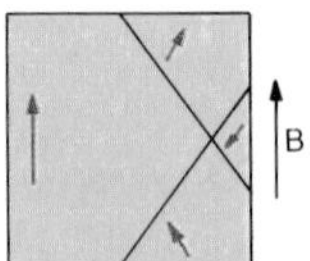

3. magnetization by domain rotation into alignment with the applied field B

ferrimagnetism (*n*) magnetic property of ferrite (↓) materials; adjacent molecular magnets are aligned anti-parallel, but have unequal strength, so that the specimen overall has strong resultant magnetization, comparable with ferromagnetism (↑). **ferrimagnetic** (*adj*).

ferrite (*n*) a material showing ferrimagnetic (↑) properties, e.g. ceramic mixed oxides of Barium (BaO) or Strontium (SrO) with iron oxide Fe_2O_3, of chemical formulae $BaO.6Fe_2O_3$ and $SrO.6Fe_2O_3$; ceramic magnetic materials are mechanically hard though brittle, inexpensive and light in weight, and electrically insulating (p.155), so that they do not suffer eddy current (p.210) energy losses; they are used for transformer cores.

magnetic flux density

normal to plane A (m^2)

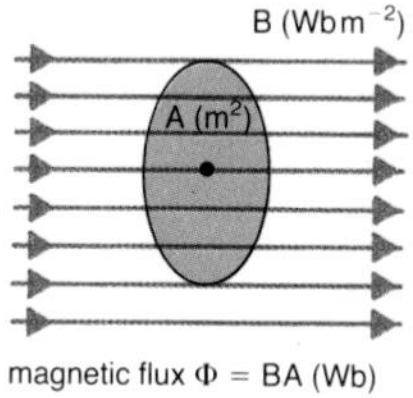

magnetic flux Φ = BA (Wb)

incident at angle θ on plane A (m^2)

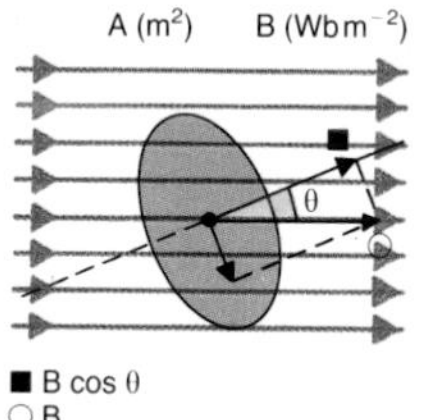

■ B cos θ
○ B

magnetic flux perpendicular to plane A (m^2) = BA cos θ (Wb)

magnetic flux density a vector quantity (p.30), conventionally defined at a specific point in a magnetic field (p.180); it gives a measure of magnetic field strength which includes both the magnetizing field (↓) and the effects of the magnetization of the medium throughout which the field forces act; denoted by B and measured in units of weber/m^2 ($Wb m^{-2}$), where weber is the unit of magnetic flux (p.204) and 1 $Wb m^{-2}$ = 1 tesla (T)(p.193); B can be measured by current balance (p.193), or by Hall probe (p.183); flux lines of an applied magnetic field converge on a ferromagnetic (↑) sample, giving high magnetic flux density in the sample due to strong magnetization effects.

magnetic induction an alternative name for magnetic flux density (↑).

magnetic field strength a vector quantity (p.30), conventionally defined at a specific point in a magnetic field (p.180), giving a measure of that part of magnetic flux density B (↑) which is independent of the medium throughout which magnetic field forces act; it is denoted by H and measured in units of ampere/metre ($A m^{-1}$); H causes magnetization of the medium, resulting in the total effect B, given by $B = \mu H$, where μ = absolute magnetic permeability (↓) of the medium.

magnetizing field an alternative name for magnetic field strength H (↑).

magnetic intensity a former name for magnetic field strength H (↑).

magnetic permeability the factor by which the presence of a medium alters the magnetic field (p.180) effects; it is denoted by μ and measured in units of weber/ampere/metre ($Wb A^{-1} m^{-1}$) or henry/metre ($H m^{-1}$) (p.207); for a ferromagnetic (↑) medium $\mu \gg 1$ and, since $B = \mu H$, the contribution of μ to total magnetic flux density B (↑) is much greater than that of magnetic field strength H (↑); but the effect of a ferromagnetic medium is to reduce the forces acting between the sources of magnetism (p.181), e.g. 2 parallel wires carrying a current (p.152); for free space or a vacuum the absolute magnetic permeability $\mu_o = 4\pi \times 10^{-7}$ $H m^{-1}$; the relative magnetic permeability $\mu_r = \mu/\mu_o$, where μ = absolute magnetic permeability for the medium; μ_r has no units.

magnetic space constant an alternative name for magnetic permeability μ_o (↑).

magnetization curve a graph showing the variation of magnetic flux density B ($Wb\,m^{-2}$) with magnetic field strength H ($A\,m^{-1}$) (p.183), for a magnetic sample, e.g. a bar magnet; for an initially unmagnetized ferromagnetic sample (p.184), the graph shape shows the variation in internalmagnetic domain structure (p.184) up to saturation magnetization (↓); for the magnetization cycle (↓) the curve is a closed hysteresis loop (↓).

saturation magnetization a condition in a magnetized ferromagnetic material in which all magnetic domains (p.184) are aligned with the applied magnetizing field H (p.185) and further increases in H do not increase the magnetization of the specimen; the specimen has saturation magnetic flux density B_s ($Wb\,m^{-2}$) (p.185).

magnetic hysteresis for a ferromagnetic sample with saturation magnetization (↑) the reduction of the applied magnetizing field H to zero leaves the sample with some residual magnetism, termed its remanence (↓); a reverse field, the coercive field (↓), is required to demagnetize the sample; changes in magnetic flux density B ($Wb\,m^{-2}$) (p.185) and magnetization of the sample show a delayed response, hysteresis, to changes in magnetizing field H ($A\,m^{-1}$) (p.185).

magnetic remanence the value of residual magnetic flux density B_r ($Wb\,m^{-2}$) (p.185) of a ferromagnetic sample (p.184), when the applied magnetizing field H ($A\,m^{-1}$) (p.185) is zero; a measure of the strength of magnetism for zero applied field.

coercive field the value H_c ($A\,m^{-1}$) of the reverse applied magnetizing field (p.185) required to completely demagnetize a ferromagnetic sample from a condition of magnetic remanence (↑) B_r ($Wb\,m^{-2}$); measure of the permanence of magnetism of the sample.

coercive force alternative name for coercive field H_c (↑).

coercivity (*n*) alternative name for coercive field H_c (↑).

magnetization cycle the process of taking an initially unmagnetized ferromagnetic sample (p.184) through the stages of magnetization to saturation (↑), demagnetization, magnetization to saturation with opposite magnetic polarity (p.181) and reverse saturation B_r, demagnetization by reversed coercive field H_c (↑) and remagnetization to saturation with initial polarity and flux density; the graph showing the variation of magnetic induction B ($Wb\,m^{-2}$) (p.185) with magnetizing field H ($A\,m^{-1}$) is a closed hysteresis

saturation magnetization

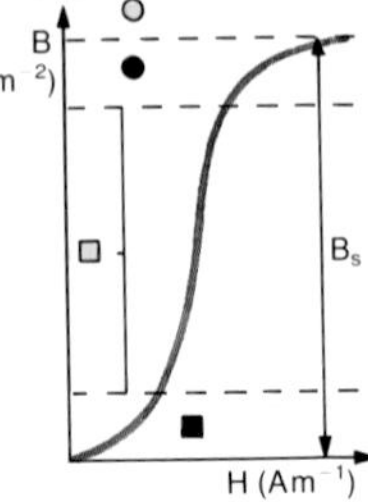

loop (↓); after following the initial magnetization curve (↑) to saturation, the specimen never returns to its initial condition at the origin O, unless demagnetized (p.181) by an a.c. current.

hysteresis loop a graph showing the variation of magnetic flux density B with magnetizing field H, through a complete magnetization cycle (↑); the loop shows symmetry about the origin O of the conditions of saturation magnetization (↑) B_s ($Wb\,m^{-2}$), residual magnetism at magnetic remanence (↑) B_r ($Wb\,m^{-2}$) and demagnetization at coercive field H_c ($A\,m^{-1}$); the loop area gives a measure of the work done in the process of taking a ferromagnetic sample through a complete magnetization cycle (↑); energy is expended in realignment of magnetic domains (p.184).

hysteresis loop
for ferromagnetic sample

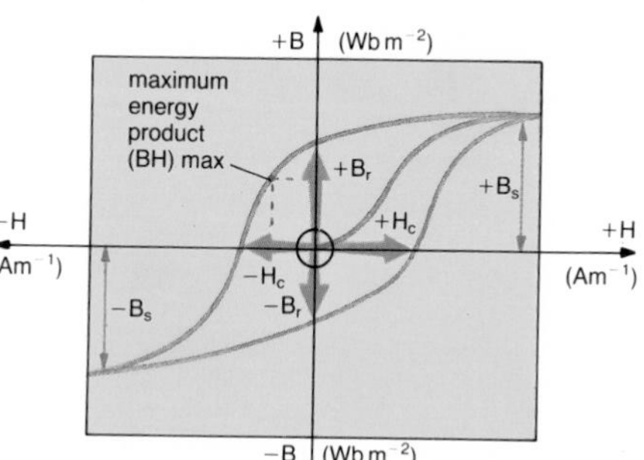

hysteresis energy losses the process of taking a ferromagnetic sample through a complete magnetization cycle (↑) requires the expenditure of energy (p.26) in the realignment of magnetic domains (p.184) twice in each cycle; the work done (p.26) per cycle is proportional to the hysteresis loop (↑) area, and the energy expended in internal friction (p.21) against domain boundary movements appears as heat in the sample; soft magnetic materials (p.188), e.g. iron, have low area loops and are used for transformer cores and a.c. generator armatures, where the sample is continuously subjected to varying magnetic flux (p.204) at a.c. mains (p.160) frequency 50 Hz, and they reduce energy wastage and heating effect to a minimum; hard magnetic materials (p.188), e.g. steel, have large area loops and are used for permanent magnets.

maximum energy product the point in the second quandrant of a hysteresis loop (p.187) for which the product BH (joule m^{-3}) is a maximum; it gives a measure of the quality of a hard ferromagnetic material (↓) for use as a permanent magnet; the units gauss-oersted (G-Oe) are also used.

soft magnetic material a ferromagnetic material (p.184) having a high magnetic remanence B_r (p.186), low coercive field H_c (p.186) and low area hysteresis loop (p.186); used in making transformer cores and a.c. generator armatures requiring high magnetic flux density (p.185), low hysteresis energy losses (p.187), and able to be readily magnetized and demagnetized when subjected to varying magnetic flux density (p.204); also used in electromagnets (p.195) and relays, in which magnetization and demagnetization should accompany the switching on and off of electric current, e.g. soft iron, magnetic alloys Mumetal (74% nickel, 20% iron, 5% copper, 1% manganese) and Stalloy (96% iron, 4% silicon).

soft magnetic material

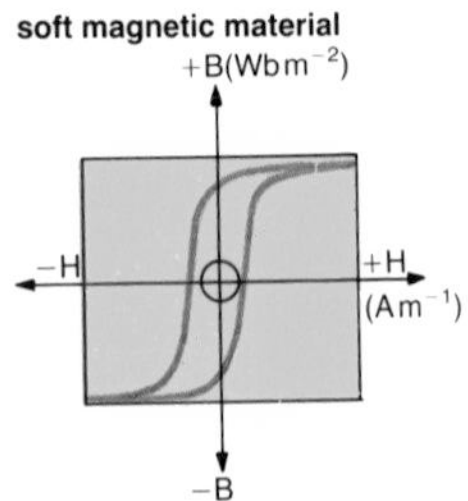

hard magnetic material

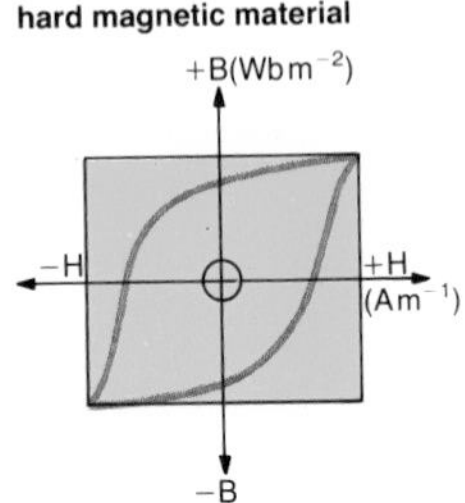

hard magnetic material a ferromagnetic material (p.184) having high magnetic remanence B_r, high coercive field H_c (p.186) and high maximum energy product BH (↑); the high area hysteresis loop (p.187) would give high hysteresis energy losses (p.187), so the material is not suitable for the same uses as soft magnetic materials (↑), but it is used for strong permanent magnets in moving-coil measuring instruments, loudspeakers, microphones and telephones, e.g. carbon steel, cobalt steel, the Alnico (iron-aluminium-nickel-cobalt) range of magnetic alloys, ferrites (p.184), and rare-earth (R)-cobalt (Co) materials of formula RCo_5 and R_2Co_{17}, such as cobalt-samarium $SmCo_5$.

Curie point the characteristic temperature for a ferromagnetic material above which its ferromagnetism (p.184) is lost due to loss of magnetic domain structure, e.g. 770°C for iron, 1120°C for cobalt, 724°C for cobalt-samarium; the Curie point should be well above the ambient temperature in which the magnet is required to operate.

geomagnetism (*n*) the study of magnetic properties of the Earth (p.189); magnetism (p.181) of the Earth as a source of magnetic field. **geomagnetic** (*adj*).

terrestrial magnetism an alternative name for geomagnetism (↑).

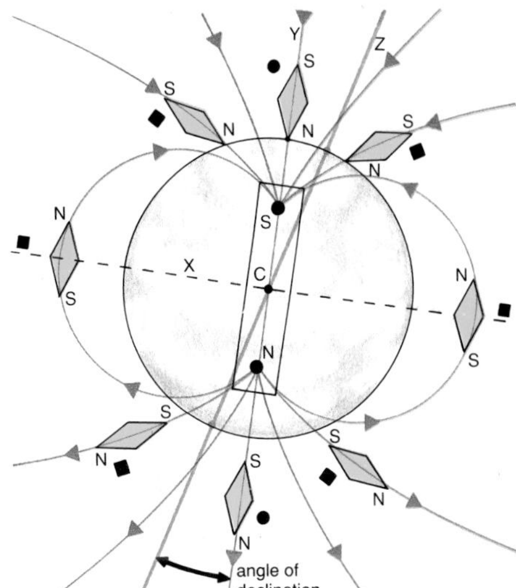

Earth's magnetic field
X magnetic equatorial plane
Y magnetic axis
Z geographic axis
● dip needle vertical
■ dip needle horizontal
◆ dip needle inclined

Earth's magnetic field the geophysical structure of the Earth indicates a partially fluid core of magnetic material as its source of magnetism; it is conveniently represented diagrammatically as a bar magnet (p.181), symmetrically placed with respect to the Earth's magnetic equator, with its S pole in the Northern hemisphere and N pole in the Southern hemisphere, giving a magnetic flux line (p.182) pattern corresponding to the Earth's external magnetic field; magnetic field strength (p.185) vectors have horizontal (p.190) and vertical (p.190) components at all except magnetic axial and equatorial points; the pole positions show slow annual variation, and palaeomagnetic measurements show past field reversals.

dip needle

vertical at magnetic N and S poles ●

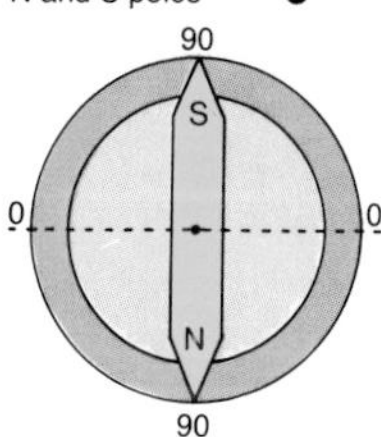

horizontal at magnetic equatorial points ■

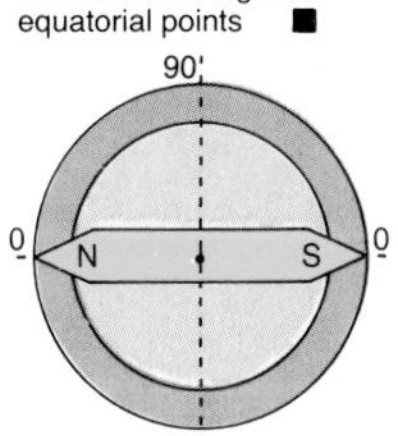

inclined at dip angle θ ◆

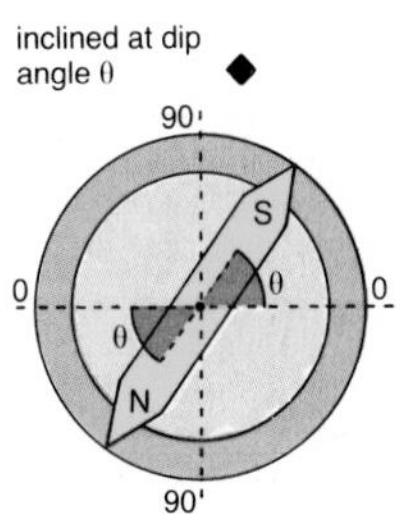

Earth's horizontal component the horizontal component of the magnetic flux density of the Earth's magnetic field (p.189) at a specific locality on the Earth's surface; the field in which a compass needle (p.182) sets when pivoted in a magnetic meridian (↓); its value in the U.K. is 2×10^{-5} tesla (T) (p.193).

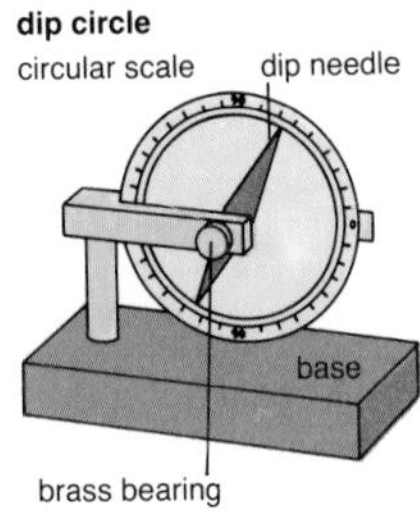

Earth's vertical component the vertical component of the magnetic flux density of the Earth's magnetic field (p.189) at a specific locality on the Earth's surface; the field in which a magnetized dip needle sets when pivoted vertically in a magnetic meridian (↓); value in the U.K. is 4×10^{-5} tesla (T) (p.193).

Van Allen radiation belts the Earth's magnetic field extends far into space, and satellite and space probe data give evidence for 2 bands of charged particles, assumed to be electrons (p.7) and protons (p.8), trapped in spiral paths around the Earth's magnetic flux lines; the nearer belt is about 2000 miles wide, with a peak radiation intensity at about 2000 miles from the Earth; the further belt is about 4000 miles wide, with its peak at about 10 000 miles from the Earth; the radiation is attenuated to zero at about 40 000 miles; the belts may be due to neutron disintegration by cosmic radiation or to particle injection from the sun; the belts would be a radiation hazard in space travel.

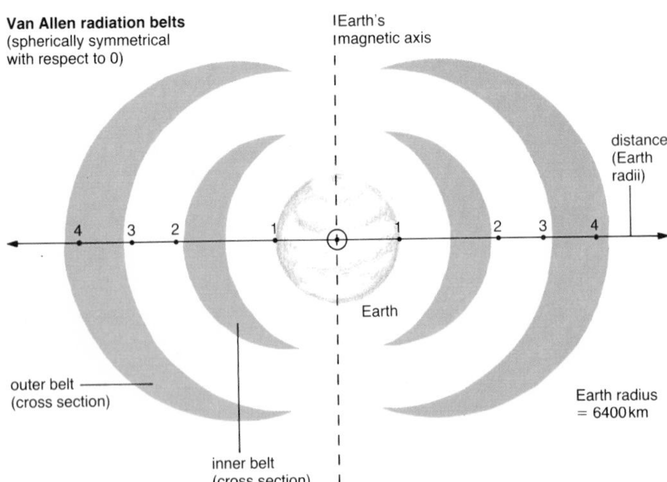

magnetic meridian any plane passing through the Earth's magnetic axis and including magnetic N and S poles.

angle of declination the angle between the magnetic meridian and the geographic meridian, passing through Earth's geographic axis, at a specific locality.

electromagnetism (*n*) refers to the study of the magnetic field (p.180) phenomena associated with electric current (p.152) or the movement of charge carriers (p.9); the property of a current, or of moving charge carriers, to generate a magnetic field in the surrounding space. **electromagnetic** (*adj*).

magnetic effects of current these are electromagnetic (↑) phenomena observable for current-carrying conductors acting as sources of magnetism (p.181), e.g. a plane coil, straight wire, solenoid; and for moving charge carriers, e.g. an electron beam (p.231).

Maxwell's Screw Rule relates the direction of a magnetic field to that of the electric current producing the magnetic effects (↑); if conventional current (p.152) flow is in the direction of forward movement of a rotating right-handed screw, the rotation direction gives the field direction; the rule can be extended to coil or solenoid, giving the axial field direction, and the N-S rule can be used to identify the magnetic polarity (p.181) of a coil or solenoid faces.

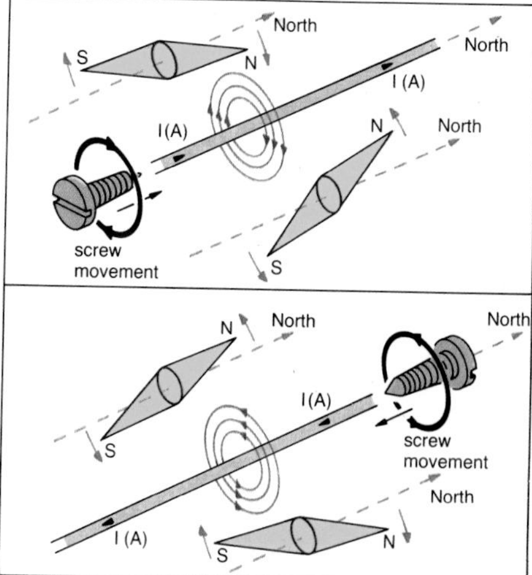

deflection of compass needle by field due to current-carrying conductor in magnetic meridian:

Maxwell's Screw Rule

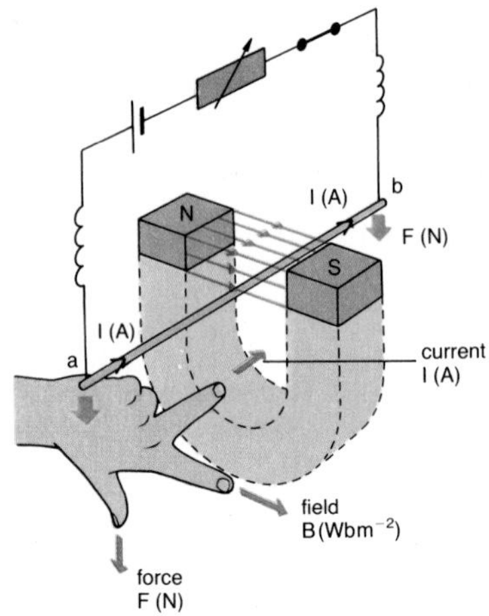

Left Hand Rule relates the mutually perpendicular directions of force on a current-carrying conductor (↓) in a magnetic field, to conventional current (p.152) flow and magnetic field direction (p.183); first finger, second finger and thumb of the left hand are held mutually perpendicular; if the first finger is now pointed in the magnetic field direction and the second finger in the conventional current direction, the thumb direction gives the direction of force; for a straight conductor carrying current, the conductor field augments the applied field above the wire, and reduces it below the wire, the difference in energy (p.26), stored in the stronger magnetic field above the wire, causes a pressure on the wire and the resulting force (p.26) displaces the wire downwards towards the region of weaker field; current reversal gives an upward displacement.

Fleming's Rule alternative name for Left Hand Rule (↑).

Left Hand Rule

force F = BIL sin α (N)
B sin α is component of B perpendicular to I
ab = conductor length L

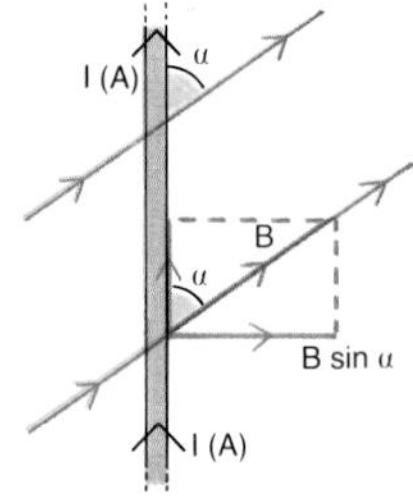

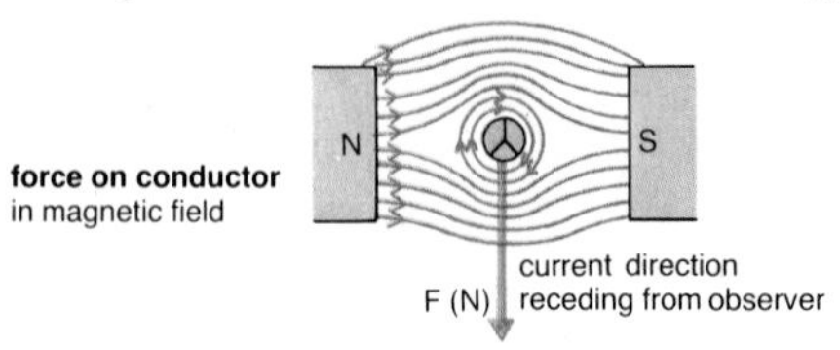

force on conductor
in magnetic field

force on current-carrying conductor interaction between the magnetic field of a current-carrying conductor and an externally applied magnetic field results in a force (p.26) acting on the conductor in the direction given by the Left Hand Rule (↑); the value of force F (newton) is directly proportional to the magnetic flux density B (weber/m^2) (p.185) of the field, electric current I (ampere) (p.152), conductor length L (m) in the field, and sin α, where α = angle between B and I, B sin α being the component of B perpendicular to I; F (N) $\propto$ BIL sin α can be verified experimentally, so F = KBIL sin α, where K = constant; when I = 1 (A), L = 1 (m), $\alpha = 90°$ and sin α = 1, then K = 1 if B is defined to have a value 1; the unit of B is tesla (T) (↓) where 1 (T) = 1 (Wb m^{-2}), so B = 1 and F (N) = BIL sin α.

tesla the unit of magnetic flux density B (p.185); defined as the magnetic flux density of a magnetic field when a conductor 1 metre long, carrying a current of 1 ampere, experiences a force (p.26) of 1 newton on the current-carrying conductor (↑); 1 tesla (T) = 1 weber/m^2 (Wb m^{-2}).

current balance an instrument, based on the principle of a magnetic force on a current-carrying conductor (↑), which may be used to measure magnetic flux density B if the current I is known, and the current I, if B is known; the primary standard apparatus used in standards laboratories to realize the practical value of the ampere (p.201) correct to one part in 10^6, from which accurate secondary standard equipment can be calibrated.

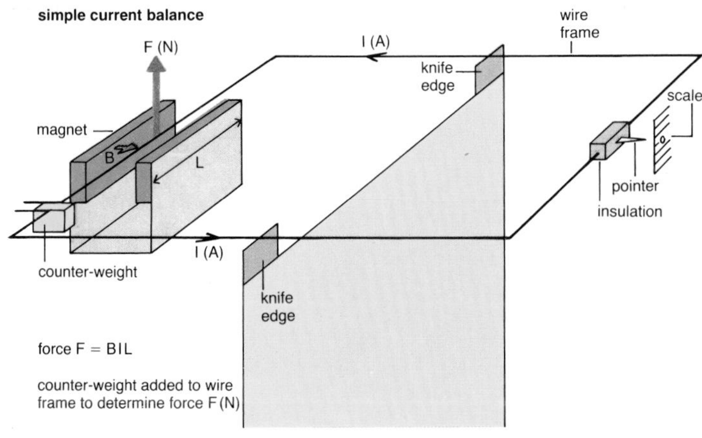

force on moving charge-carriers charge-carriers (p.9) moving through air, a vacuum or a transmitting medium, e.g. electrons (p.7) in a conducting wire or a vacuum tube, electrons or protons (p.8) in high energy particle accelerators (↓), electrons and holes (p.223) in semiconductor devices (p.222), are equivalent to a conventional electric current (p.152), and experience a force, as on a current-carrying conductor in a magnetic field; N particles each carrying electric charge q (coulomb) (p.165), travel distance L (m) in time t (s) at velocity v ($m s^{-1}$) (p.30) along the charge-carrier beam, crossing perpendicular to a field of magnetic flux density B ($Wb m^{-2}$) (p.185); $t = L/v$ and the equivalent conventional current I (A) = $Nq/t = Nqv/L$;
force F′ (N) = $BIL = BNqv$ on N particles, so the force/charge-carrier $F = F'/N = Bqv$ (newton); for electrons in a cathode ray tube (p.230), having electronic charge e, (p.7) force/electron $F = Bev$ (N).

centripetal force on moving charge the force Bqv (↑) acting on a particle of mass m (kg) carrying charge q (C) moving at velocity v perpendicular to magnetic field B acts in a direction perpendicular to v and hence is a centripetal force (p.47) resulting in the charge moving in a circle of radius r, where $Bqv = mv^2/r$.

fine-beam tube cathode-ray tube (p.230) containing a small quantity of hydrogen which is ionized by electrons, thus making the beam visible; Helmholtz coils (p.200) may be used to deflect the electron beam into a circular path whose radius may be measured; knowing the field B (T) between the Helmholtz coils, the accelerating voltage V (volt) of the electron gun and the radius of the circular beam r (m) the specific charge on the electron (e/m) may be calculated; $Bev = (mv^2)$ and $eV = \frac{1}{2}mv^2$; $e/m = 2V/B^2r^2$.

mass spectrometer an instrument used for separating streams of positive ions of different masses using magnetic and electric fields; it is calibrated to measure the mass of the ions.

particle accelerator a machine for increasing the kinetic energy of charged particles such as protons or electrons by accelerating them in a suitable electric field arrangement.

cyclotron a particle accelerator (↑) in which charged particles spiral at right angles to a magnetic field and are accelerated by an alternating electric field applied between D-shaped conductors.

mass spectrograph

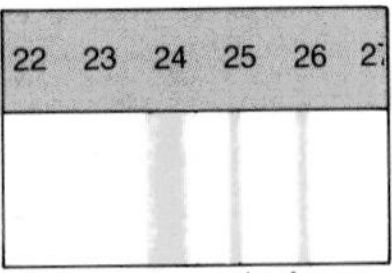

mass spectrograph of magnesium

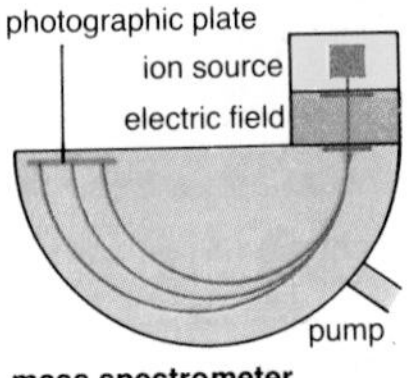

mass spectrometer

electromagnet

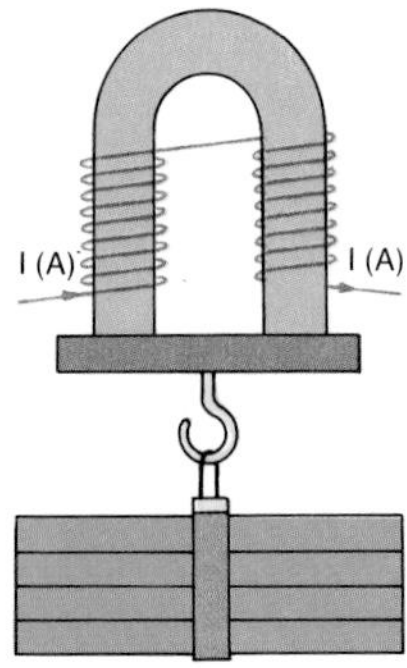

electromagnet (*n*) a horseshoe-shaped magnet of soft magnetic material (p.188), magnetized by current-carrying field-windings and demagnetized (p.181) on switching off the current; the magnet is very strong due to its high permeability (p.185) but it has low remanence (p.186).

magnetic relay a low-current switching circuit, based on a readily magnetized and demagnetized soft iron core responding to electric current pulses; with current on, the soft iron becomes magnetized and attracts a rocking contact, causing circuit make in another, higher-current circuit controlled by the relay.

couple on current-carrying coil refers to a plane rectangular coil of N turns, length L (m) × breadth b (m) = area A (m^2), carrying an electric current I (ampere) and suspended vertically or pivoted horizontally in a field of magnetic flux density B (Wbm^{-2}) (p.185) with its long sides L being perpendicular to the field and experiencing opposite forces F (N) = BIL on the current-carrying conducting sides L; the couple (p.35) Fx (Nm) causes rotation of the coil about its axis of suspension; x = b cos Φ, where Φ = angle between coil plane and magnetic field, so Fx = BIL cos Φ = BIA cos Φ (Nm), and for N turns, Fx = BINA cos Φ (Nm); for a radial magnetic field (p.182), the coil is wound on a cylindrical soft iron armature, suspended or pivoted between the cylindrical pole faces of a permanent magnet of hard magnetic material (p.188), giving Φ = 0, cos Φ = 1 and a constant couple Fx = BINA (Nm) on the coil; coil rotation is opposed by the torsional couple of the suspension.

magnetic relay

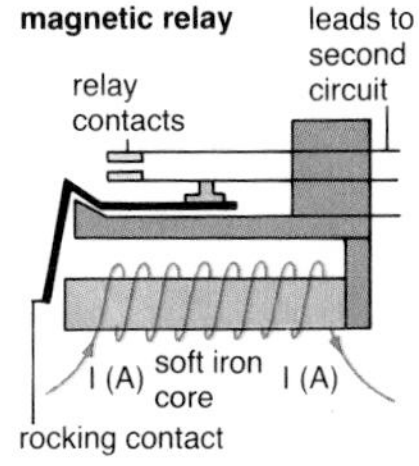

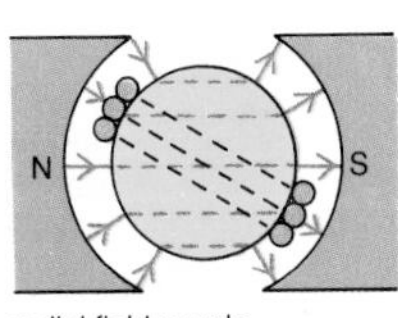

radial field couple
Fx = BINA (Nm)

radial magnetic field

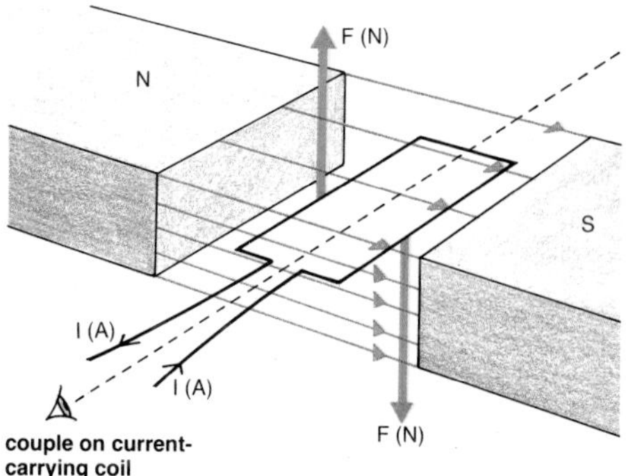

couple on current-carrying coil

moving coil galvanometer a sensitive electric current (p.152) measuring instrument in which the couple on a current-carrying coil (p.195) in a radial magnetic field is opposed by a torsional couple resisting twist, the angle of twist β (rad) being directly proportional to the current I (A); the deflecting couple BINA (N m) = kβ, so I = kβ/BNA, and I ∝ β since k, B, N and A are constant; a suspended coil (↓) instrument is in a vertical plane, a pivoted coil (↓) instrument is in a horizontal or vertical plane; the coil has many turns of fine insulated copper wire and moves between the cylindrical pole faces of a strong permanent magnet; β is observed from deflection of a light spot over a centre-zero scale in an optical lever arrangement (p.64).

current sensitivity for a moving coil galvanometer (↑) the value β/I is constant, giving the value of coil deflection/ampere (rad A^{-1}) as a measure of its sensitivity as a current-measuring instrument; for a suspended coil (↓) instrument with a mirror, lamp and scale in optical lever (p.64) arrangement the sensitivity is expressed in mm/microampere for a scale 1 m distant; for a pivoted coil (↓) instrument with pointer and graduated scale arrangement, the sensitivity is expressed in ohm/volt.

suspended coil galvanometer a sensitive moving coil galvanometer (↑) suspended in a vertical plane by a phosphor bronze wire of low torsional constant; the coil deflection in a radial magnetic field is measured by reflection of a narrow light beam from a mirror fixed to the suspension on to a linear mm scale in optical lever (p.64) arrangement; for steady current measurement, the coil moves to the appropriate position and remains there; for transient current, the coil may move rapidly to the appropriate position and then oscillate (p.49) about its rest position, operating in ballistic mode (↓), or it may return slowly to zero, operating in dead-beat mode (↓), depending on the amount of damping (↓) of the coil movement.

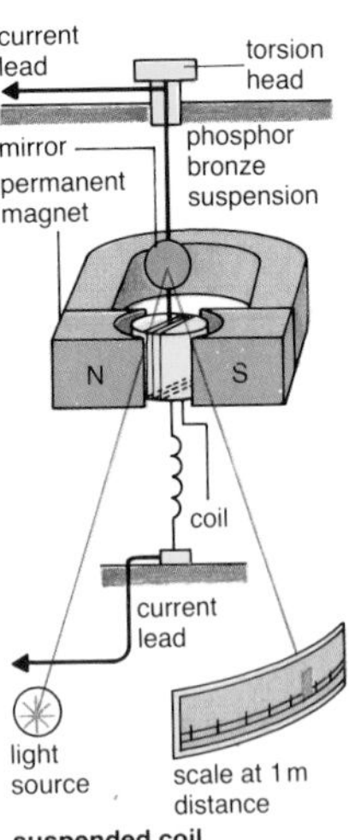

suspended coil galvanometer

damping (*n*) for an oscillating (p.49) mechanical system, e.g. a pendulum, galvanometer coil (↑) on passage of a transient current, the return to its rest position occurs after many oscillations, whose number can be reduced by the action of forces applied to slow down the coil movement and dissipate its kinetic energy (p.27); mechanical damping uses air frictional forces acting on a vane, moving inside a dashpot and attached to the end of the pointer of a pivoted coil galvanometer (↓); electromagnetic damping uses eddy currents (p.210) induced in the metal former on which the pivoted coil is mounted, or in the copper ring inside the former; for a suspended coil galvanometer (↑) operating in ballistic mode (↓), electromagnetic damping is provided by currents induced in the oscillating coil itself, opposing the coil movements causing them, according to Lenz's Law (p.205). **damp** (*v*), **damped** (*adj*), **damping** (*adj*).

dead-beat operation refers to operation of a moving coil galvanometer (↑) under conditions such that, when a transient current passes, the deflected coil returns slowly to its rest position and the pointer or light-spot indicator returns slowly to the zero of the scale due to mechanical and electromagnetic damping (↑).

ballistic operation refers to operation of a suspended coil galvanometer (↑) under conditions such that, when a transient current passes, the deflected coil oscillates about its rest position and the light-spot indicator oscillates about the zero of the scale.

pivoted coil galvanometer a moving coil galvanometer (↑) pivoted in a horizontal or vertical plane by jewelled bearings; the coil deflection in a radial magnetic field is opposed by phosphor bronze hairsprings of low torsional constant and is measured by the movement of a pointer over a centre-zero linear scale calibrated in milliampere (mA) or microampere (μA).

pivoted coil galvanometer

coil pivot bearing
current lead
zero adjuster
linear scale (0 – 10 mA)
pointer
N
S
coil
permanent magnet
phosphor bronze control hair spring
coil pivot bearing
current lead

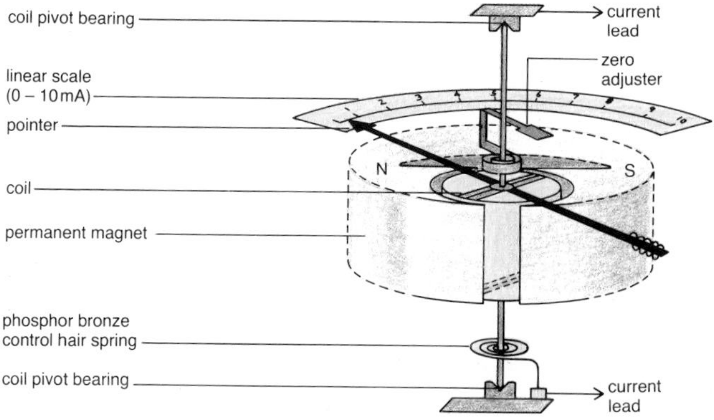

full scale deflection for an ammeter (↓) or voltmeter (↓) type of measuring instrument, full scale deflection is the value of the measured quantity causing the pointer to move from the zero of an end-zero scale to the maximum measurable deflection at the opposite end, e.g. voltmeter reading e.m.f. on open circuit (p.154).

ammeter (*n*) an electric current (p.152) measuring instrument with a pivoted coil galvanometer (p.197) connected in parallel with a low resistance shunt (↓) to carry the major proportion of the circuit current; the instrument may be modified for any current range by suitable choice of shunt value; the ammeter is connected into a circuit in series with other components.

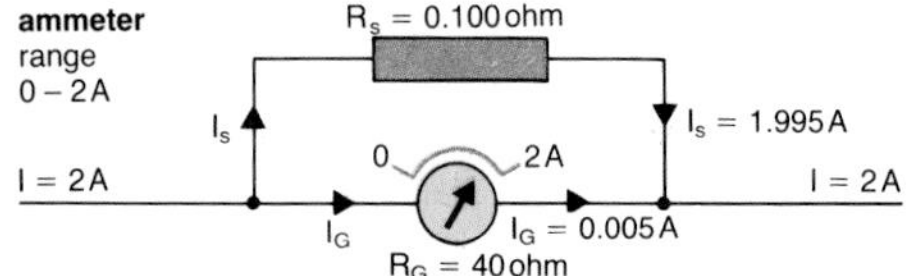

shunt (*n*) a resistor (p.156) of calculated low value; connected in parallel, to adapt a moving coil galvanometer to operate as an ammeter (↑) of specified range; it is made from metal of low temperature coefficient of resistivity (p.158), e.g. manganin. **shunt** (*v*), **shunted** (*adj*), **shunting** (*adj*).

voltmeter (*n*) an instrument measuring electric p.d. (p.170) with a pivoted coil galvanometer (p.197) connected in series with a high resistance multiplier (↓), across which the major proportion of the measured p.d. is applied; the instrument may be modified for any voltage range by suitable choice of multiplier value; the voltmeter is connected in parallel with the circuit component across which the p.d. is being measured and must take a small amount of the main circuit current

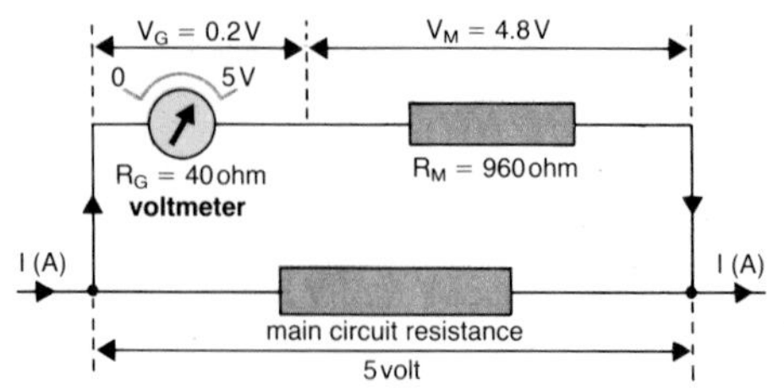

voltmeter
range 0–5V

in order to operate; the figure of merit of the instrument (ohm/volt) is high when its current is low, e.g. for 1000 ohm voltmeter of 5 volt range, 200 ohm/volt (A^{-1}); for a low figure of merit, the current taken may have to be accounted for.

multiplier (*n*) a resistor (p.156) of calculated high value, connected in series to adapt a moving coil galvanometer to operate as a voltmeter (↑) of specified range.

multi-range meter a multi-purpose meter, incorporating both shunt (↑) and multiplier (↑) selection ranges for a.c. or d.c. voltage and current measurement; an internal dry battery (p.151) is also included for resistance (p.155) measurement, the ohm scale (0–∞) being graduated in the reverse direction, the current decreasing with external resistance increase.

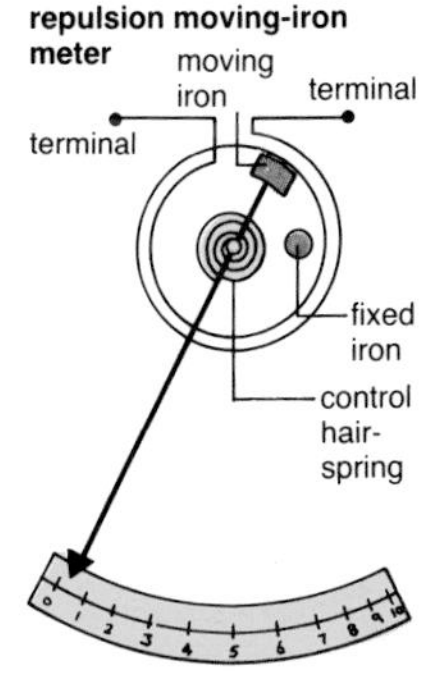

repulsion moving-iron meter an electric current (p.152) measuring instrument, based on the repulsive magnetic forces acting between 2 soft iron bars magnetized (p.180) with like magnetic polarity (p.181) by a surrounding current-carrying coil; one bar is fixed and the other moveable, carrying a pointer over a non-linear scale; the instrument can be adapted for use as an ammeter (↑) or voltmeter (↑) of chosen range using selected shunts (↑) or multipliers (↑).

attraction moving-iron meter an electric current (p.152) measuring instrument, based on the attractive magnetic forces acting between a current-carrying coil and a soft iron bar, magnetized (p.180) by induction (p.182) with unlike magnetic polarity (p.181) to the coil; the iron moves towards the coil, carrying a pointer over a non-linear scale; adaptations for use as an ammeter or voltmeter of chosen range is as for the repulsion moving-iron meter (↑).

wattmeter (*n*) an instrument measuring electrical power VI (watt) (p.159) generated in a conducting component R, as the product of physical quantities directly proportional to p.d. V (volt) and electric current I (ampere); a moving coil, carrying current I′, is connected in parallel with R and suspended between 2 fixed coils carrying the supply current I, which causes a field of magnetic flux density B ($Wb\,m^{-)}$ (p.185) to act on the moving coil; the couple ∝ VI and the moving coil deflection indicates electrical power in R; it also measures the average power in a cycle of a.c. voltage and current (p.214).

current element a very small part of a conducting wire, length δL, in a circuit carrying electric current I (ampere) (p.152).

Biot-Savart Law a law giving the value of the small contribution δB made to magnetic flux density B ($Wb\,m^{-2}$) (p.185) at point P due to a current element δL (↑) distant r (m) from P; $\delta B \propto I\delta L \sin\alpha/r^2$, where I = electric current (ampere), so $\delta B = k.I\delta L \sin\alpha/r^2$, where $k = \mu_o/4\pi$ and the magnetic space constant $\mu_o = 4\pi \times 10^{-7}$ ($H m^{-1}$) (p.185); from the law $\delta B = \mu_o I\delta L \sin\alpha/4\pi r^2$.

Ampere-Laplace Rule an alternative approach to the Biot-Savart Law (↑), using the current element idea, giving the same practical results.

current-carrying coil the magnetic flux density B ($Wb\,m^{-2}$), at the centre of a plane circular coil of N turns of radius r (m), is calculated using the Biot-Savart Law (↑) for $2\pi rN/\delta L$ elements with $\alpha = 90°$: $B = \int_0^{2\pi rN} \delta B = \mu_o/4\pi \int_0^{2\pi rN} I\delta L \sin 90°/r^2 = \mu_o NI/2r$ ($Wb\,m^{-2}$).

Helmholtz coils 2 identical flat coils of N turns of radius r (m), mounted coaxially distance r (m) apart, have a uniform field between them of $B = 0.72\mu_o NI/r$ ($Wb\,m^2$), the same current I (ampere) flowing in both coils.

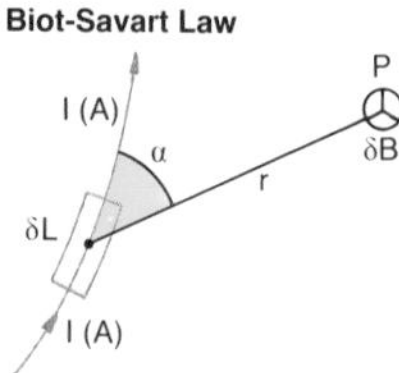

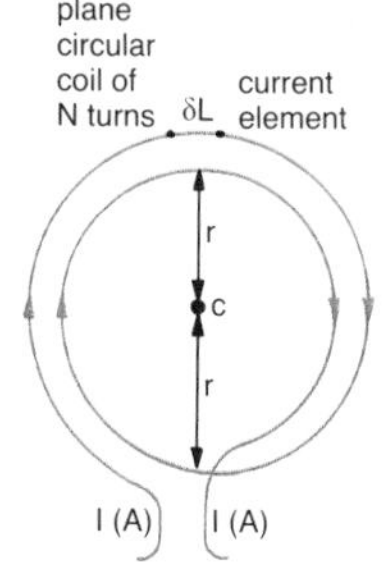

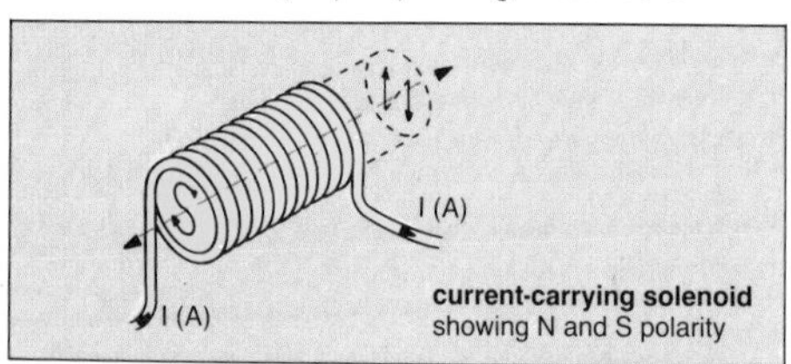

current-carrying solenoid showing N and S polarity

current-carrying solenoid the magnetic flux density B ($Wb\,m^{-2}$) along the axis of a solenoid of n turns/metre carrying electric current I (ampere) can be calculated using the Biot-Savart Law (↑); for a straight solenoid of infinite length, or an endless solenoid wound on a ring, $B = \mu_o nI$ ($Wb\,m^{-2}$) in free space; the solenoid faces have a polarity given by Maxwell's Screw Rule (p.191).

magnetic circuit the completely closed path enclosing a given set of lines of magnetic flux (p.182).

reluctance of a solenoid considering the magnetic flux Φ (Wb) flowing through a solenoid (↑) of NI (ampere-turns) the reluctance of the solenoid is given by $NI \div \Phi$ ($henry^{-1}$); reluctance in a magnetic circuit can be compared with electrical resistance (p.155) in a current circuit.

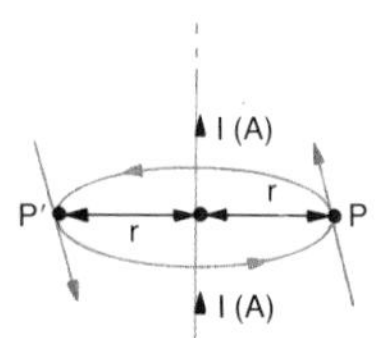

At P and P′: $B = \mu_o I/2\pi r$
current-carrying straight wire

current-carrying straight wire the magnetic flux density B ($Wb\,m^{-2}$) at points in the magnetic field of a long straight current-carrying wire of negligible cross-section can be calculated using the Biot-Savart Law (↑); for a point P close to a long wire, the wire subtends almost 180° at P; $B = \mu_o I/2\pi r$ ($Wb\,m^{-2}$), where r (m) = perpendicular distance between P and the wire.

parallel current-carrying wires for 2 parallel current-carrying straight wires (↑), each is in the magnetic field of the other and experiences a force F = BIL (newton) as a current-carrying conductor (p.152) of length L (m), with electric current I (ampere) (p.152), in a field of magnetic flux density B ($A\,m^{-2}$) (p.185); for parallel currents I_1 (A) and I_2 (A) in the same direction forces between the wires are attractive; magnetic flux density B_1 due to I_1 at $I_2 = \mu_o I_1/2\pi r$ ($A\,m^{-2}$), so force F_1/metre of $I_2 = B_1 I_2 = \mu_o I_1 I_2/2\pi r$ ($N\,m^{-1}$); similarly, the magnetic flux density B_2 due to I_2 at $I_1 = \mu_o I_2/2\pi r$ ($Wb\,m^{-2}$), so force F_2/metre of $I_1 = B_2 I_1 = \mu_o I_1 I_2/2\pi r$ ($N\,m^{-1}$) and $F_1 = F_2$ ($N\,m^{-1}$); for parallel currents I_1 and I_2 in opposite directions, the magnetic fields reinforce each other between the wires, giving a stronger field, so that forces between the wires are repulsive; for $I_1 = I_2$, the force/metre of each wire $= \mu_o I^2/2\pi r$ ($N\,m^{-1}$).

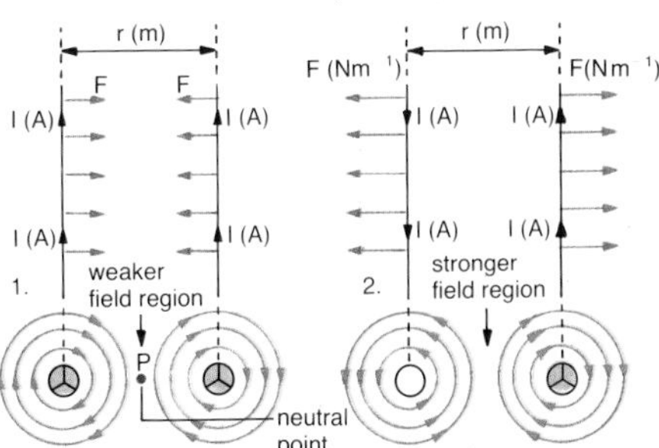

parallel current-carrying wires
1. currents in same direction
2. currents in opposite direction

ampere (A) the unit of electric current (p.152) defined from forces acting between parallel current-carrying wires (↑) as: that current which, flowing in each of 2 infinitely long straight parallel wires of negligible cross-sectional area, situated 1 metre apart in a vacuum, produces forces between the wires of value 2×10^{-7} newton/metre; since, for equal currents, the force F/metre of each wire $= \mu_o I^2/2\pi r$, when I = 1 (A), r = 1 (m) and $F = 2 \times 10^{-7}$ ($N\,m^{-1}$), $\mu_o = 4\pi \times 10^{-7}$ ($H\,m^{-1}$).

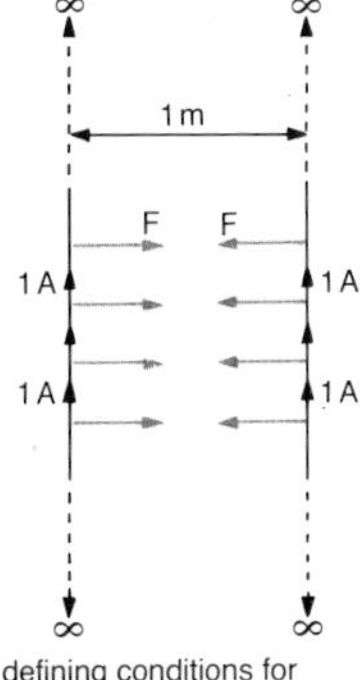

defining conditions for **ampere (A)**
$F = 2 \times 10^{-7}$ ($N\,m^{-1}$)

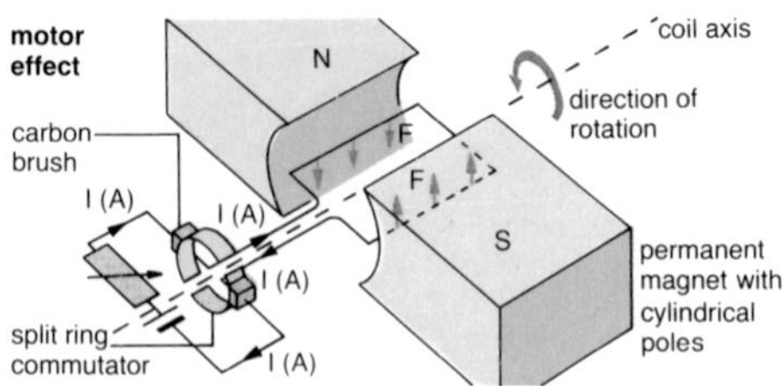

motor effect the force experienced by a current-carrying conductor (p.155) in a magnetic field.

electric motor a machine designed to convert electrical energy (p.28) input to rotational mechanical energy (p.28).

motor armature the central component of an electric motor (↑) with a cylindrical soft iron core, laminated to minimize eddy current energy losses (p.210), carrying the current supply windings of many turns of copper wire wound in slots around the core, and connected to the commutator (↓) segments, ensuring constant maximum rotational couple in a radial magnetic field (p.182).

field winding the armature (↑) of an electric motor rotates in a radial magnetic field between the cylindrically shaped pole pieces of a soft iron field magnet, magnetized by the current-carrying field coil winding, drawing current from the same supply source as the armature, in series, shunt or compound winding; a series-wound motor has a very high starting torque (p.32), used in electrically powered lifting gear and electrically driven gear of high rotational inertia (p.51); a shunt-wound motor has a steady running speed and is used where speed control is essential; a compound-wound motor combines series and shunt properties for use in the traction of heavy loads.

motor armature
laminated construction (with separating insulating laminations)

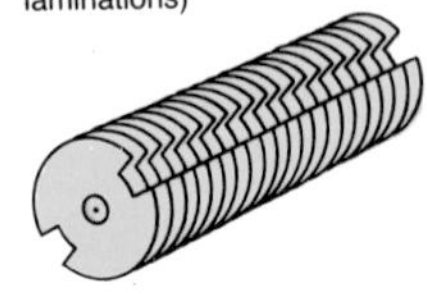

field windings
for motor series wound

field winding
I (A)
armature
I (A)
line voltage (V volt)

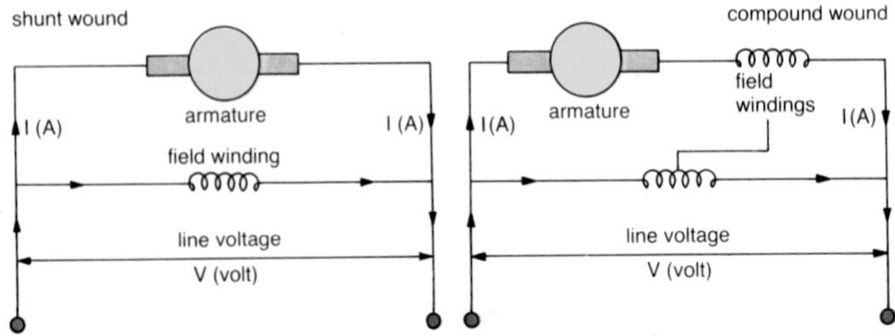

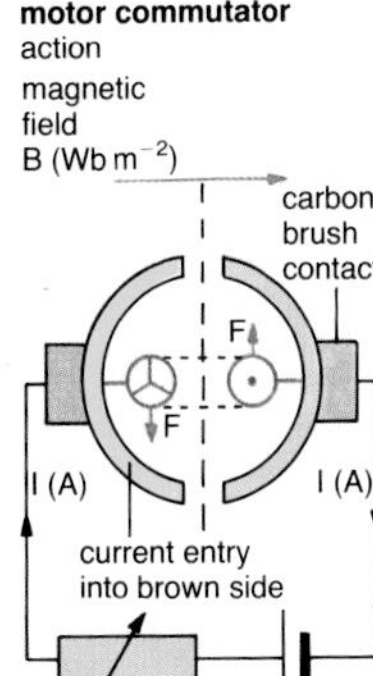

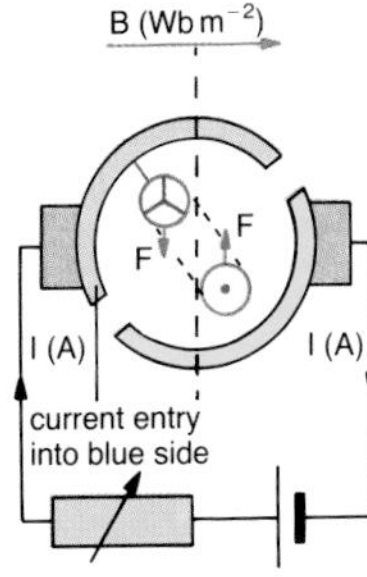

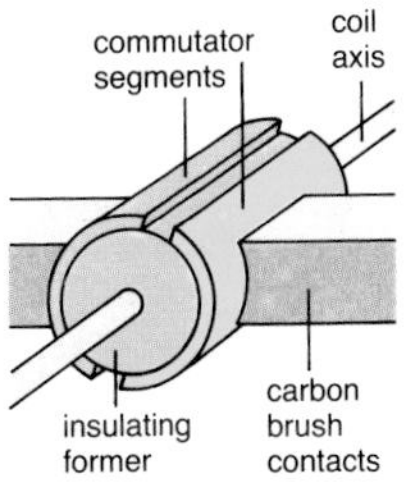

motor commutator a segmented copper ring mounted on the armature (↑) shaft of an electric motor to reverse the supply current direction through the armature every half-cycle of rotation, giving a unidirectional rotational couple on the armature current windings; a commutator has 2 segments for current reversal through each current winding, so for a practical electric motor with many (N) current supply windings, it has twice as many (2N) segments, connecting with the current supply circuit through carbon brush contacts.

back e.m.f. in motor the rotating armature (↑) of a working electric motor generates an induced back e.m.f. (p.207) by interaction between the rotating armature coils and the radial magnetic field; the e.m.f. e (volt) has its direction given by Lenz's Law (p.205) and opposes that of the line supply voltage V (volt) driving the motor, so $V - e = Ir$, where I = armature current (ampere) and r = internal armature resistance (ohm) of the motor; also $VI - eI = I^2r$, where VI (watt) = electrical power (p.159) input to the motor, I^2r (watt) = rate of generation of heat in the armature by Joule heating effects (p.159), and eI (watt) $= VI - I^2r$ = mechanical power developed by the motor = back e.m.f. × armature current; a working motor acts as a dynamo generating a constant d.c. voltage e (V).

d.c. motor an electric motor (↑) designed to operate on d.c. current; the armature (↑) and field windings (↑) are supplied with current from the same d.c. source, in series, shunt or compound winding.

a.c. motor an electric motor (↑) designed to operate on a.c. current; the armature (↑) and field windings are supplied with current from the same a.c. source (p.151) in series winding, so that the a.c. voltage and a.c. current are in phase (p.59); the design can also be as for a 3-phase alternator (p.214) operating as a motor, using 3-phase a.c. current supply, with the motor rotating at supply frequency 50 Hz requiring an auxiliary starter motor to attain this speed; also an a.c. squirrel-cage induction motor has an inner rotor (p.214) with a cage-like structure for the armature conducting bars, surrounded by an outer stator (p.215) carrying the field windings run by a 3-phase a.c. supply, the rotating magnetic field causing large induced currents (p.205) in the armature bars, so that the rotor accelerates towards, but never reaches, the frequency of the a.c. supply.

electromagnetic induction the generation of an electromotive force (p.149) in a conducting circuit by appropriate relative movement between the circuit and a magnetic field, or by variation in magnetic flux linkage (↓) with the circuit; if the conducting circuit is closed, the induced e.m.f. (↓) will cause an induced electric current (↓) to flow in a direction given by Lenz's Law (↓).

magnetic flux denoted by Φ weber (Wb) (↓), it is the product of magnetic flux density B ($Wb m^{-2}$) and area A (m^2) through which the flux passes perpendicular to B.

magnetic flux linkage magnetic flux density B ($Wb m^{-2}$) passing through a coil or solenoid of N turns of area A (m^2), gives magnetic flux linkage $\Phi = BAN$ (weber-turns); for magnetic flux cutting the coil plane obliquely, only the component perpendicular to the coil plane contributes to flux linkage; a coil or solenoid carrying current I (ampere) is its own source of flux linkage, and for varying current I (As^{-1}), it generates an induced e.m.f. and induced current by electromagnetic induction (↑), using its own property of self-induction.

Faraday's Laws of electromagnetism the results of Faraday's induction experiments, experiments with the Faraday ring transformer and experiments on the generation of induced e.m.f. (↓) by the cutting of magnetic flux can be summarized: 1. Whenever the magnetic flux linkage (↑) of a closed conducting circuit is changing, an induced current (↓) will flow in the circuit; 2. The magnitude of the induced e.m.f. (↓) e (volt) is directly proportional to the rate of change of magnetic flux linkage $d\Phi/dt$: $e \propto d\Phi/dt$.

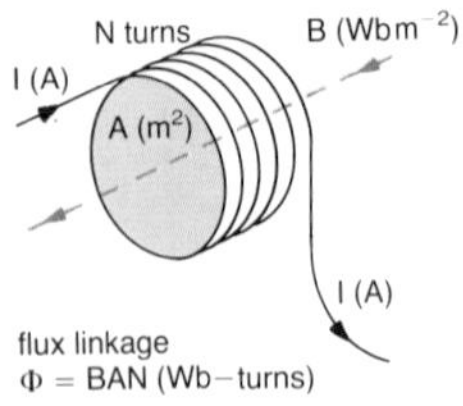

magnetic flux linkage

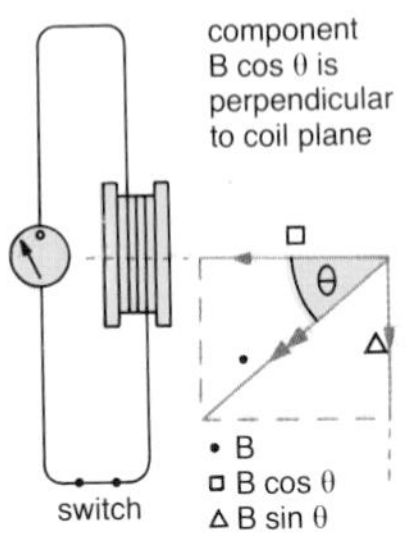

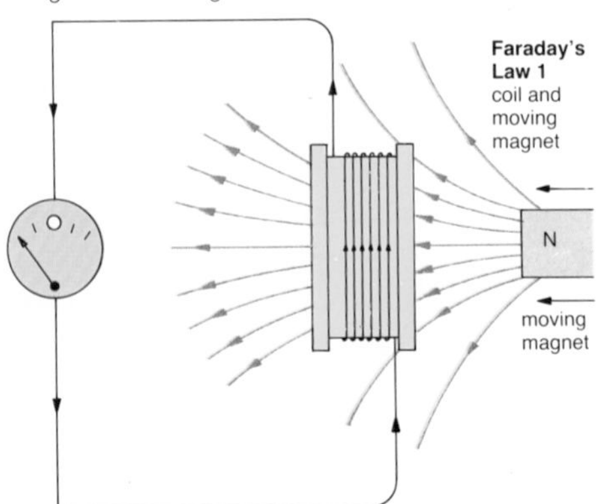

Faraday's Law 1
coil and moving magnet

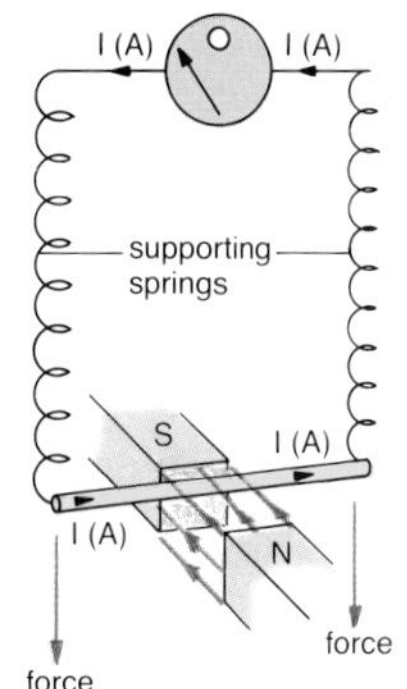

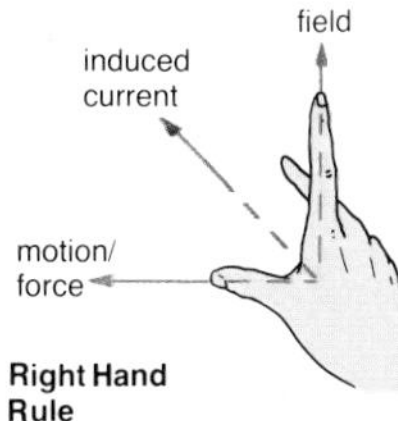

Right Hand Rule

induced e.m.f.
e (volt)

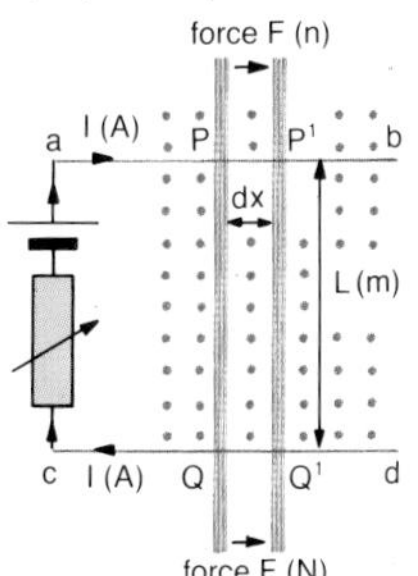

magnetic field B (Wbm^{-2}) perpendicular into diagram plane

Lenz's Law an induced current generated in a circuit by electromagnetic induction (↑) always flows in such a direction as to oppose the change causing it; a magnet moving towards a coil will be repelled, an induced current giving the coil a similar polarity to the approaching magnet; on withdrawal of the magnet it will be attracted to the coil, the induced current giving the coil an opposite polarity to the receding magnet; when a straight bar conductor is moved by an applied force across the flux lines of a magnetic field, the direction of the induced current, given by the Right Hand Rule (↓), is such that its flow in the magnetic field causes an opposing force, acting on the current-carrying conductor according to the Left Hand Rule (p.192).

Right Hand Rule relates mutually perpendicular directions of applied force and movement of a conductor in a magnetic field, to the magnetic field direction (p.183), and to the direction of the induced current and e.m.f. generated by electromagnetic induction (↑); first finger, second finger and thumb of the right hand are held mutually perpendicular; if the first finger is now pointed in the magnetic field direction and the thumb in the direction of movement, the second finger gives the direction of induced current.

induced e.m.f. if a straight movable bar conductor PQ of length L (m), is moved perpendicular to magnetic field of flux density B (Wbm^{-2}) (p.185) with velocity v (ms^{-1}), the e.m.f. e (volt) induced across PQ = B.L.v = B × (area swept out/second) = B.dA/dt = dΦ/dt; since induced e.m.f. opposes the change producing it, we write $e = -d\Phi/dt$.

induced current when an electromotive force is generated in a closed conducting circuit by electromagnetic induction (↑), an induced current flows in a direction given by Lenz's Law (↑); for an induced e.m.f. e (volt) (↑), in a circuit of total electrical resistance R (ohm) (p.155) having no other source of e.m.f., the induced current I (ampere) = e/R (volt ohm^{-1}).

weber (Wb) the unit of magnetic flux Φ (↑); induced e.m.f. (↑) e (volt) = −dΦ/dt, so when e = 1 (V), and dt = 1 (s), then dΦ = 1 (Wb); also when dΦ/dt = 1 (Wbs^{-1}), then e = 1 (V) and 1 volt = 1 weber/second.

Neumann's Law an independent statement of experimental findings essentially confirming Faraday's Law 2. of electromagnetism (↑); induced e.m.f. (↑) e (volt) = −dΦ/dt; Lenz's Law (↑) gives − sign for dΦ/dt.

mutual induction refers to electromagnetic induction (p.204) between 2 circuits with magnetic flux linkage (p.204); when magnetic flux linkage in primary coil P is varied by varying the primary current I (ampere) at a rate dI/dt (As^{-1}), an induced e.m.f. e (volt) (p.205) is generated by the accompanying variation in flux linkage in the secondary coil S; the induced e.m.f. $e \propto dI/dt$ and $e = -M\,dI/dt$, where M = coefficient of mutual induction, or mutual inductance M (↓), between the 2 circuits; according to Lenz's Law (p.205), the induced e.m.f. e (V) would be in such a direction as to oppose the variation of the primary flux causing it; it would create its own magnetic flux associated with the induced current, tending to oppose the growth of primary flux caused by increasing primary current, or to oppose the fall of primary flux caused by decreasing primary current; the primary coil P experiences self-induction (↓) due to its own varying flux.

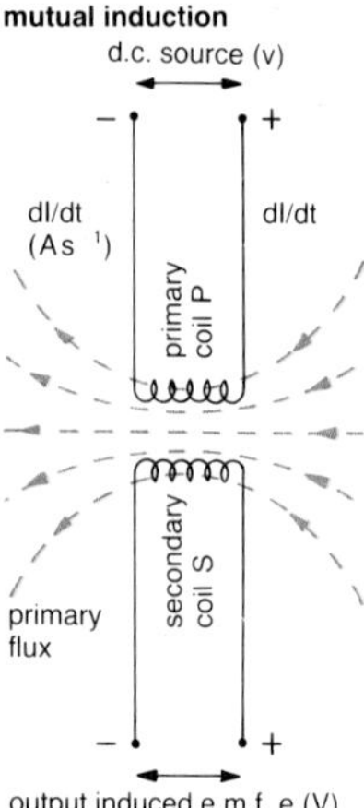

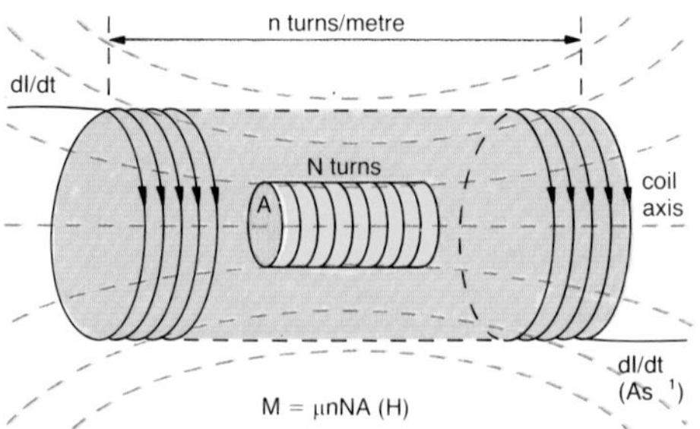

mutual inductance
of a small coil inside a large one

mutual inductance mutual induction effects (↑) between a particular pair of coils are expressed by the equation: induced e.m.f. e (V) = $-M\,dI/dt$, where dI/dt (As^{-1}) = rate of change of primary coil current and M = mutual inductance between the pair of coils; M is a constant for a particular coil arrangement with a surrounding medium of uniform magnetic permeability (p.185) and is measured in units of the henry (H) (↓); for practical purposes the millihenry (mH) is used.

self-induction refers to electromagnetic induction (p.204) occurring in a current-carrying coil or solenoid due to the variation in its own magnetic flux linkage (p.204), accompanying the variation in its own current I (ampere); current variation at a rate dI/dt (As^{-1}) causes an induced e.m.f. e (volt) (p.205) to be generated in the coil itself; the induced e.m.f. $e \propto dI/dt$ and $e = -L\,dI/dt$, where L = coefficient of self-induction, or self-inductance L (↓), of the coil; according to Lenz's Law (p.205), the induced e.m.f. e (V) would be in such a direction as to oppose the flux variation causing it; it would create its own magnetic flux, associated with induced current, tending to oppose the growth of flux caused by increasing current I (A) or to oppose the fall of flux caused by decreasing current I (A); the induced e.m.f. e (V) acts as a back e.m.f. (↓) in the coil itself, and opposes the applied e.m.f., which must overcome e in order for current I to flow.

self-inductance self-induction effects (↑) in a current-carrying coil or solenoid are expressed by the equation: induced e.m.f. $e\ (V) = -L\,dI/dt$, where $dI/dt\ (A\,s^{-1})$ = rate of change of coil current and L = self-inductance of the coil; it is a constant for a particular coil arrangement with surrounding medium of uniform magnetic permeability (p.185); L is measured in units of henry (H) (↓) and, for practical purposes, millihenry (mH).

henry (H) the unit of mutual inductance M (↑) and self-inductance L (↑), defined from the equations: induced e.m.f. $e\ (V) = -M\,dI/dt$, where $dI/dt\ (A\,s^{-1})$ = rate of change of primary coil current, and $e\ (V) = -L\,dI/dt$, where $dI/dt\ (A\,s^{-1})$ = rate of change of coil current; when the induced e.m.f. e = 1 volt for dI/dt = 1 ampere/second the mutual inductance M = 1 henry (H) for the specified pair of coils, and the self-inductance L = 1 H for the specified coil; 1 millihenry (mH) = 10^{-3}.

back e.m.f. an induced e.m.f. (p.205) generated in a conducting circuit through which magnetic flux linkage (p.204) is varying; this electromagnetic induction effect (p.204) causes an induced current to flow in such a direction as to oppose the process causing it, according to Lenz's Law (p.205); mutual induction and self-induction effects (↑) generate a back e.m.f., which opposes current growth and decay in an inductive circuit (p.208); the back e.m.f. in a motor, driven at constant speed with frequency of rotation f (s^{-1}), is a steady d.c. voltage $e\ (V) = BNA\omega$, where B = magnetic induction $(Wb\,m^{-2})$, N = number of turns of area A (m^2) on the motor coils and $\omega = 2\pi f\ (s^{-1})$, the motor acting as a dynamo working against the line supply voltage.

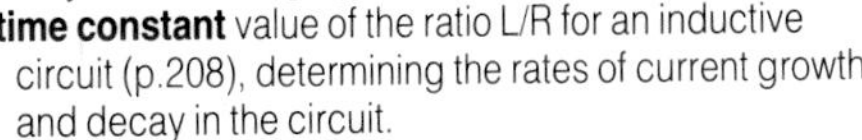
time constant value of the ratio L/R for an inductive circuit (p.208), determining the rates of current growth and decay in the circuit.

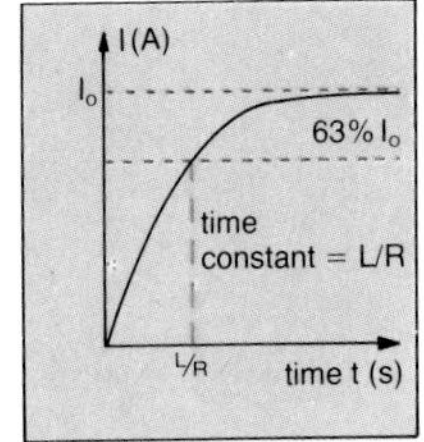

current growth in an inductive circuit

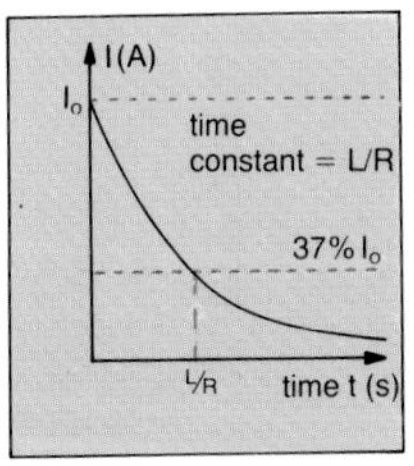

current decay in an inductive circuit

inductor (*n*) a conducting circuit component having the property of self-induction (p.206) and a specified value of self-inductance L (H) (p.207); it is usually a coil of a few turns of wire of low resistance R (ohm) (p.158), so that resistive energy losses (p.159) are minimal; an inductor demonstrates inductive reactance (p.219) in a.c. circuits. **inductance** (*n*), **inductive** (*adj*).

inductor energy for an inductor (↑) of self-inductance L (henry), carrying a current I_o (ampere), the work done in establishing the magnetic flux $= \int_o^t e\,I = \int_o^t LI.dI/dt.dt = \int_o^{I_o} LI.dI = \frac{1}{2}LI_o^2$ (joule), e and I being back e.m.f. and current respectively at time t; this energy is stored in the magnetic field and released into the circuit on circuit break, when I_o falls rapidly to zero.

search coil a conducting coil used in series with an oscilloscope to measure magnetic flux density B ($Wb\,m^{-2}$) (p.185) of an alternating field; the search coil plane should be placed parallel to the magnetic field B and rotated to give a magnetic flux linkage parallel to the coil axis; a search coil may be used to investigate the strength of an alternating magnetic field by joining it to an oscilloscope and holding the coil in the alternating field; the magnitude of the r.m.s. (p.218) induced voltage (p.205) across the coil, as given by the oscilloscope, is proportional to the r.m.s. value (p.218) of the field B at the point.

inductive circuit electrical circuit with inductance (↑) only, power being supplied by an a.c. or d.c. source (p.150).

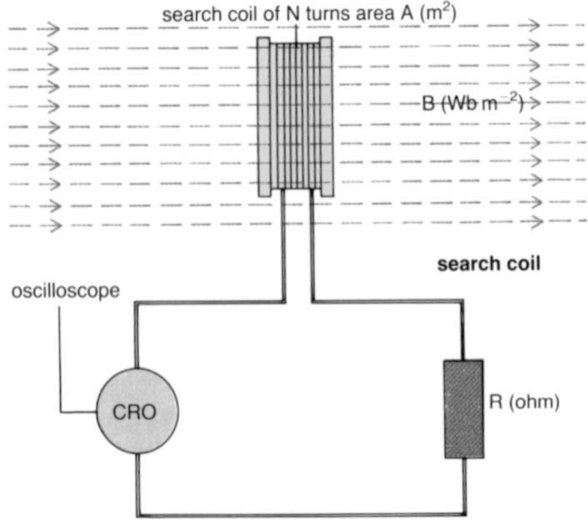

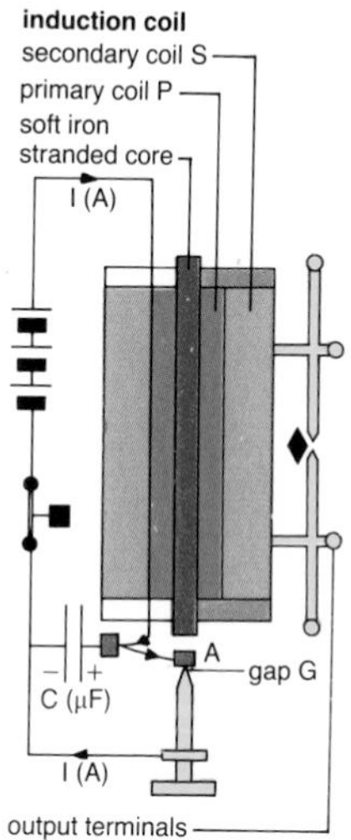

Rogowski spiral a series of search coils (↑) in series used for investigating magnetic flux around a closed loop; according to Ampere's theorem the net flux will be zero unless the closed loop encloses a current-carrying conductor.

induction coil a step-up voltage transformer operating on d.c. current, designed to produce high voltage d.c. output pulses across the secondary coil terminals; the input current is interrupted by a rapid make-and-break on the primary; there are far more turns on the secondary, so that the secondary output is a continual and rapid succession of high voltage d.c. pulses; the induction coil principle is used in car ignition systems (↓); it was formerly used in operating X-ray tubes (p.236) and gas discharge tubes.

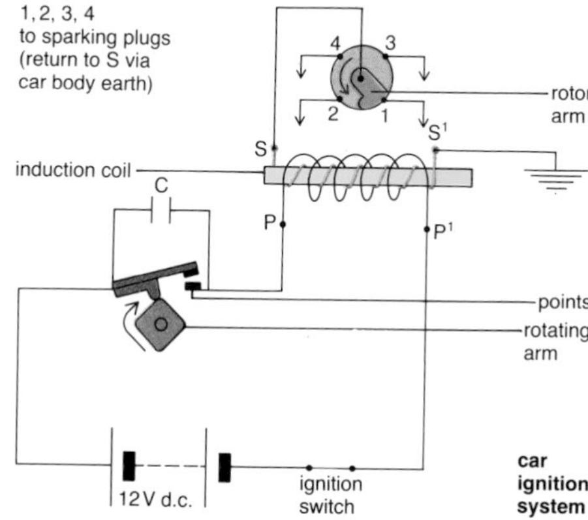

car ignition system an induction coil (↑) operated by a 12 volt d.c. battery to produce 12 kilovolt d.c. output pulses in precisely timed sequence, igniting the vaporized petrol and air mixture in the 4-cylinder internal combustion engine (p.146) of a car; the primary circuit make-and-break is controlled by a rotating cam, which opens and closes the tungsten contact points 4 times in each sequence, providing 4 high voltage pulses to the rotor arm of the distributor, firing the sparking plugs in the cylinder heads in sequence 1–3–4–2 to give engine balance and smooth running; the cam and rotor arm are driven by a common shaft from the camshaft, synchronizing pulses and distribution.

eddy currents any electrically conducting material near to a source of varying magnetic flux linkage (p.204) is itself subjected to the varying magnetic flux linkage, and an induced e.m.f. (p.205) is generated in the material, in a direction tending to oppose the flux variation causing it; the accompanying induced current is called an eddy current; eddy currents usually represent energy wastage from a circuit as resistive energy losses (p.159) and attempts are made to prevent their formation or minimize their effects, e.g. motor and alternator armatures (p.202) and transformer cores have laminated construction, with many, thin laminations cut perpendicular to the eddy current flow direction and separated from each other by thin layers of insulating varnish or oxide film; ferrimagnetic materials (p.184) are themselves good electrical insulators with minimal eddy current losses, while retaining the high remanence and low coercivity of soft magnetic materials (p.188), and so they replace laminated soft iron.

shaded-pole principle if part of the pole of an electromagnet carrying a.c. is covered by a plate of electrically conducting material, eddy currents (↑) are induced in the conductor; the magnetic field produced by eddy currents in the conductor (the shaded pole) is out of phase with the field of the electromagnet; it is as though the magnetic field moves from the unshaded pole to the shaded pole so that a nearby metal disc will respond by rotating.

eddy currents

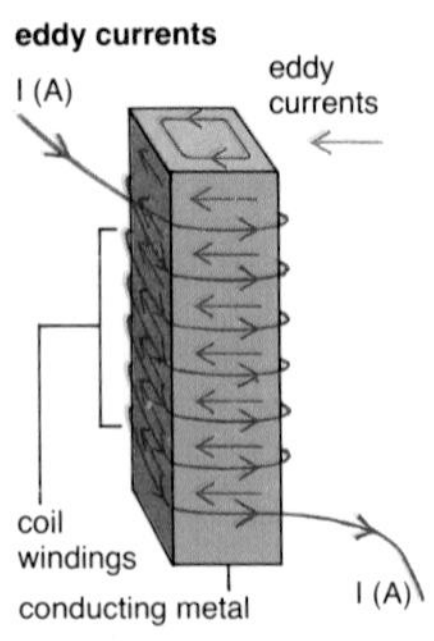

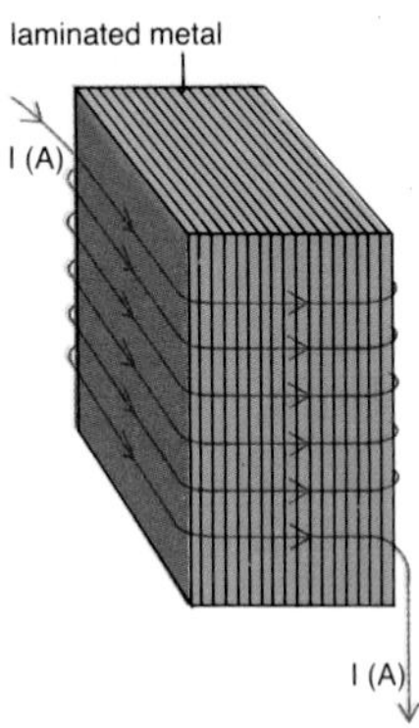

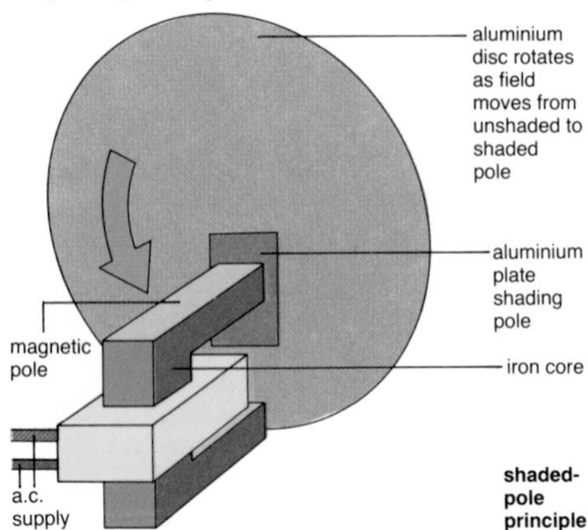

shaded-pole principle

shaded-pole induction motor this employs the shaded-pole principle (↑). The rotor (p.214) is a 'squirrel-cage' with a laminated iron core; the laminated stator (p.215) core is shaded by 2 (or more) welded copper loops; these cause the alternating magnetic flux to be delayed in the parts of the core they enclose and the rotor turns towards these; both efficiency and torque of this motor are low, but it is quiet running and mechanically very reliable; it is used in record players, small fans and aquarium pumps.

shaded-pole induction motor

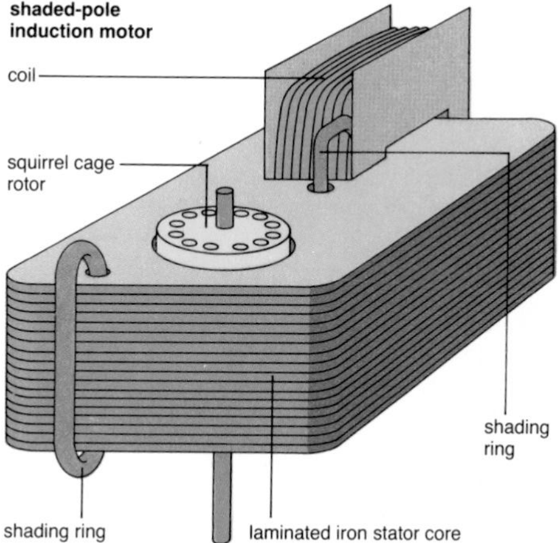

induction heating the generation of heat in a sample of conducting material using eddy current (↑) heating effects; high frequency a.c. current (p.214) is induced by a radio frequency (r.f.) a.c. voltage source, e.g. 500 kHz, as efficiency of energy transfer from source to sample increases with supply frequency; temperature rise due to eddy current heating is greater when the sample is heated in vacuum, and processes requiring continuous heating over a long period are carried out in a suitable vacuum system, e.g. the growth of a single crystal (p.12) of semiconducting material (p.222), from a melt containing a controlled amount of impurity content, for use in semiconductor device (p.227) manufacture; gases adsorbed on the metal structures inside electron tubes, e.g. a cathode ray tube (p.230), can be released on eddy current heating by a surrounding r.f. coil.

a.c. mains transformer a device for converting a low voltage input supplying high current to a high voltage output at low current, or vice versa, with minimized energy losses; it is designed to operate on a.c. current (↓) drawn from the a.c. mains (p.160); a secondary coil S is wound coaxially around the primary coil P on a laminated soft iron yoke core minimizing hysteresis losses (p.187), providing a closed path for magnetic flux linkage (p.204) and minimizing eddy currents (p.210); for a.c. mains voltage V_p, the secondary voltage V_s is given by $V_s/V_p = N_s/N_p$,
where N_p = number of turns on primary and N_s = number of turns on secondary; neglecting power losses $I_s.V_s = I_p.V_p$, where I_p = primary current and I_s = secondary current; thus $I_s/I_p = V_p/V_s = N_p/N_s$.

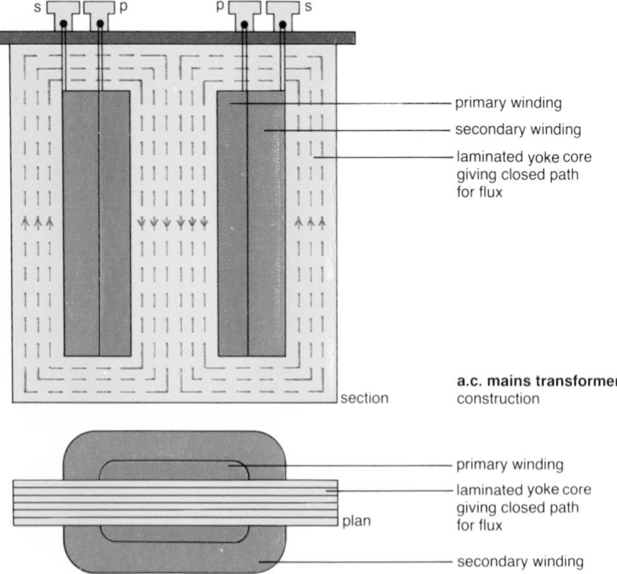

a.c. mains transformer construction

generator (*n*) a machine or physical system designed to produce electrical energy (p.28) from some other form of energy, e.g. mechanical energy (p.28) in a dynamo (↓) or alternator (p.214); solar energy in solar power devices; it can be designed to generate a.c. (↓) or d.c. voltage. **generate** (*v*), **generating** (*adj*).

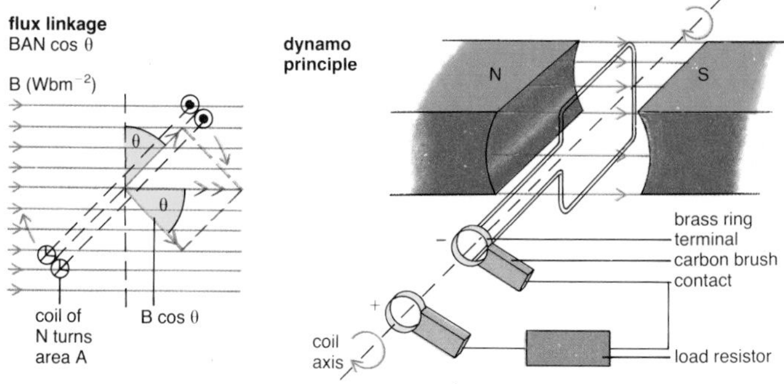

dynamo principle is based upon electromagnetic induction (p.204) in a conducting coil, rotating at constant angular velocity ω (rad s^{-1}) (p.46) in a magnetic field of flux density B (Wbm^{-2}) (p.185); at time t the e.m.f. induced across the coil is given by $E = E_o \sin \omega t$ where E_o = peak value of the e.m.f.; for coil of area A (m^2) and with N turns, $E_o = BAN\omega$ (volt); during each cycle of alternation (↓) the e.m.f. varies through 0, $+E_o$, 0, $-E_o$ to 0, giving an alternating voltage (↓).

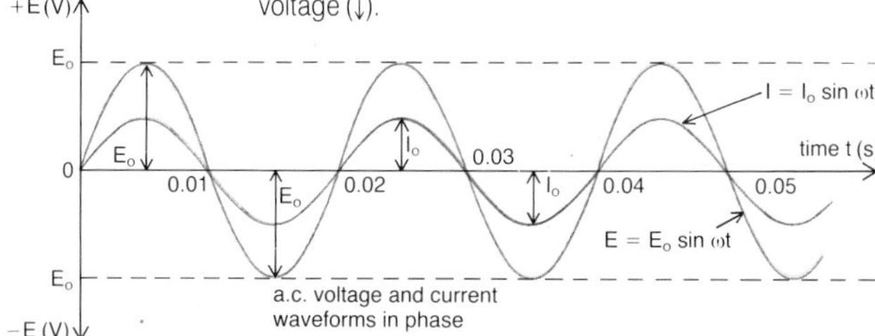

cycle of alternation

cycle of alternation the pattern of variation with time t (s) of the alternating voltage (↓) output E (volt) from a dynamo (↑), during 1 rotation of the dynamo coil.

alternating voltage the voltage output from a dynamo (↑) or alternator (p.214) of simple harmonic waveform (p.53) $E = E_o \sin \omega t$.

alternating current the current output from a dynamo (↑) or alternator (p.214) of simple harmonic waveform (p.53) $I = I_o \sin \omega t$, where I = instantaneous value at specific time instant t, and I_o = peak value (p.214); in an a.c. circuit with resistive impedance (p.219) only, alternating voltage (↑) and current are in phase.

a.c. voltage an alternative name for alternating voltage (p.213).
a.c. current an alternative name for alternating current (p.213).
peak value maximum value or amplitude (p.49) of a.c. voltage (p.213) or current (p.213).
alternator (*n*) an alternating voltage generator (p.212) operating on the dynamo principle (p.213), designed to give an alternating voltage output of maximum peak value (↑); the principal feature of design is that the magnetic field windings are wound on a rotating rotor (↓), requiring a relatively small operating current, while the voltage-generating coils are wound on a fixed stator (↓), so that the very large currents generated can be carried by fixed output connections; variations in the couple exerted on the rotor can be reduced by using a multi-pole armature construction with 3-phase wiring (↓) arrangements; a.c. voltage (↑) is generated with a waveform $V = V_0 \sin \omega t$ and a.c. current (↑) has a waveform $I = I_0 \sin \omega t$. **alternation** (*n*), **alternate** (*v*), **alternating** (*adj*).

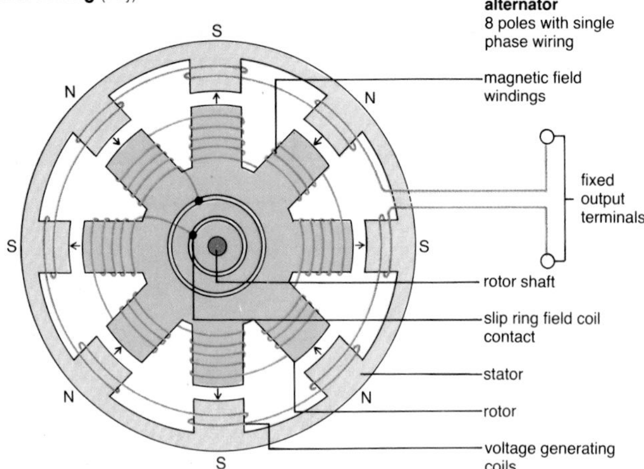

a.c. generator an alternative name for alternator (↑).
rotor (*n*) the central rotating core of an alternator (↑) armature, carrying the magnetic field windings; a small exciter dynamo, on the common shaft of the turbo-alternator set, provides a relatively small d.c. current (p.150) to energize the field windings.

stator (*n*) the outer fixed core of an alternator (↑) armature, surrounding the rotor (↑) and carrying the voltage-generating coils; the very large currents generated are carried by fixed output connections.

alternator armature the central component parts of an alternator (↑) including the rotor (↑) and stator (↑) with windings on soft iron laminated core structures, giving reduced eddy current energy losses (p.210).

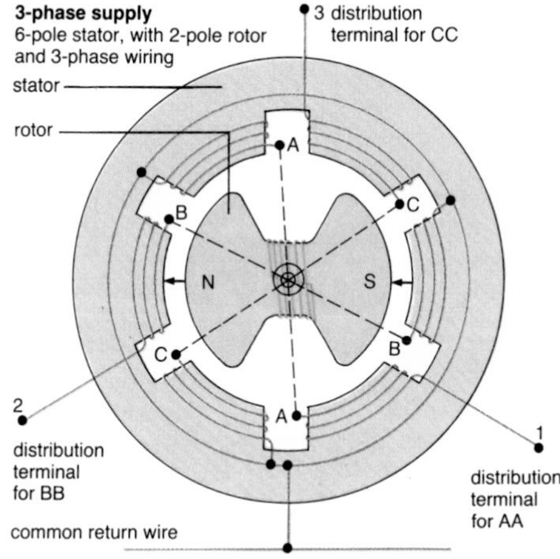

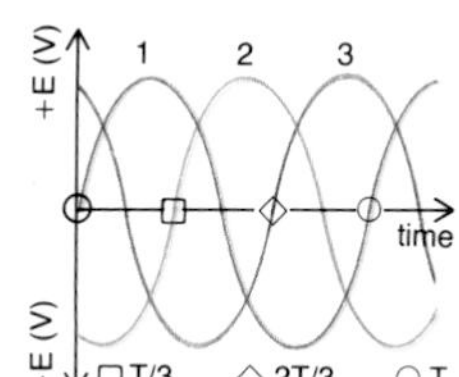

a.c. currents in 3-phase system: always zero current in common return wire

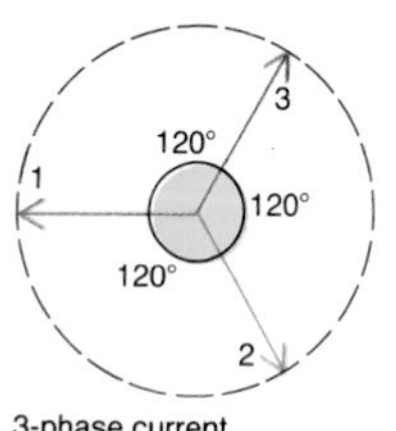

3-phase current vectors

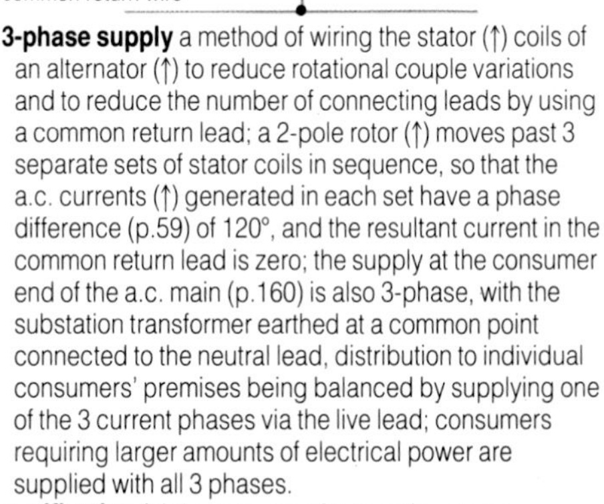

3-phase supply a method of wiring the stator (↑) coils of an alternator (↑) to reduce rotational couple variations and to reduce the number of connecting leads by using a common return lead; a 2-pole rotor (↑) moves past 3 separate sets of stator coils in sequence, so that the a.c. currents (↑) generated in each set have a phase difference (p.59) of 120°, and the resultant current in the common return lead is zero; the supply at the consumer end of the a.c. main (p.160) is also 3-phase, with the substation transformer earthed at a common point connected to the neutral lead, distribution to individual consumers' premises being balanced by supplying one of the 3 current phases via the live lead; consumers requiring larger amounts of electrical power are supplied with all 3 phases.

rectification (*n*) the process of converting an a.c. voltage or current (↑) to a d.c. voltage or current (p.150) using an appropriate device, e.g. d.c. generator, metal rectifier, vacuum diode or semiconductor diode (p225).

half-wave rectification rectification (p.215) of a.c. voltage using a single rectifying device; the rectifier conducts only on alternate half-cycles of alternation (p.213) of the a.c. input, and the rectified output is a series of periodic unidirectional d.c. voltage pulses with corresponding pulsating current (p.150).

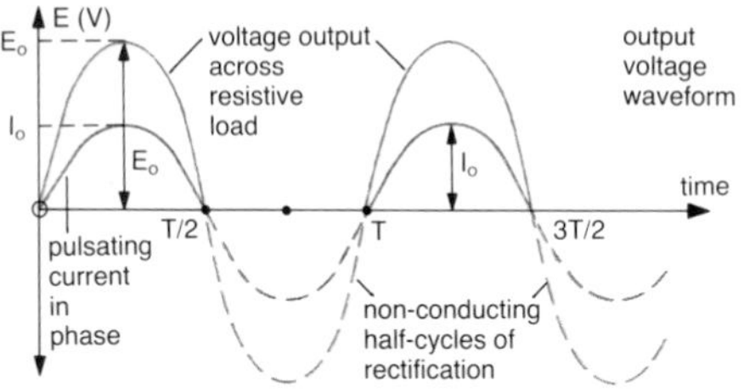

half-wave rectification

a.c. mains voltage input

\+ −

resistive load

metal rectifier

voltage transformer

full-wave rectification rectification (p.215) of a.c. voltage in which the positive and negative half-waves of the single phase a.c. input wave are both effective in delivering unidirectional current to the load; the rectified output is a continuous series of unidirectional d.c. voltage pulses with corresponding continuous pulsating current (p.150); filter circuits, using the a.c. reactance (p.219) conducting properties of inductors (p.208) and capacitors (p.171), can be used to remove the a.c. components, giving a smoothed d.c. output voltage and current.

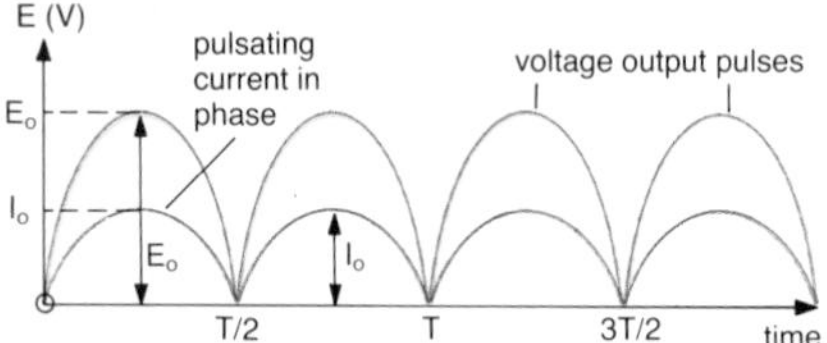

full-wave rectification output voltage waveform: unidirectional pulsating current in phase

bridge rectifier a full-wave (↑) rectifier consisting of a bridge with a rectifier in each arm.

centre-tap full-wave rectifier a full-wave (↑) rectifier whose output is taken from the output of a transformer with a centre tapping; two rectifying devices, each conducting only on alternate half-cycles are used.

rectified voltage the d.c. voltage output resulting from half-wave (↑), full-wave (↑) or 3-phase rectification.

rectified current the d.c. current (p.150) output resulting from half-wave (↑), full-wave (↑) or 3-phase rectification.

bridge rectifier

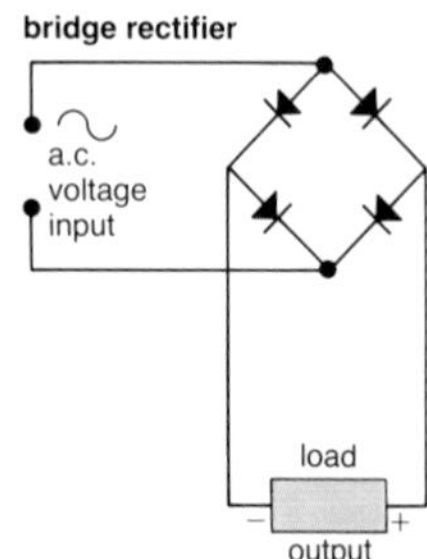

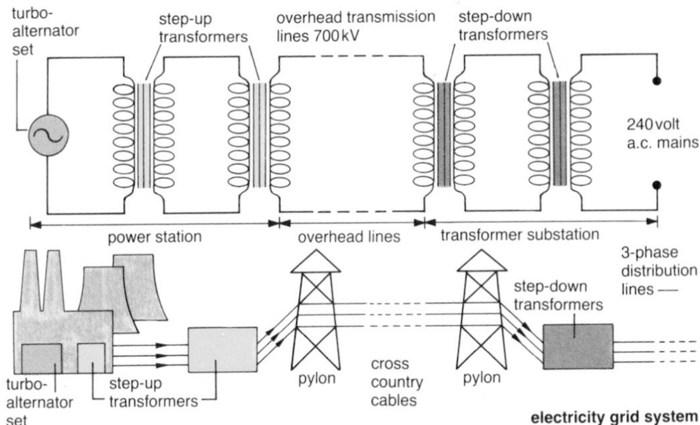

electricity grid system

electricity grid system a method of distribution of electrical power over large regions; it includes power stations, using step-up transformers to supply the long-distance overhead transmission lines with electrical power at high voltage, and step-down transformers at local substations (at the consumer supply points), to provide 3-phase supply (p.215) at reduced voltage for industrial and domestic consumption; the U.K. National Grid System generates megawatt output with a.c. voltage (p.214) of frequency 50 hertz (p.54) at up to 32 kilovolt, with step-up transformers raising this to 270 kV or 400 kV for long distance power transfer on the overhead transmission lines; voltage reduction at consumer supply points is by step-down transformers in stages, to 132 kV, then 33 kV and 11 kV for industrial consumers, and 415 V or 240 V for local and domestic use; U.S.A. and Canada operate overhead lines at 700 kV, U.S.S.R. at 750 kV and at 1500 kV.

a.c. circuit a conducting circuit including an a.c. source (p.151) and components capable of conducting alternating current, e.g. inductive and capacitative reactances (p.219), resistance.

a.c. circuit

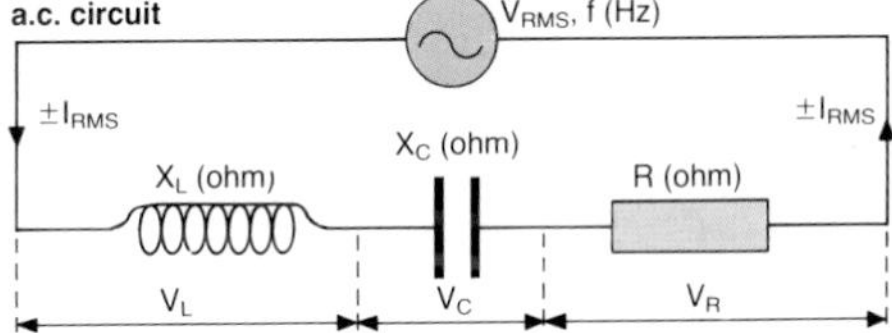

root-mean-square (r.m.s.) value the effective value of an a.c. voltage, current or other periodic quantity; for an a.c. current (p.214) of waveform $I = I_o \sin \omega t$, with amplitude I_o and frequency f (hertz) (p.54) given by $\omega = 2\pi f$, it is defined as equal in value to the d.c. current (p.150) I (ampere) having the same Joule heating effect (p.159) in a specified resistor of electrical resistance R (ohm) (↓); the average heating effect per cycle of alternation (p.213) of a.c. current = $\frac{1}{2}I_o^2R$ (watt), so I^2R (W) = $\frac{1}{2}I_o^2R$ (W), and $I = I_{RMS} = I_o/\sqrt{2} = 0.707I_o$ (A); similarly for a.c. voltage of waveform $V = V_o \sin \omega t$, V^2/R (W) = $\frac{1}{2}V_o^2/R$ (W), and $V = V_{RMS} = V_o/\sqrt{2} = 0.707V_o$ (volt); current and voltage values for a.c. circuits are always given as r.m.s. values; they are measured by moving-iron meters, which give a unidirectional pointer deflection for both positive (+) and negative (−) half-cycles of a.c. current flow, deflection being $\propto$ (current)^2; the average electrical power (p.159) generated per cycle = $\frac{1}{2}V_oI_o = V_o/\sqrt{2} \times I_o/\sqrt{2} = V_{RMS} \times I_{RMS}$ (W) and is measured by a wattmeter (p.199).

effective value an alternative name for r.m.s. value (↑) of voltage or current.

a.c. vector representation a.c. voltage and current (p.214) of the same frequency in an a.c. circuit (p.217) usually differ in phase (p.53), and their waveforms do not overlap unless the circuit has a purely resistive impedance (↓); they are vector quantities (p.30) and are represented diagrammatically by vectors drawn radially with respect to the centre of a reference circle, having an angle equal to the phase difference (p.59) between them; vectors can be added by the vector parallelogram law (p.33).

rotating vector a name given to a.c. vector representation (↑).

rotating vector representation of 2 a.c. quantities of the same frequency but differing in phase by 45° ($\pi/4$)

Φ phase angle
● current vector
□ voltage vector

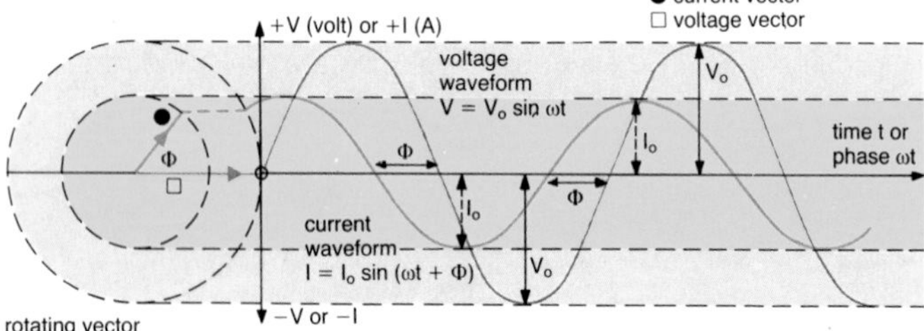

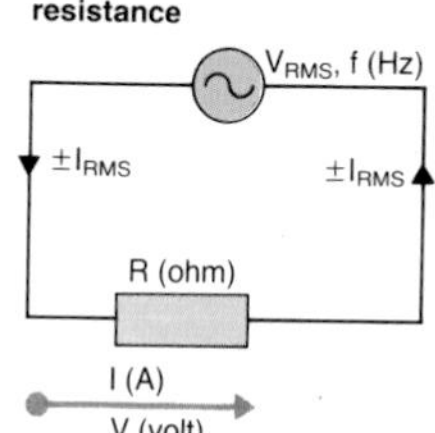

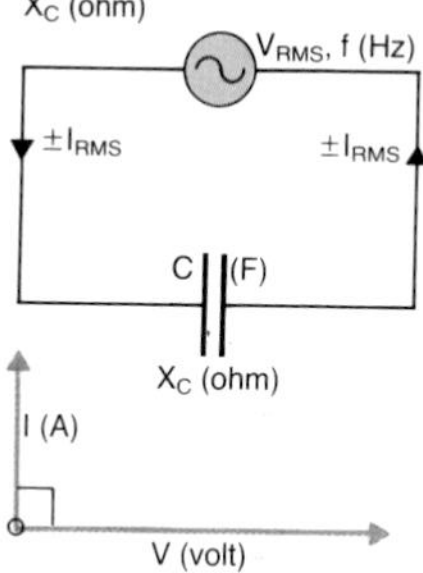

phase angle the angle between the rotating vectors (↑) representing resultant a.c. voltage V (volt) and a.c. current I (ampere) in an a.c. circuit; denoted by Φ.

resistance in a.c. circuit a.c. voltage V (volt) and a.c. current I (ampere) are in phase, with their a.c. vectors (↑) parallel; from Ohm's Law (p.155), $R = V/I$, where V and I are r.m.s. values (↑), so a.c. resistance is identical with d.c. resistance R (ohm).

capacitative reactance the a.c. current I (ampere) leads the a.c. voltage V (volt) by a phase difference of 90° ($\pi/2$) and their a.c. vectors (↑) are perpendicular; the capacitative reactance X_c (ohm) is defined to be the ratio of r.m.s. voltage to r.m.s. current for a.c. flowing through the capacitor (p.171); for capacitance C (farad) and a.c. of frequency f, $X_c = 1/\omega C$, where $\omega = 2\pi f$.

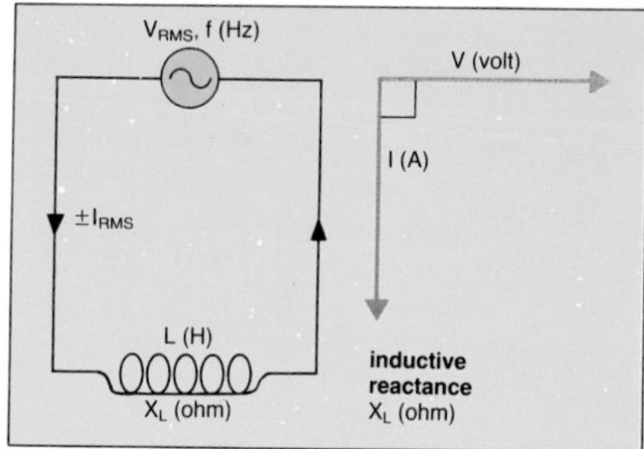

inductive reactance the a.c. current I (ampere) lags behind the a.c. voltage V (volt) by a phase difference of 90° ($\pi/2$) and their a.c. vectors (↑) are perpendicular; the inductive reactance X_L (ohm) is defined to be the ratio of r.m.s. voltage to r.m.s. current for a.c. flowing through the inductance (p.208); for inductance L (henry) and a.c. of frequency f, $X_L = \omega L$, where $\omega = 2\pi f$.

impedance (*n*) the electrical resistance offered to the flow of a.c. current (p.214) in an a.c. circuit (p.217) by the combined effects of all its components, e.g. capacitative (↑) and inductive reactances (↑), resistance (↑); denoted by Z, it is defined to be the ratio of r.m.s. voltage to r.m.s. current; its value for an a.c. series circuit is $Z = \sqrt{R^2 + (X_L - X_C)^2}$ (ohm).

a.c. series resonant circuit a.c. circuit (p.217) showing minimum impedance (↑) at resonant frequency f_o (Hz), forming a detector (p.221) circuit for a radio receiver.

voltage magnification in an a.c. series resonant circuit (p.219), voltage across the inductor L is increased by factor $Q = \omega L/R$ where $\omega = 2\,\pi f_o$.

a.c. parallel resonant circuit an a.c. circuit (p.217) including an inductor of negligible resistance and capacitor in parallel arrangement, giving inductive and capacitative reactance (p.219) contributions to the total circuit impedance Z (ohm) (p.219); the a.c. current I_C (ampere) in the capacitor branch leads the applied a.c. voltage V (volt) by 90° ($\pi/2$) and the a.c. current I_L (A) in the inductor branch lags behind V by 90° ($\pi/2$); I_C, I_L and V are in r.m.s. values (p.218); I_C and I_L are in antiphase and, if the a.c. supply frequency f (hertz) is varied, the condition for current resonance is fulfilled at resonant frequency f_o (Hz), when $I_C = -I_L$; since $I_C = V/X_C$ and $I_L = V/X_L$, $X_C = X_L$ at resonance, and $1/\omega C = \omega L$, so $\omega^2 = 1/LC$, where $\omega = 2\pi f$, giving $f_o = 1/2\pi\sqrt{LC}$ (Hz); the frequency-response curve (p.102) for the circuit shows minimum current I (A) at f_o for a total circuit impedance Z (ohm) approaching infinity (∞) and the circuit appears to reject current of frequency f_o; but currents I_C and I_L remain in exact antiphase in the 2 branches of the circuit, which forms an oscillatory circuit continuing to oscillate without any further energy input from the supply; in practice there is always some resistance present and oscillations need to be sustained, e.g. transistor oscillator (p.227); this circuit forms the basis of the signal generator (p.108) oscillator (↓) and, with the use of an aerial, of the radio transmitter (↓).

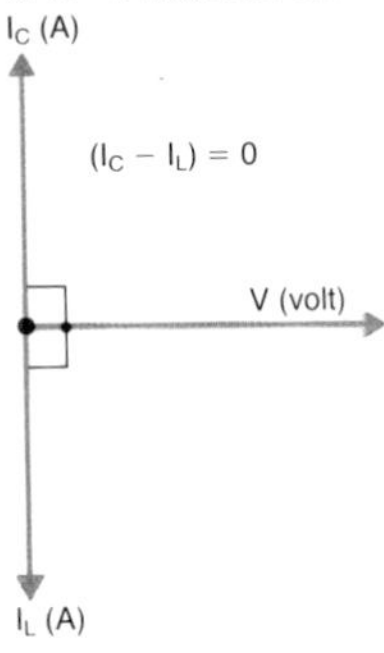

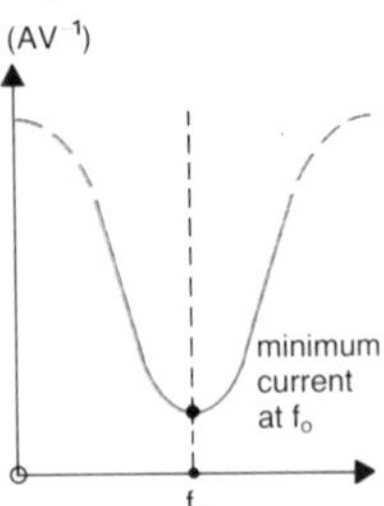

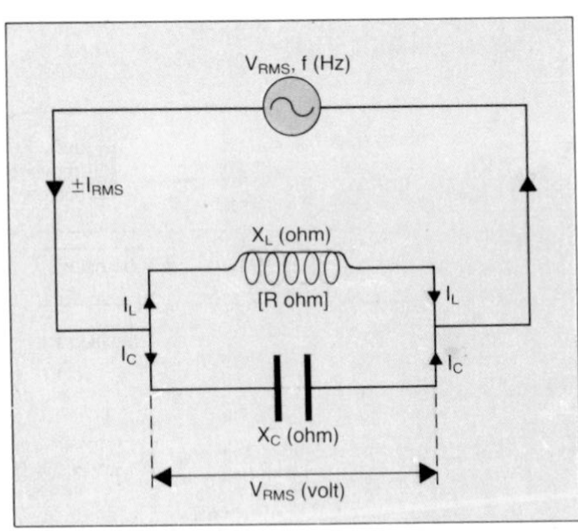

a.c. parallel or current resonant circuit

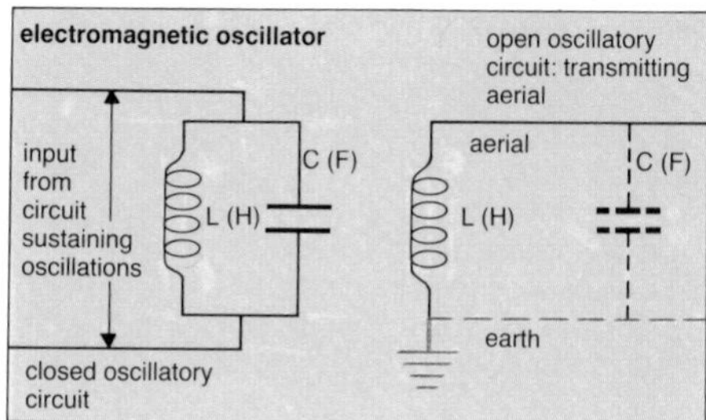

electromagnetic oscillator an a.c. parallel resonant circuit (↑) used as a means of generating electromagnetic oscillations, e.g. transistor oscillator (p.227).

electromagnetic tuned circuit an oscillatory circuit containing inductance and capacitance which has been adjusted to resonate at the frequency of an applied signal.

radio transmitter a circuit designed to generate electromagnetic waves of radio frequency (p.101), using an electromagnetic tuned circuit (↑), and to transmit electromagnetic energy into free space, by means of a transmitting aerial; the waves are modulated to act as information carriers of radio and T.V. signals.

radio receiver a circuit designed to receive radio waves; it consists of an aerial linked to an electromagnetic tuned circuit and a detector (↓).

detector a circuit used to extract a signal from a carrier; in the case of a radio receiver it separates the audio-frequency signal from the radio frequency signal and commonly contains a semiconductor diode (p.225).

radio receiver

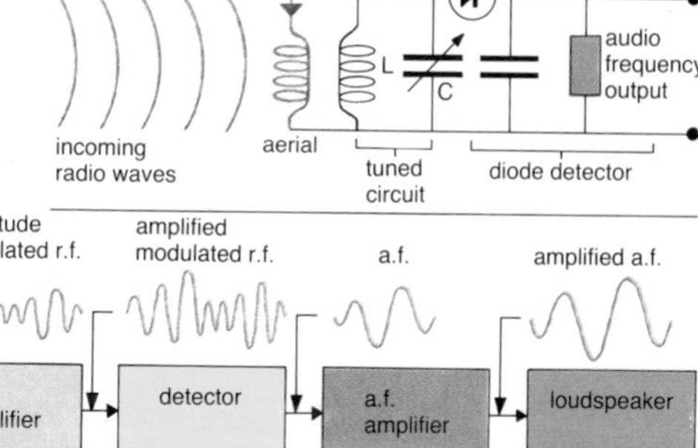

electronic systems any complex electronic equipment may be considered to be made up of a number of basic building blocks each of which performs a specific task, e.g. amplifier, oscillator, switch, counter etc.; designing circuits in terms of these basic blocks is known as the systems approach; a system is a combination of blocks put together to perform a particular task.

semiconductor (*n*) a solid with electrical resistivity (p.158) intermediate in value between that of electrical conductors and electrical insulators, e.g. germanium 0.47 ohm m and silicon 3×10^3 ohm m at 27°C; the semiconducting elements are included in the Group 4 elements (p.7) of the Periodic Table, and so have 4 valence electrons, e.g. germanium, silicon. **semiconducting** (*adj*).

intrinsic semiconductor a semiconductor (↑) of high purity having equal numbers of electrons (p.7) and holes (↓) as charge carriers (p.9), 1 in 10^{10} of the atoms in its crystal lattice (p.12) contributes charge carriers; its electrical conductivity (p.155) rises rapidly with temperature as more hole-electron pairs are produced.

extrinsic semiconductor a semiconductor (↑) whose electrical conductivity (p.155) is greatly enhanced by doping (↓) with trace impurity (↓) in precisely controlled amounts, e.g. 1 impurity atom in 10^7 crystal lattice atoms (p.12); either electrons (p.7) or holes (↓) are present in excess as majority charge carriers (↓).

doping (*n*) refers to the introduction of a very small and precisely controlled amount of impurity into the crystal lattice (p.12) of a single crystal of pure semiconductor (↑) during crystal growth from the molten state, e.g. 1 impurity atom in 10^7; doping enhances the electrical conductivity (p.155) of a pure semiconductor. **dope** (*v*), **doped** (*adj*).

donor impurity Group 5 elements of the Periodic Table, e.g phosphorus, arsenic, antimony, possessing 5 valence electrons; when introduced as trace impurity into the crystal lattice (p.12) of a pure Group 4 semiconducting (↑) element, excess electrons are donated as negative charge carriers.

acceptor impurity Group 3 elements of the Periodic Table, e.g. indium, gallium, possessing 3 valence electrons; when introduced as trace impurity into the crystal lattice (p.12) of a pure Group 4 semiconducting (↑) element, excess holes (↓) are generated as positive charge carriers.

doping
1 impurity atom in 10^7 atoms of intrinsic semiconductor (germanium)

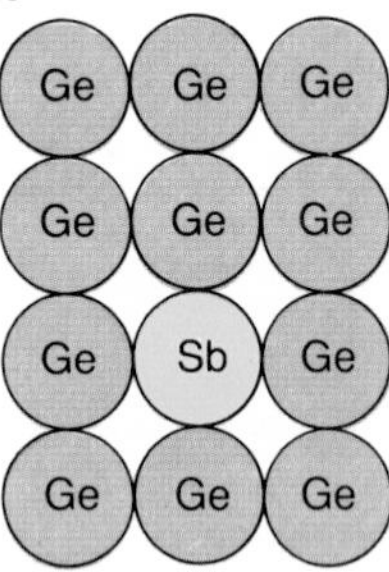

donor impurity (antimony)

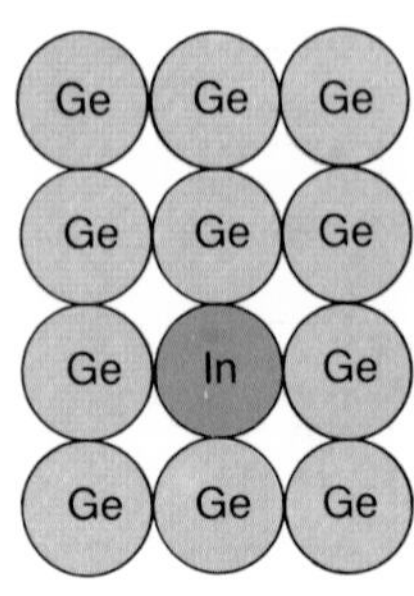

acceptor impurity (indium)

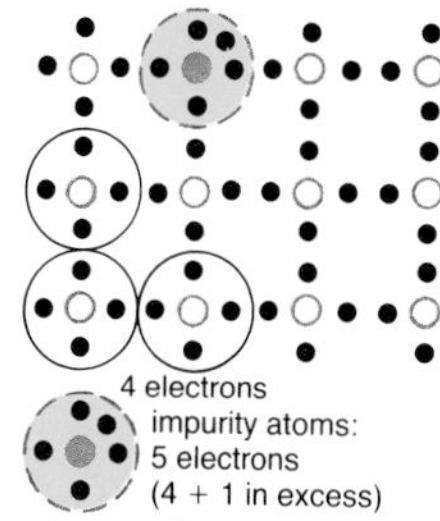

n-type semiconductor

hole (*n*) a small location in the crystal lattice (p.12) where the number of electrons (p.7) available is insufficient to satisfy the valence bonding requirements of nearby atoms, thus creating a small site of excess positive charge called a hole; a hole can migrate through the crystal lattice, as electrons move in successively from adjacent atoms to fill the vacant site; a hole acts as a positive charge carrier of lower mobility than an electron; holes are majority charge carriers (p.224) in p-type semiconductor (↑) material and minority carriers in n-type semiconductor material.

n-type semiconductor pure Group 4 element, silicon or germanium, doped (↑) with Group 5 element impurity, e.g. phosphorus, arsenic, antimony, so that negative (n) charge carriers (p.9), electrons (p.7), are in excess.

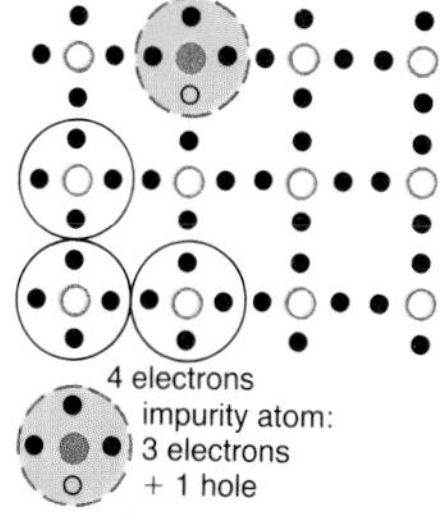

p-type semiconductor

p-type semiconductor pure Group 4 element, silicon or germanium, doped (↑) with Group 3 element impurity, e.g. gallium, indium, so that positive (p) charge carriers (p.9), holes (↑), are in excess.

energy band model for solids the sharply defined discrete energy levels (p.7) for isolated atoms broaden out into much wider distinct energy bands when atoms are at close proximity in the crystal lattice (p.12) of a solid; the bands overlap in electrically conducting metals; electrons exist in these broad energy bands, rather than in discrete energy levels.

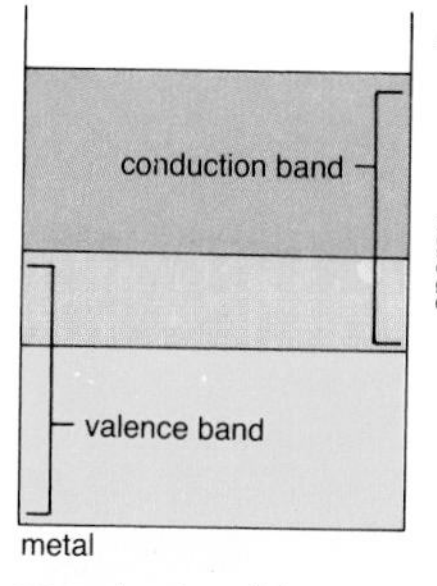

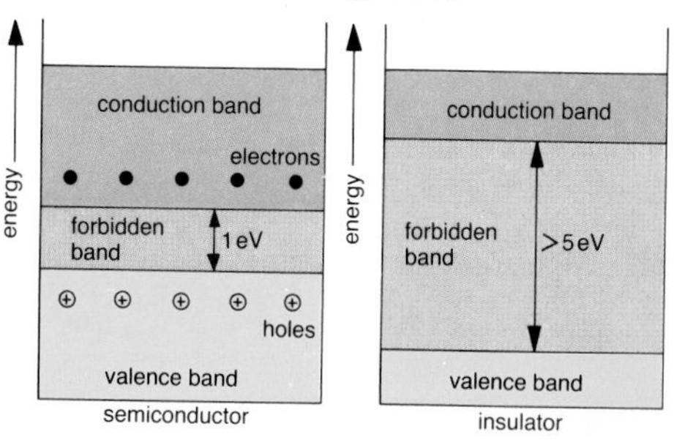

energy band model for solids

valence band the energy band containing the valence electrons of atoms in a particular crystal lattice (p.12).

conduction band the energy band containing the conduction electrons (p.9) of atoms in a particular crystal lattice (p.12); in metals the conduction band overlaps with the valence band (↑).

forbidden band the energy gap between the valence (p.223) and conduction bands (p.223) in a semiconductor (p.222) or electrical insulator; semiconductors have a small gap approximately 1 eV (p.228) at room temperature (p.139), so that electrons can cross from the valence band into the conduction band by thermal energy (p.28) alone; electrical insulators have a larger gap, 5 eV or more at room temperature, so that electrons can only cross the gap when a strong electric field is applied, causing breakdown of the insulation properties of the material.

forbidden zone an alternative name for forbidden band (↑).

majority carriers the charge carriers (p.9) present in the greatest numbers in extrinsic semiconductors (p.222), e.g. electrons in n-type (p.223) materials, holes in p-type materials (p.223).

minority carriers the charge carriers (p.9) present in the lowest numbers in extrinsic semiconductors (p.222), e.g. holes in n-type material (p.223), electrons in p-type material (p.223).

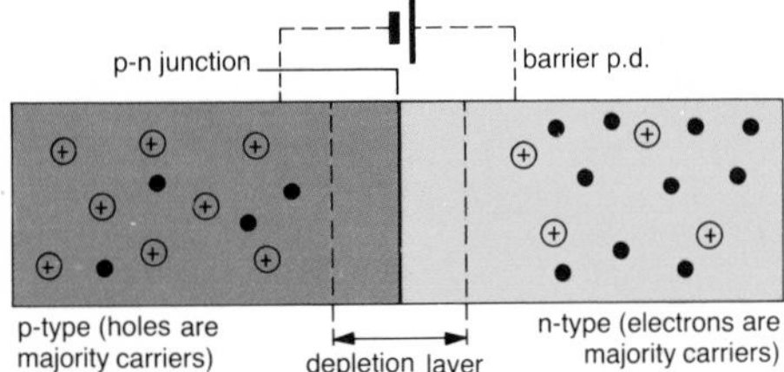

p-n junction with **barrier potential difference**

barrier potential difference charge carriers (p.9), drifting across a p-n junction (↓) into the region of opposite charge excess, tend to repel the drift of further charge carriers of the same sign, thus acting like a small potential difference (p.154) to prevent further flow of charge carriers across the p-n junction; this is about 0.1 V for germanium and about 0.6 V for silicon.

junction voltage an alternative name for barrier potential difference.

p-n junction the interface separating regions of n-type (p.223) and p-type (p.223) semiconductor material within the same crystal of germanium or silicon; the regions are separated by a narrow depletion layer (↓), free of charge carriers, across which the barrier potential difference (↑) acts.

depletion layer the narrow region on either side of a p-n junction (↑) in which the recombination of electrons with holes (p.223) leaves the region free of charge carriers.

barrier layer an alternative name for depletion layer (↑).

diode an electronic device with two electrodes; most commonly used as a rectifier (p.215).

junction diode a p-n junction (↑) in series with an external source of potential difference, e.g. a battery, to provide forward (↓) or reverse bias (↓), so that the junction can act as a rectifier (p.215).

semiconductor diode an alternative name for a p-n junction diode (↑).

photodiode a p-n junction (↑) with a transparent window through which light can pass when used in reverse bias (↓); the leakage current is proportional to the amount of light incident on the device.

junction diode
– circuit symbol

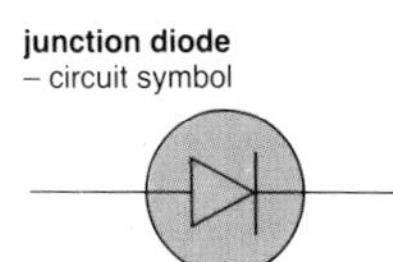

light-emitting diode (LED) a junction diode (↑) made from the semiconductor (p.222) gallium arsenide phosphide; it emits light when forward biased (↓).

Gunn diode a diode formed from n-type (p.223) gallium arsenide; produces coherent microwaves (p.101) when a large electric field (p.165) is applied across it.

vacuum diode 2 electrode vacuum tube, producing electrons by thermionic emission (p.228); functions as a rectifier.

forward bias the application of an external source of potential difference, e.g. a battery, to a p-n junction (↑), in opposition to the barrier potential difference (↑), to increase current flow across the junction.

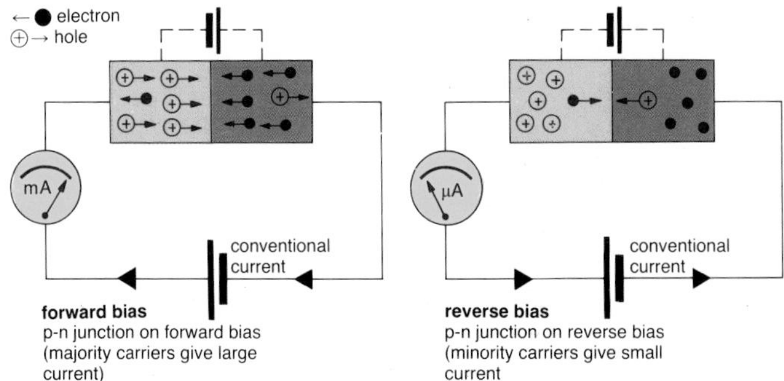

forward bias
p-n junction on forward bias (majority carriers give large current)

reverse bias
p-n junction on reverse bias (minority carriers give small current

reverse bias the application of an external source of potential difference, e.g. a battery, to a p-n junction (↑), in the same direction as the barrier potential difference (↑), to decrease current flow across the junction.

junction diode characteristic graphs of forward and reverse current against forward and reverse bias (p.225) voltage, usually plotted on the same axes.

Zener voltage the voltage marking a sudden increase in reverse current as the junction diode (↑) resistance decreases on reverse bias (p.225).

breakdown voltage an alternative name for Zener voltage (↑).

Zener diode a junction diode (↑), operating at its Zener voltage (↑) which remains constant over a wide range of currents; the diode is used as a voltage stabilizer.

junction transistor a 2-junction semiconductor device (↓) capable of current, voltage and power amplification; it can have 2 configurations: p–n–p and n–p–n, and it has 3 electrodes referred to as the emitter, base and collector; since electrons are more mobile than holes, devices depending mainly on electron flow are faster in operation than those depending on the flow of holes, thus n–p–n transistors are most commonly used and the semiconductor material is usually silicon.

phototransistor a junction transistor (↑) whose base is directly illuminated as an alternative to making an electrical connection to it; collector/emitter current is then governed by the light radiation falling on the base; the device is 100 times more sensitive than the photodiode (p.225).

transistor amplifier since a small change in base current in a junction transistor (↑) can cause a large change in current in the emitter/collector circuit, transistors can be used to amplify current, voltage or power.

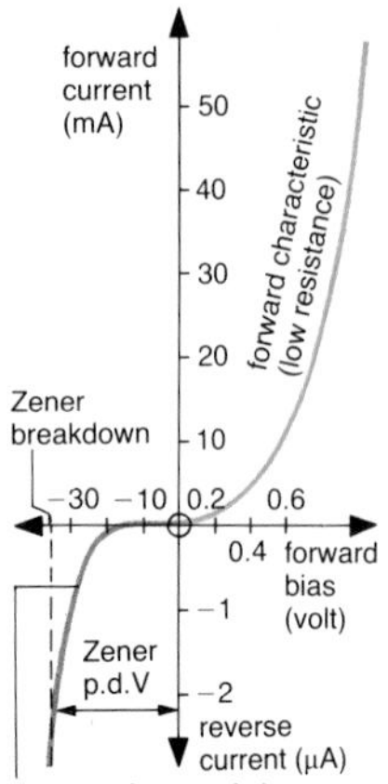

junction diode characteristic (silicon)

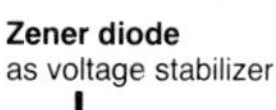

Zener diode as voltage stabilizer

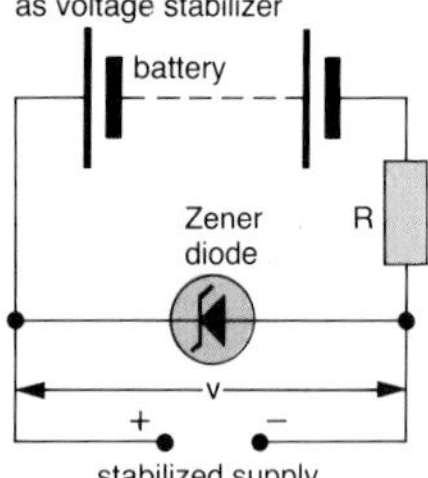

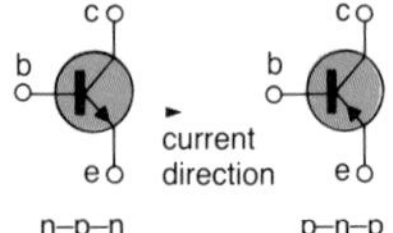

junction transistors

n-p-n junction transistor as common emitter amplifier

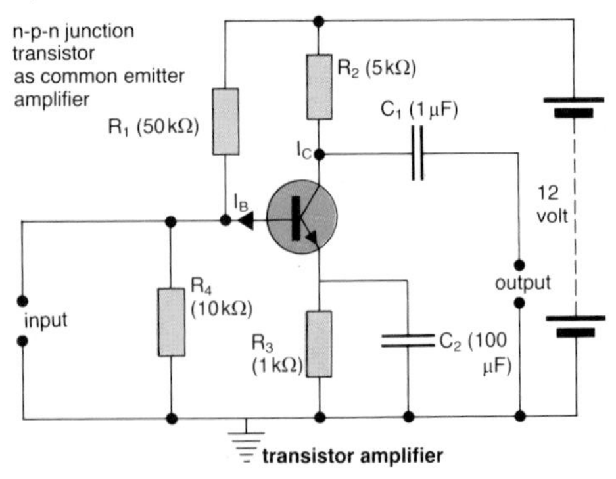

transistor amplifier

$I_C \simeq \beta I_B$
β amplification current factor
C_2 by-pass capacitor
R_3 emitter stabilizing resistor

operational amplifier
as an integrator

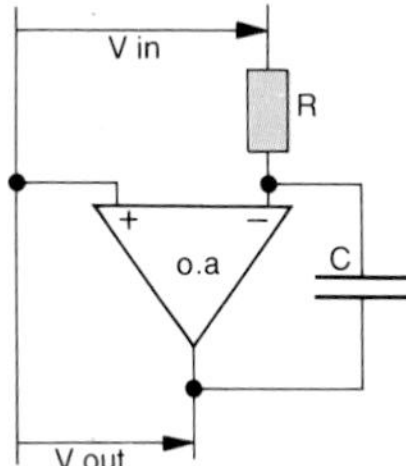

o.a operational amplifier

transistor oscillator

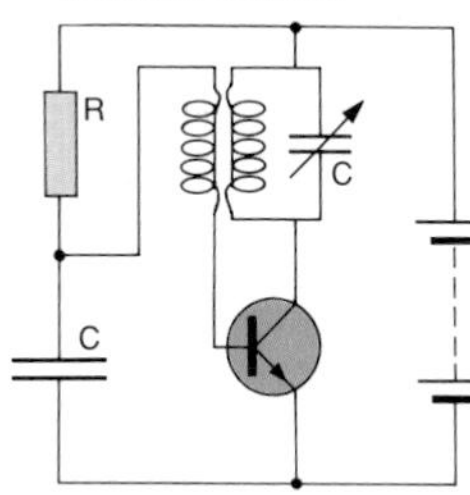

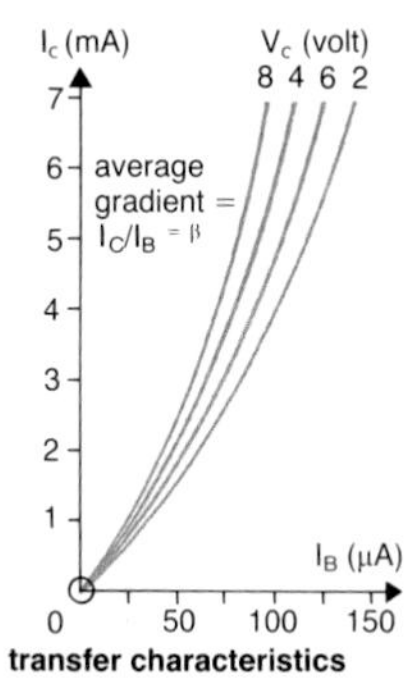

transfer characteristics

operational amplifier a high gain, high input impedance d.c. amplifier normally used with considerable external feedback (↓) which completely determines its characteristics; can be used to perform mathematical operations such as differentiating and integrating, has a wide range of applications.

transistor oscillator a junction transistor (↑) operating so that part of the amplifier output is fed back (↓) in phase with the input, and electromagnetic oscillations (p.221) are established at the circuit resonant frequency (p.102).

feedback the process of returning part of an output signal back to the input of a device; if feedback increases the input signal it is called positive feedback; if feedback decreases the input signal it is negative feedback.

current amplification factor the ratio $I_C \div I_B$ for a junction transistor (↑), where I_C = collector current and I_B = base current; denoted by β; β may be between 30 and 75, but a typical value is about 50.

transfer characteristics graphs of collector current I_C against base current I_B for various values of collector-emitter voltage V_C, plotted for a junction transistor (↑); the average gradient gives the current amplification factor (↑).

solid state device a small-scale silicon manufactured device using the electrical properties of a p-n junction (p.224).

semiconductor device an alternative name for solid state device (↑).

integrated circuit a very small-scale solid state device (↑), manufactured using a silicon chip of size approximately 1.5 mm square and 0.2 mm thick, and containing the many transistors, diodes, capacitors and resistors required to make up a complex electronic circuit.

silicon chip an alternative name for integrated circuit (↑).

large-scale integration describes an integrated circuit (↑) manufactured on a silicon chip (↑) 2 mm square or smaller, containing 1000 to 10000 separate electronic components; the density of components is 25000/cm^2 or more.

very large-scale integration describes an integrated circuit containing 10000–1 million components on the silicon chip (↑).

cathode rays high velocity electrons (p.7) emitted from the negative cathode electrode of a vacuum discharge tube at pressure below 10^{-2} torr and electric field 100 volt cm^{-1}; they form a non-luminous discharge travelling in straight line paths along the tube to the positive anode electrode; they can be detected by the kinetic energy (p.27) they transfer to objects in their path and by their fluorescent effects, e.g. on the screen of a T.V. tube; they can cause the emission of X-rays from the target of an X-ray tube (p.236).

electron energy an electron (p.7) of mass m (kg), having negative charge e (coulomb), is accelerated by the attractive forces of an electric field (p.165) and acquires velocity v ($m s^{-1}$) and kinetic energy $\frac{1}{2}mv^2$ (joule) from the work done upon it by the field forces; in an X-ray tube (p.236) or vacuum discharge tube, with electric potential difference V (volt) (p.154) between its electrodes, electrons emitted from the cathode as cathode rays (↑) with zero velocity acquire velocity v on reaching the anode; the work done on an electron = eV (joule) = kinetic energy $\frac{1}{2}mv^2$ (J); so $v = \sqrt{2eV/m}$ ($m s^{-1}$).

electron-volt (eV) the unit of kinetic energy for electrons and nuclear particles, e.g. protons (p.8); it is the energy (↑) acquired by an electron when accelerated through p.d. V = 1 volt; for electronic charge $e = 1.60 \times 10^{-19}$ (coulomb): $1\ eV = 1.60 \times 10^{-19}$ (joule); for very high energy particles: $1\ MeV = 10^6\ eV$ and $1\ GeV = 10^9\ eV$.

thermionic emission the emission of electrons from a solid surface when sufficient energy is provided, by direct or indirect heating of the surface, to overcome the work function (↓) energy; these free electrons are usually produced from specially prepared surfaces, e.g. mixed barium and strontium oxides, of low work function, in vacuum tube conditions, as in a cathode-ray tube (p.230).

photoelectric emission the emission of electrons from a solid surface when sufficient energy is provided, by the energy of photons (p.234) in a light beam, to overcome the work function (↓) energy; these free electrons are produced from low work function surfaces by the photoelectric effect (↓) in vacuum tube conditions, e.g. in a photoelectric cell (p.229).

work function the energy (p.26) required to remove an electron from its containing solid surface; denoted by W and measured in units of electron-volt (eV) (↑) or joule (J); the value of W is characteristic of a particular solid

cathode rays
vacuum discharge tube (cathode-ray tube)

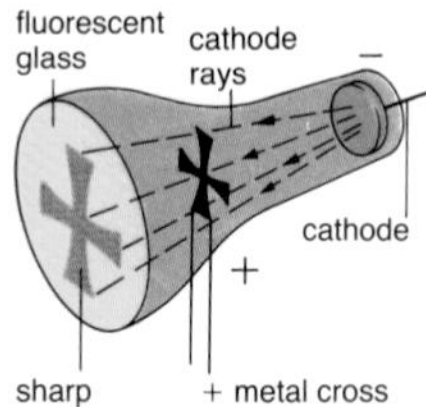

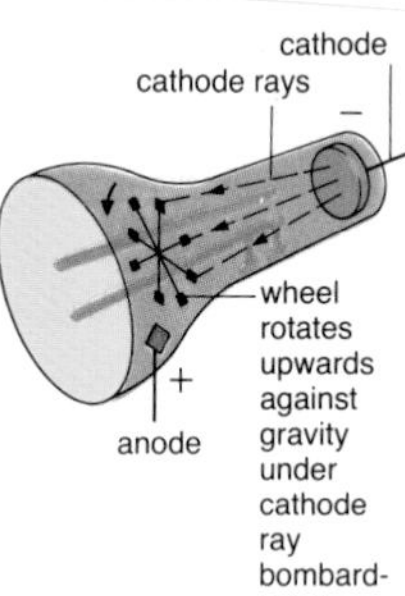

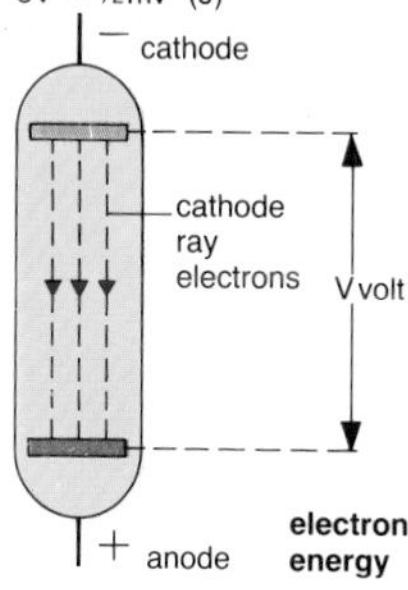

electron energy

surface; the work function is overcome by energy from an external source in thermionic (↑) or photoelectric emission (↑) giving free electrons.

photoelectric effect refers to the production of free electrons by photoelectric emission (↑); photons (p.234) of energy E = hf (joule), where f (hertz) = incident radiation frequency (p.54) and h (J s) = Planck's constant (p.235), irradiate a surface of work function W (J) (↑); when hf < W, there is no emission of electrons from the surface; when hf = W, electrons are emitted from the surface with zero excess energy; when hf > W, electrons are emitted with excess kinetic energy $\frac{1}{2}mv^2$ (J) (p.27); the emission of a photoelectron occurs by a quantized energy transfer from a photon in the incident light beam to an electron in the irradiated surface; the energy quantum (p.234) must be sufficient to overcome the work function energy; the threshold condition for photoelectric emission is $W = hf_o$, where f_o (Hz) is the threshold frequency (↓) characteristic of the emitting surface; for $f > f_o$: $hf = W + \frac{1}{2}mv^2 = hf_o + \frac{1}{2}mv^2$, so kinetic energy of an emitted photoelectron $\frac{1}{2}mv^2 = h(f - f_o)$ (J); this energy remains unchanged until the incident photon energy hf is changed; increase in the intensity of illumination (↓) of the surface increases the number of electrons emitted by increasing the number of incident photons, but has no effect on the energy of emitted photoelectrons; the photoelectric effect is evidence for the photon theory of the nature of light, proposed by Einstein in 1905; it occurs with suitable photo-sensitive surfaces under irradiation by visible light, ultraviolet radiation or infrared radiation (p.73), and can occur when metals are irradiated with X-rays (p.100).

threshold frequency the minimum frequency f_o (hertz) for which the photoelectric effect (↑) can occur for a specific surface; it is characteristic of the surface, and $hf_o = W$, where W (joule) is the characteristic work function (↑) of the surface and h (J s) = Planck's constant (p.235).

photoelectric cell a light-sensitive device capable of detecting variation in the intensity of illumination (↓) of its light-sensitive surface and recording the variation as electric current, e.g. photoemissive, photovoltaic cells (p.230).

intensity of illumination light energy received/s/m^2 at normal incidence (p.63) on an illuminated surface.

photoemissive cell a photoelectric cell (p.229) whose sensitivity to the variation in intensity of illumination (p.229) of its light-sensitive surface causes a corresponding variation in the number of electrons emitted from the surface by the photoelectric effect (p.229); a low work function surface is composed of mixed silver and caesium or potassium oxides for the detection of the wavelengths present in white light (p.73); it shows saturation current in the characteristic curve above a certain minimum operating voltage, when all emitted photoelectrons are collected by the anode electrode; this saturation current varies with intensity of illumination of the surface and acts as a measure of illumination when recorded as electric current; the emitting surface and anode can be contained in a vacuum tube, giving a very accurate response to illumination changes and direct proportionality between measured current and illumination; alternatively the tube can contain an inert gas, e.g. argon, at reduced pressure of a few torr (p.39), giving increased current due to ionization by collision and avoiding saturation.

photovoltaic cell a photoelectric cell (p.229) whose sensitivity to the variation in intensity of illumination (p.229) of its light-sensitive surface causes an e.m.f. (p.149) to be generated within the surface itself, giving a corresponding current proportional to the surface illumination; one type of photosensitive surface has an oxidized copper layer on a copper base, coated with a very thin transparent gold film; alternatively, an iron base is coated with selenium and a transparent gold film; the photovoltaic cell is less sensitive than the photoemissive cell, but sufficiently accurate for practical use as an industrial or photographic light meter.

photoconductive cell a photoelectric cell (p.229) whose sensitivity to the variation in intensity of illumination of its light-sensitive surface causes a decrease in electrical resistance (p.155) of the cell and a corresponding increase in electric current; the semiconductor cadmium sulphide is commonly used.

light dependent resistor (LDR) an alternative name for a photoconductive cell (↑).

cathode-ray tube (CRT) describes an evacuated tube with cathode rays (p.228) in transit between cathode and anode electrodes, e.g. X-ray tube (p.236), cathode-ray oscilloscope (p.232), T.V. tube (p.232).

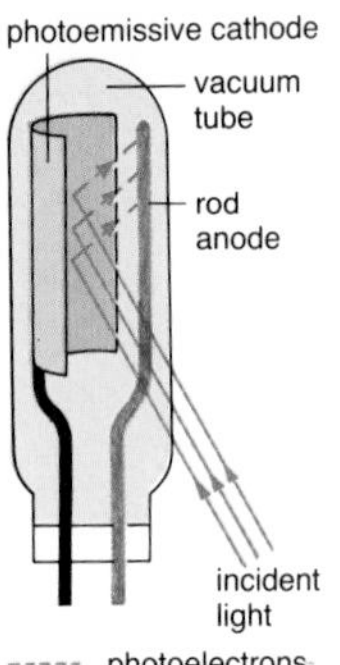

photoemissive cell

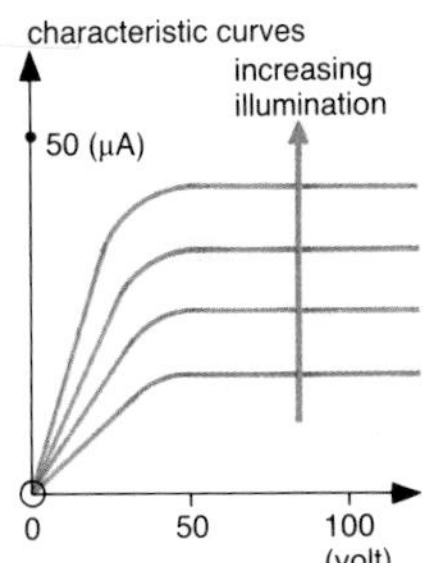

characteristic curves

photovoltaic cell

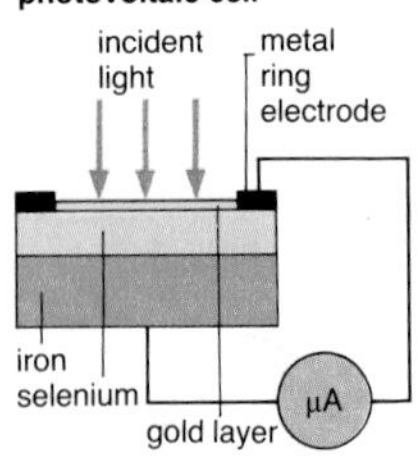

electron beam electrons in transit in a CRT (↑).

electron gun the component parts of a cathode-ray oscilloscope (p.232) from which high energy accelerated cathode-ray electrons emerge; it includes an indirectly heated cathode from which electrons are emitted by thermionic emission (p.228), a surrounding grid at negative electric potential controlling the number of electrons leaving the cathode region and so the brightness of the spot on the fluorescent screen (↓), and an arrangement of anode electrodes at high positive electric potential which both accelerate and focus (↓) the electron beam (↑).

electron focusing system the component parts of a cathode-ray oscilloscope (p.232) acting to converge the electron beam (↑) as it passes through the electron gun (↑); negatively charged electrons tend to diverge the beam by their own repulsive forces due to their like electric charges (p.165); the equipotential surfaces (p.171) of the electric field between the arrangement of accelerating anodes oppose this tendency and converge the beam to a sharply focused spot on the fluorescent screen (↓).

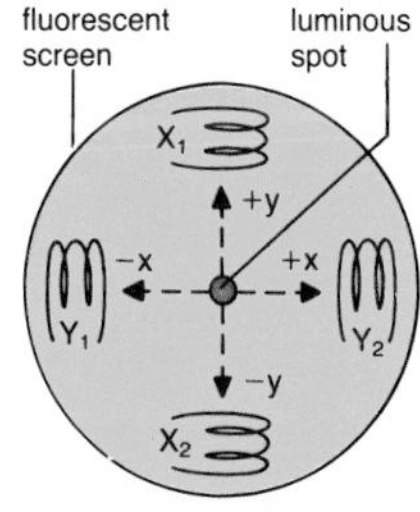

magnetic deflection of electrons
X_1X_2 coils give horizontal deflection
Y_1Y_2 coils give vertical deflection

electric deflection of electrons deflection of the electron beam (↑) in a cathode-ray oscilloscope (p.232) by the attractive and repulsive forces exerted by the electric field (p.165) between 2 pairs of electric field plates placed symmetrically round the beam; the X-plates impart a horizontal and the Y-plates a vertical deflection to the beam, the resultant deflection giving a trace on the fluorescent screen (↓).

magnetic deflection of electrons 2 pairs of magnetic field coils placed vertically and laterally with respect to the electron beam (↑) give X and Y deflections.

fluorescent screen the cathode-ray oscilloscope (p.232) or T.V. tube (p.232) opens out at its end into a circular or rectangular plane screen of fluorescent material coated over the tube glass; the screen receives the impact of the accelerated electron beam (↑) from the electron gun (↑) and magnetic or electric deflection (↑) system, showing a sharply focused fluorescent spot, of a colour characteristic of the screen material, e.g. zinc sulphide-green, as kinetic energy (p.27) is transferred to it from the electrons; the screen exhibits the resultant trace of the deflected beam; electrons are conducted away from the screen by a graphite coating on the tube.

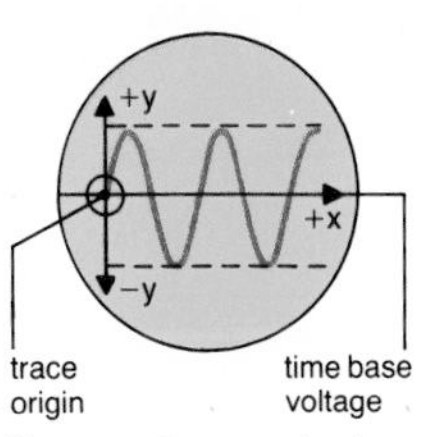

fluorescent screen showing oscilloscope voltage trace

time-base voltage a voltage (p.155) increasing linearly with time for a specific time interval, at the end of which it falls very rapidly to zero, before repeating this cycle of linear increase followed by rapid flyback; the voltage has a saw-tooth waveform and, when applied to the X-plates or X-coils of the magnetic or electric deflection system (p.231) of a cathode-ray oscilloscope (↓), it gives a lateral trace across the fluorescent screen (p.231) with rapid flyback to the trace origin, at the point where the electron beam (p.231) initially impinges; this point is given a lateral bias away from the screen centre to a point near the screen edge, so that the electron beam is deflected across the full width of the screen; a T.V. tube (↓) has 2 time-base voltages scanning the screen simultaneously in horizontal and vertical directions.

cathode-ray oscilloscope a cathode-ray tube (p.230) incorporating an electron gun, electric deflection system and fluorescent screen (p.231) for the visual display and measurement of voltage-dependent quantities.

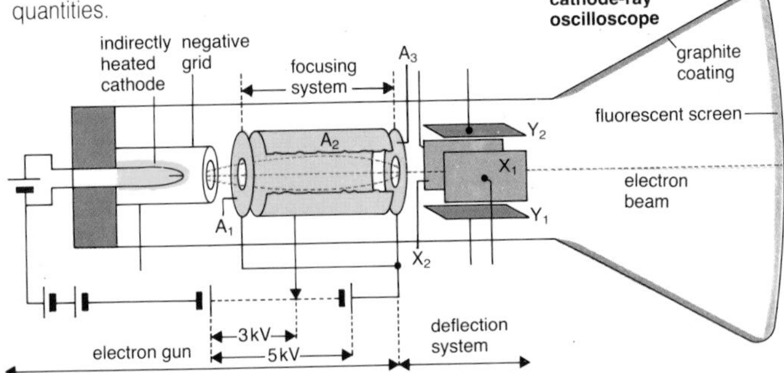

television T.V. tube a cathode-ray tube (p.230), with cathode-ray oscilloscope (↑) construction, and associated receiving circuitry and aerial employed to detect electromagnetic radio waves generated by a radio or T.V. transmitter; the T.V. signals are detected with their carrier waves, which are then demodulated to separate the sound and picture signals; these are then passed to audio- and video-amplifiers, before applying audio-signals to a loudspeaker for conversion into speech or music, and video-signals to the T.V. tube for conversion into a black and white, or coloured, picture; the video-signal imposes its variations in intensity on

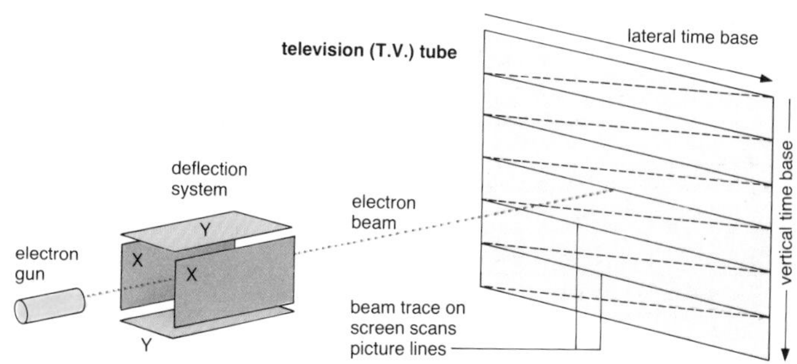

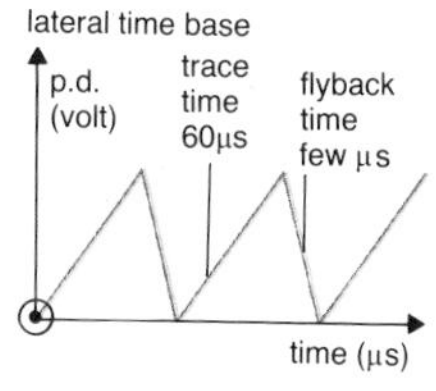

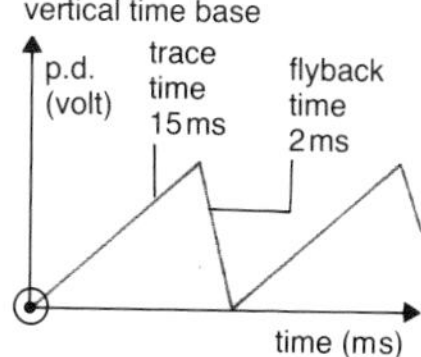

the intensity of the electron beam emerging from the electron gun (p.231) in the T.V. tube, resulting in variation in brightness of the trace scanning the fluorescent screen (p.231) and varying contrast which reveals detail as a black and white picture is built up.

colour T.V. tube a T.V. tube (↑) with 3 electron guns, each carrying a different colour signal: red, blue and green; the fluorescent screen is covered by a shadowmask perforated sheet with thousands of tiny holes, each hole revealing a composite phosphor dot, comprising 3 individual closely spaced red, blue and green-sensitive phosphorescent dots, located with extreme precision, in exactly the same positions relative to each other and the hole, for each shadowmask hole; the 3 guns are accurately aligned in the T.V. tube so that each electron beam stimulates only its appropriate colour-phosphor dot in each shadowmask hole; the combined colour signal variations on a composite dot build up a colour picture as line scanning of the screen area is carried out by the time-base voltages.

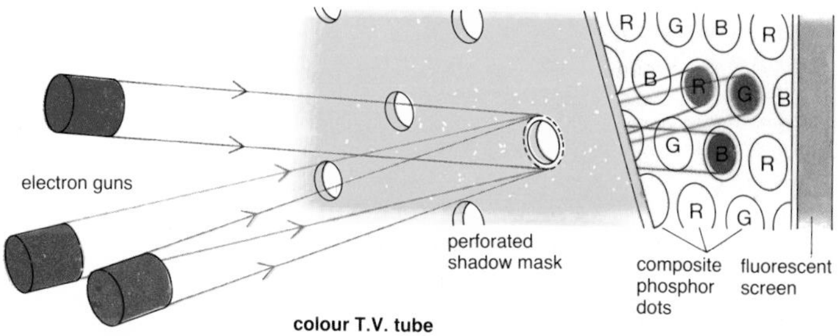

colour T.V. tube

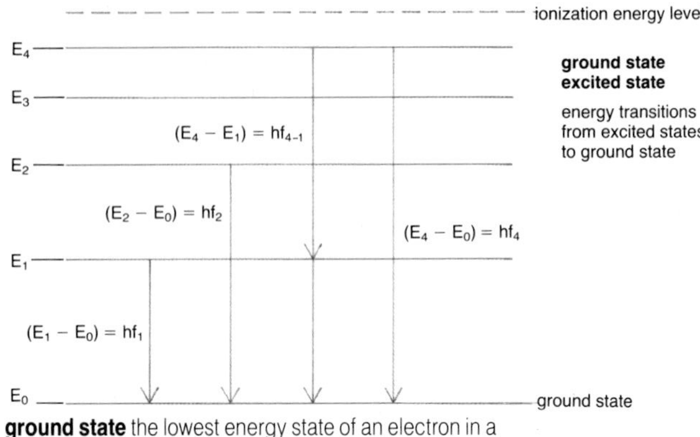

ground state the lowest energy state of an electron in a particular energy level (p.7) in an atom.

excited state the energy state of an electron which has gained enough energy to raise it to a higher energy level (↓) above the ground state (↑). **excitation** (*n*).

discrete energy level an orbital energy level (p.7) in which an electron orbiting an atomic nucleus (p.7) can remain without losing energy as electromagnetic radiation (p.55); electron shells K, L, M, N represent such orbitals and an electron in one of these will have a unique energy characteristic of that shell.

hyperfine level an energy level (p.7) observed when a magnetic field (p.180) is applied.

quantized electron transition electrons orbiting an atomic nucleus can only reside in discrete energy levels (↑); an electron can make a jump or transition between one energy level and another by gaining or losing a specific amount of energy equal to the energy difference between the 2 levels.

energy quantum the discrete or unique amount of energy gained or lost by an electron in making a quantized transition (↑) within an atom. **quantization** (*n*), **quantize** (*v*), **quantized** (*adj*).

quantum theory the principle that all electron transitions (↑) between energy levels (↑) within an atom must be quantized (↑); on transition from a higher energy state to a lower energy state, emitted photons (↓) are quanta (↑) of specific characteristic energy.

photon (*n*) the energy pulse emitted from an atom when an electron in an excited energy state (↑) returns to a lower or ground energy state (↑).

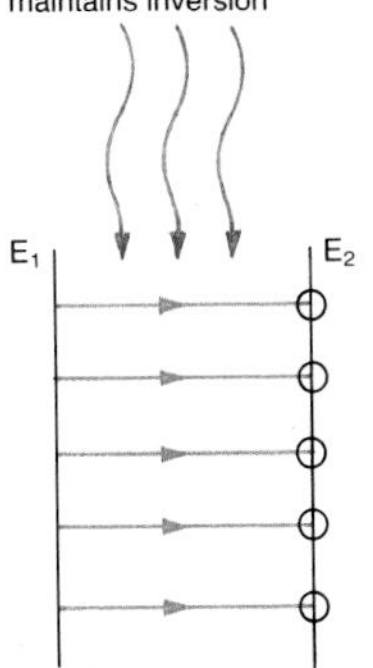

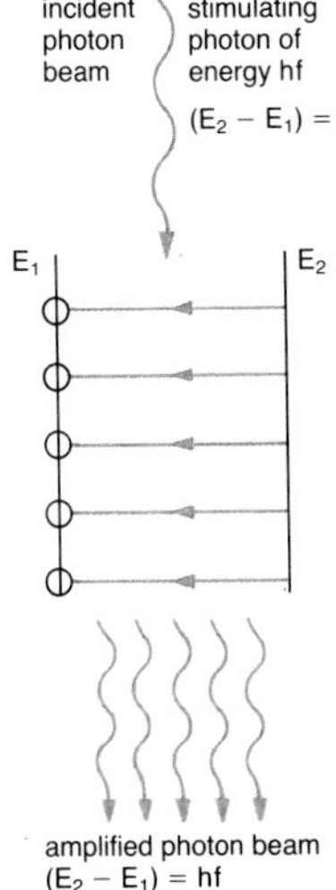

photon energy the energy E (joule) of an emitted photon (↑) is directly proportional to the frequency f (hertz) of electromagnetic radiation emitted from an atom; $E \propto f$ and $E = hf$, where h = Planck's constant (↓); if E_0 = ground energy state (↑) and E_1 = excited energy state (↑), then $(E_1 - E_0) = E = hf$.

Planck's constant the proportionality constant in the expression $E = hf$ for photon energy (↑); $h = 6.62 \times 10^{-34}$ joule second (J s).

stimulated emission a process whereby an incoming photon (↑) of energy hf can stimulate an electron in a high energy state E_2 to jump to a lower energy state E_1, where $hf = E_2 - E_1$; the photon resulting from this process has the same frequency $(E_2 - E_1)/h$ as the stimulating photon and travels in the same direction; if there are sufficient electrons in the high energy level the stimulated photons can cause further stimulated emission and a narrow beam of monochromatic (p.74) radiation results, the intensity of which increases exponentially; the beam is coherent (p.91) and has a very high energy density as in a laser (↓).

laser acronym for Light Amplification by Stimulated Emission (↑) of Radiation; the laser operates by producing a large number of electrons in a particular high energy state; this condition, called population inversion, is a non-equilibrium state and power must be fed into the system to maintain the inversion; materials that are used to make lasers include ruby, helium-neon and carbon dioxide; laser light is collimated (p.88) monochromatic (p.74), coherent (p.91) and has a high energy density.

ionization energy the energy required to remove an electron from its normal ground state (↑) to the ionization energy level (p.7) where it is considered to be no longer associated with that particular atom.

X-rays a continuous beam of electromagnetic radiation (p.55) is emitted from the target (↓) of an X-ray tube (p.236) under continuous bombardment with high energy cathode-ray electrons (p.228); its wavelength (p.54) range is approximately 10^{-12}–10^{-14} m and the photon energy (↑) range is approximately 30–50 keV; characteristic high energy peaks occur for the specific target (↓) material. **X-radiation** (*n*).

target (*n*) refers to a heavy metal, e.g. tungsten, anode of an X-ray tube (p.236); it emits X-rays (↑) on bombardment with high energy electrons.

X-ray tube vacuum cathode-ray tube (p.230) with sealed-in cathode and anode; operated at high voltage, e.g. 50 kV; the cathode acts as an electron source by thermionic emission (p.228); accelerated cathode-ray electrons (p.228) are focused on to the tungsten target (p.235) embedded in the copper anode, which rapidly conducts away heat to the cooling fins; approximately 99.5% of the cathode-ray kinetic energy is converted to heat on target impact, only 0.5% causing the production of X-rays (p.235); the X-ray beam is given an approximate direction by the target alignment to the tube axis.

X-ray diffraction refers to 3-dimensional diffraction (p.61) of electromagnetic radiation of X-ray wavelengths by the 3-dimensional array of crystal unit cells (p.12) in a single crystal; constructive interference (p.58) between wavetrains diffracted by adjacent parallel crystal lattice planes gives the condition defined in Bragg's Law (↓), from which the characteristic dimensions of the crystal unit cell can be calculated.

Bragg's Law defines the constructive interference condition (p.58) observed in X-ray diffraction (↑) by the relationship $2d \sin \theta = n\lambda$, where n = integral number and λ (m) = wavelength of the characteristic X-ray (p.235) spectrum line used; λ is previously determined using a single crystal of known dimensions and θ is measured for known n, so that d (m) can be calculated; the technique is used in crystallography for identification of substances of simple structure in single crystal (p.12) form.

X-ray powder photography a technique for obtaining an X-ray diffraction (↑) photograph of a polycrystalline specimen or a crystalline powder; X-rays directed at the specimen are diffracted by many crystal planes oriented at the Bragg angle to the incident beam; as a result, the diffracted beams form the surface of cones which intersect X-ray film to produce lines from which the internal structure of the crystals may be calculated.

electron density measurements the modern method of analysis of crystal structure replacing Bragg's Law (↑) as the method for identification of substances of complex crystalline structure (p.12), e.g D.N.A., insulin, penicillin; diffraction patterns are recorded photographically and calculations of complex molecular structure are carried out by computer.

X-ray tube

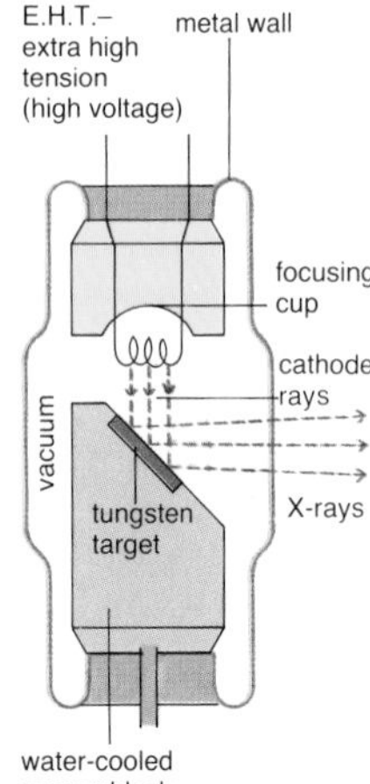

Braggs Law
for X-ray diffraction

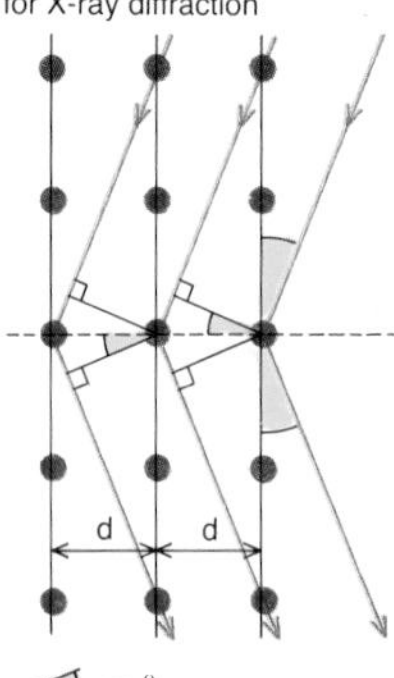

nuclear model of the atom the model of the atom proposed by Rutherford in 1911 in which the atom consists mainly of empty space but contains a minute, massive, positively charged nucleus (p.7) surrounded by electrons; the results of the Geiger and Marsden scattering experiment (↓) were as predicted by this model.

Geiger and Marsden scattering experiment the experiment in which alpha-particles (p.239) were directed towards thin metal foils and the number of alpha-particles scattered at different angles counted; since most of the alpha-particles passed straight through the foil undeflected, and only a few were scattered through large angles, it was assumed that atoms were mainly empty space but had small positively charged nucleii.

Geiger and Marsden scattering experiment

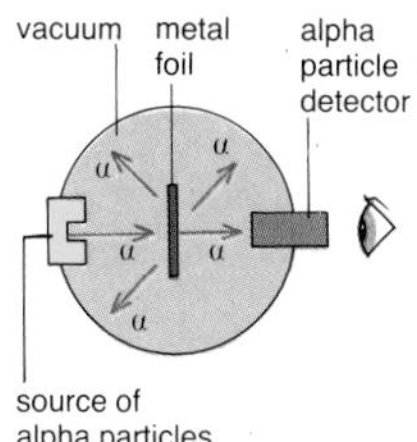

Rutherford scattering law the relationship deduced by Rutherford for the number of alpha-particles scattered through a given angle by a metal foil in terms of the proton number (p.8) of the metal and initial kinetic energy of the alpha-particles, assuming an inverse square law of repulsion between an alpha-particle and a nucleus.

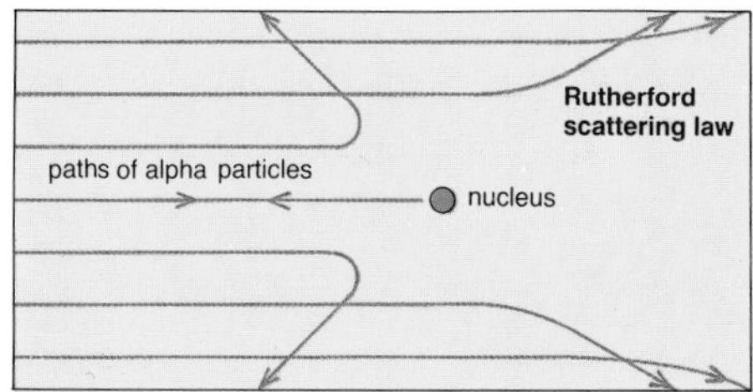

stable nucleus the nucleus (p.7) of an atom in which the proportion of neutrons to protons is such that nuclear attractive forces (p.8) exceed electrostatic repulsive forces (p.166) between similarly charged protons and the nucleus does not disintegrate.

unstable nucleus the nucleus (p.7) of an atom in which the proportion of neutrons to protons is such that electrostatic repulsive forces (p.166) between similarly charged protons exceed nuclear attractive forces (p.8) and the nucleus tends to disintegrate spontaneously into smaller and more stable fragments; the nuclide (p.8) shows spontaneous radioactivity (p.239).

nuclear disintegration the natural tendency of an unstable nucleus (p.237) to break up spontaneously into smaller and more stable fragments emitting certain radioactive radiations (↓). **disintegrate** (*v*).

nuclear fission the breaking up of a stable or unstable nucleus (p.237) into smaller fragments by bombardment with heavy particles, e.g. neutrons in a nuclear reactor; the fission process releases large amounts of energy held in the nucleus in association with nucleons (p.7) bound together by a binding energy (↓) which depends on the mass number (p.8); fission is accompanied by the emission of gamma-radiation (↓) from the nucleus, and violent fission, as in a nuclear fission bomb, by large amounts of heat and blast shock waves (p.56). **fission** (*v*).

binding energy the total mass of the individual nucleons (p.7) comprising a heavy nucleus, e.g. uranium-235 ($^{235}_{92}U$), is greater than their mass when bound together by nuclear forces (p.8) in a nucleus; this mass defect or binding energy is considered to be shared between all nucleons in the nucleus and is partly released as energy on nuclear fission (↑); the heavy nucleus breaks into 2 or more fragments whose binding energy per nucleon is greater than in the parent nucleus and this greater mass defect results in the release of fission energy.

nuclear fusion 2 light nuclei, e.g. hydrogen $^{1}_{1}H$ or deuterium $^{2}_{1}H$, can fuse together to form a single nucleus of greater binding energy (↑) per nucleon, e.g. helium $^{4}_{2}He$; the greater mass defect results in the release of fusion energy as gamma-radiation (↓), and violent fusion, as in the nuclear fusion hydrogen bomb, gives heat and blast shock waves (p.56); fusion occurs in plasma (p.9) in a torus, e.g. JET (p.29).

radioactive nucleus (radionuclide) an unstable nucleus (p.237) capable of emitting radioactive radiations (↓) on nuclear disintegration (↑).

radioisotope (*n*) a radioactive (↑) isotope (p.8) usually found in trace quantities in a naturally occurring mixture of stable isotopes, e.g. carbon-14 in natural carbon compounds, tritium ($^{3}_{1}H$) in water (H_2O) (p.8); radioisotopes can be formed artificially by neutron bombardment in a nuclear reactor.

radioactive radiations the single name given to alpha- (↓) and beta-particles (↓) and gamma-radiation (↓), emitted as products of nuclear disintegration (↑).

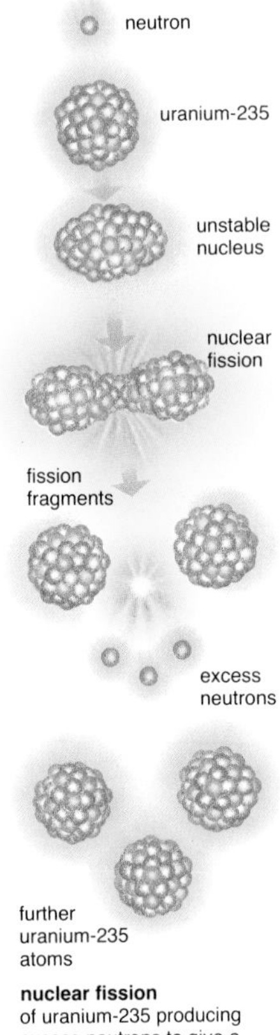

nuclear fission of uranium-235 producing excess neutrons to give a neutron chain reaction

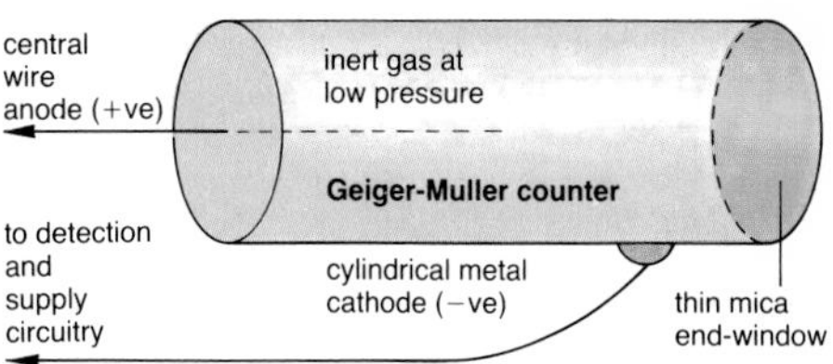

Geiger-Muller counter a radiation detector (p.241) consisting of a gas-filled tube containing two electrodes with a high voltage between them; incoming radiation ionizes the gas causing a pulse of charge to flow between the electrodes and be counted by a scaler (p.241) circuit.

background radiation the low intensity radiation (↑) resulting from cosmic rays and from naturally occurring radioisotopes (↑) in rocks, soil, air, building materials, etc.

alpha (α)-particle refers to a helium nucleus ($^4_2$He), emitted from certain nuclei undergoing spontaneous nuclear disintegration (↑) or nuclear fission (↑), it has a very short range (↓) and is readily absorbed by air, paper or the surface of the skin of the body.

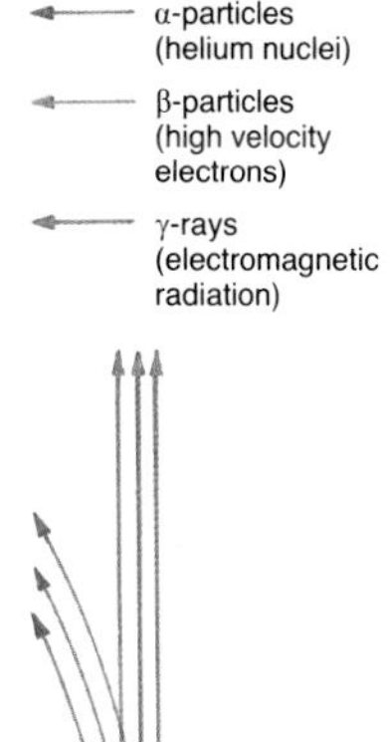

radioactivity
deflection of α- and β-particles by a magnetic field; γ-rays are undeflected

beta (β)-particles electrons of high kinetic energy (p.27) emitted from certain nuclei undergoing spontaneous nuclear disintegration (↑) or nuclear fission (↑); they have varying energies and ranges (↓), e.g. 10–100 cm in air, and are absorbed by thin metal; some high energy β-particles show a relativistic (p.38) variation of mass with velocity.

gamma (γ)-radiation high energy photons (p.234) above 50 keV, emitted as short wavelength (p.54) (10^{-14} m) electromagnetic radiation (p.55) from most nuclei undergoing spontaneous nuclear disintegration (↑) or nuclear fission (↑); γ-rays are highly penetrating and have no definite range (↓), being effectively absorbed only by thick lead or concrete according to an exponential absorption law as for X-rays.

particle range the distance a particle of specific kinetic energy (p.27) can travel in a specific medium before all its energy is absorbed, e.g. α-particles (↑) have a range of only a few cm in air.

radioactivity (*n*) the emission of one or more radioactive radiations (↑) from a substance whose atoms have radioactive nuclei (↑). **radioactive** (*adj*).

radioactive decay the continual loss of radioactive nuclei (p.238) from a sample of radioactive substance by the process of spontaneous nuclear disintegration (p.238). **decay** (*v*).

radioactive disintegration rate the number of radioactive nuclei (p.238) disintegrating per second in a sample of radioactive substance; units are curie (↓) and becquerel (↓).

curie (Ci) a unit of activity (↓); 1 curie = 3.7×10^{10} disintegrations/second.

becquerel (Bq) a unit of activity (↓); 1 bequerel = 1 disintegration/second.

activity (*n*) an alternative name for radioactive disintegration rate (↑).

law of radioactive decay an exponential law relating the number of radioactive nuclei N_0 (p.238), initially present in a sample of radioactive (p.239) substance at time $t = 0$, to the number N remaining after time t and given by $N = N_0 \exp(-\lambda t)$, where λ = decay constant (↓) for the specific radioactive nucleus; experimentally the energy intensity I (Wm^{-2}) (p.57) of the emitted radioactive radiation (p.238) is measured as being directly proportional to N at time t (s) so that $I = I_0 \exp(-\lambda t)$; the law implies zero activity (↑) after infinite time.

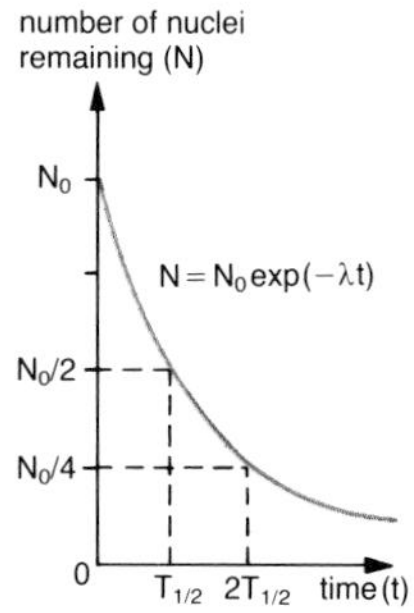

$T_{1/2}$ = radioactive half life
law of radioactive decay

decay constant a constant representing the probability of a radioactive nucleus (p.238) of a specified substance undergoing disintegration at any instant of time; denoted by λ with units of s^{-1}.

radioactive half-life the time taken for the number of radioactive nuclei N (p.238), present at time t (s) in a sample of radioactive (p.239) substance, to decay to half its value N/2; denoted by $T_{1/2}$; units are μs, ms, s, h, y; time $T_{1/2}$ is characteristic of the specific radioactive substance, e.g. for iodine-131 ($\beta\gamma$), 8 days; for plutonium-239 (α), 24400 years.

radioactive dating a method of determining the age of archeological and fossil remains, rocks, etc. by measuring the abundance (p.9) of one or two specific radioisotopes (p.238) contained in the sample; carbon-dating, for measuring the age of organic matter up to 10000 years old, is based on measuring the abundance of ^{14}C isotope, a radioisotope of half-life (↑) 5730 years, with that of ^{12}C, a stable isotope; potassium-argon dating is used for dating rocks up to 10^7 years old; rubidium-strontium dating is used for dating rocks up to several billion years old.

neutron chain reaction nuclear fission (p.238), by a single neutron (p.8), of a nucleus of uranium-235 ($^{235}_{92}U$), gives fission fragments and one or more neutrons, each of which is capable of producing a further nuclear fission; the chain reaction is self-sustaining when each successive fission by a neutron yields one other neutron; it becomes uncontrolled when excess neutrons are yielded; in the controlled and moderated process nuclear fusion energy (p.238) is released in a nuclear reactor (↓).

nuclear reactor a specially designed large-scale commercial plant in which the heat energy released by controlled nuclear fission (p.238) in a neutron chain reaction (↑) is removed from the core by a coolant and used to generate steam for a conventional turbo-alternator set at a nuclear power station.

radiation detector a device which takes energy (p.26) from ionizing (p.235) radiation for the purpose of measurement.

scaler an electronic device for counting pulses, e.g. in association with a radiation detector (↑) such as a spark counter (↓).

spark counter a radiation detector (↑) consisting of a wire mesh close to a fine wire with high voltages between them such that ionizing (p.235) radiation passing through sets off sparks; particularly suitable for detecting alpha-particles (p.239).

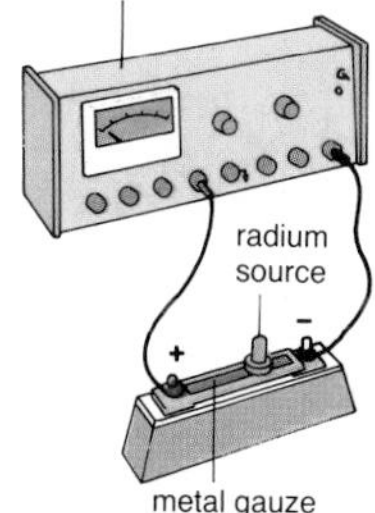

cloud chamber a radiation detector (↑) containing supersaturated vapour, e.g. alcohol or water, which condenses as ions are formed by the incoming radiation; liquid condensing on these ions produces visible tracks which may be photographed.

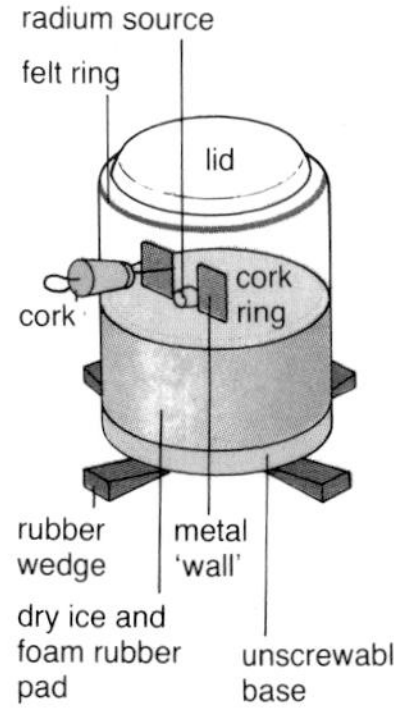

bubble chamber a radiation detector (↑) containing a liquid such as hydrogen or helium which is about to boil; incoming radiation ionizes atoms in the liquid, the energy causing the liquid to form bubbles of vapour along the track of the radiation; these tracks are therefore visible.

Bohr theory of atom states that an electron in an atom occupies one of a number of fixed orbits and that when in that orbit it does not radiate energy; energy is emitted when it jumps from one orbit to another.

Quantum theory the theory based on Planck's ideas that physical systems can only possess certain properties, such as energy (p.26) or angular momentum (p.5) in certain amounts (quanta).

Units and measurement

physics the study of energy (p.26) and the properties (↓) of matter and of the relationships between them. **physical** (*adj*).

physical property something we can observe about the physical appearance or behaviour of a body or substance, e.g. physiological colour (p.77), elasticity (p.19).

physical quantity a physical property (↑) of matter or energy (p.26) which we can observe and measure, e.g. mass (p.7), force (p.26), electric current (p.152).

physical field a volume of space throughout which forces (p.26) can act and in which energy (p.26) is available, e.g. gravitational (p.42), electric (p.165) and magnetic fields (p.180).

physical standard a standard amount of a specified physical quantity (↑) with which other amounts of that quantity can be compared, e.g. standard metre, kilogram, second, ampere on the SI system of units (↓); it must be practically realized in a precisely agreed and reproducible arrangement of equipment or apparatus from which accurate measurements can be made.

physical unit a precisely defined amount of a specified physical quantity, related directly to physical standards (↑) of measurement, e.g. milliampere ($1\,mA = 10^{-3}A$) for electric current (p.152), newton for force (p.26) ($1\,N = 1\,kg\,m\,s^{-2}$); a measurement of a physical quantity is stated as a number and a unit, e.g. peak mains voltage (p.155) is 250 volt.

SI system of units (Système International d' Unités) the internationally accepted system of basic units (↑) of measurement with which all other units can be compared and from which all others can be derived.

calibration (*n*) comparison of a specified but uncertain amount of a physical quantity (↑) with a physical standard (↑), e.g. an electric lamp of specified luminous intensity (p.229) with a standard lamp; comparison of a scale of measurement of doubtful accuracy with physical standards at 2 or more points on the scale, e.g. thermocouple characteristic (p.164), calibration of a voltmeter (p.198) or ammeter (p.198) using a calibrated laboratory potentiometer (p.162), calibration of a mercury-in-glass thermometer (p.119) on the Celsius temperature scale (p.119).

linear scale equal changes in the value of the physical quantity (↑) being measured are indicated by equal changes on the scale of the measuring instrument, e.g. graduations on a mercury-in-glass thermometer (p.119) and on a moving coil galvanometer (p.196) scale.

non-linear scale equal changes in the value of the physical quantity (↑) being measured are indicated by unequal changes on the scale of the measuring instrument, e.g. graduations on the scale of a moving iron (p.199) ammeter or voltmeter.

graph (*n*) an accurate representational display of the variation in value of one physical quantity (↑) (a dependent variable), with changes in the value of another related physical quantity (a dependent variable); corresponding pairs of values each give a graph point with respect to co-ordinate axes, and joining these points to give a straight line or curve shows the continuous variation in value of one variable with changes in the other; a linear graph has characteristic gradient and intercept; for a relationship of direct proportionality the graph passes through the origin, e.g. variation of volume of a fixed mass of gas with pressure according to Boyle's Law (p.137).

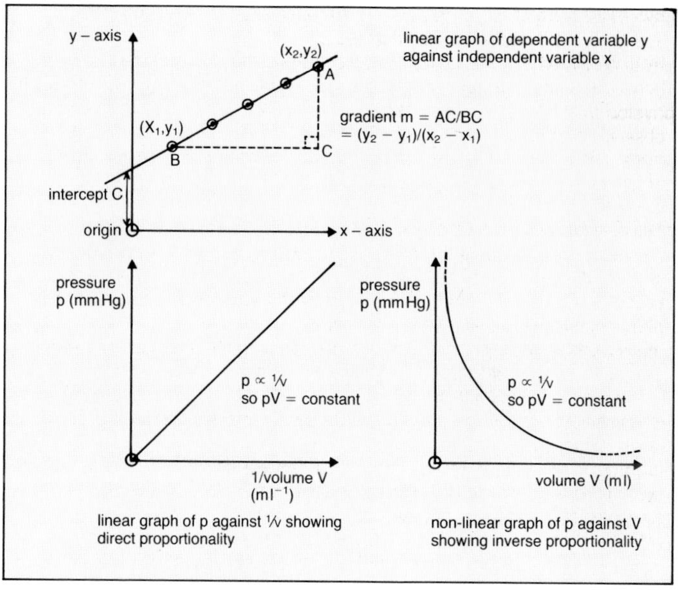

International System of Units (SI)

Prefixes for SI units

MULTIPLE	FIGURE	PREFIX	SYMBOL
10^{12}	1 000 000 000 000	tera	T
10^{9}	1 000 000 000	giga	G
10^{6}	1 000 000	mega	M
10^{3}	1 000	kilo	k
10^{-3}	0.001	milli	m
10^{-6}	0.000 001	micro	μ
10^{-9}	0.000 000 001	nano	n
10^{-12}	0.000 000 000 001	pico	p
10^{-15}	0.000 000 000 000 001	femto	f
10^{-18}	0.000 000 000 000 000 001	atto	a

Basic units

The SI (Système International d'Unités) system of units has 7 basic units from which all derived units are obtained. Multiples and sub-multiples of the basic units may be used with approved prefixes.

1. metre *(unit of length)* *symbol:* m The metre is the length equal to 1 650 763.73 wavelengths (p.54) in vacuum corresponding to the quantized electron transition (p.234) between energy levels $2p_{10}$ and $5d_5$ of the krypton-86 atom.

2. kilogram *(unit of mass)* *symbol:* kg The kilogram is the unit of mass (p.7) equal to the mass of the international prototype kilogram kept at Sèvres, France.

3. second *(unit of time)* *symbol:* s The second is the duration of 9 192 631 770 periods of the radiation corresponding to the quantized electron transition (p.234) between the 2 hyperfine levels (p.234) of the ground state (p.234) of the caesium-133 atom.

4. ampere *(unit of electric current)* *symbol:* A The ampere is that constant electric current (p.152) which, if maintained in 2 straight parallel conductors of infinite length, of negligible cross-section and placed 1 metre apart in vacuum, would produce between these conductors a force equal to 2×10^{-7} newton/metre (p.201).

5. kelvin *(unit of temperature)* *symbol:* K The kelvin, unit of thermodynamic temperature (p.145), is the fraction 1/273.16 of the thermodynamic temperature of the triple point of water (p.146).

6. candela *(unit of luminous intensity)* *symbol:* cd The candela is the luminous intensity (p.62), in a perpendicular direction, of a surface 1/600 000 metre^2 of a black body at the freezing point of platinum at a pressure (p.39) of 101 325 newton/metre^2.

7. mole *(unit of amount of substance)* *symbol:* mol The mole (p.11) is the amount of substance containing as many elementary units as there are carbon atoms in 0.012 kilogram of carbon-12. The elementary unit may be an atom, molecule, ion or electron.

Two supplementary units are also used:

radian *(unit of plane angle)* *symbol:* rad The radian is the unit of measurement of angle and is the angle subtended at the centre of a circle by an arc equal in length to the circle radius.

steradian *(unit of solid angle)* *symbol:* sr The steradian is the unit of measurement of solid angle and is the solid angle subtended at the centre of a circle by a spherical cap equal in area to the square of the circle radius.

Some useful physical constants

PHYSICAL CONSTANT	SYMBOL	VALUE
electron mass	m	9.11×10^{-31} kg
electronic charge	e	1.60×10^{-19} C
specific charge	e/m	1.76×10^{11} C kg^{-1}
atomic mass unit	a.m.u.	1.660×10^{-27} kg
proton mass	1.007 a.m.u.	1.673×10^{-27} kg
neutron mass	1.009 a.m.u.	1.675×10^{-27} kg
Avogadro constant	N_o	6.02×10^{23} mol^{-1}
Faraday constant	F	9.65×10^{4} C mol^{-1}
acceleration due to Earth's gravity	g	9.81 m s^{-2}
universal gravitational constant	G	6.67×10^{11} N m^2 kg^{-2}
velocity of light, electromagnetic waves in free space	c_o	2.99×10^{8} m s^{-1}
velocity of sound waves in air at 0°C	—	330 m s^{-1}
Stefan's constant	σ	5.67×10^{-8} W m^{-2} K^{-4}
electric permittivity of free space	ε_o	8.85×10^{-12} F m^{-1} (C^2 N^{-1} m^{-2})
magnetic permeability of free space	μ_o	$4\pi \times 10^{-7}$ H m^{-1} (Wb A^{-1} m^{-1})
electron – volt	eV	1.60×10^{-19} J
Planck's constant	h	6.62×10^{-34} J s
Angstrom Unit	Å	10^{-10} m

Some common physical quantities and their units

PHYSICAL QUANTITY	SYMBOL	S.I. UNIT	SYMBOL
acceleration, deceleration	a	metre/second^2 kilometre/hour/second	ms^{-2} $kmh^{-1}s^{-1}$
angular velocity	ω	radians/second	$rads^{-1}$; s^{-1}
capacitance	C	farad (coulomb/volt)	F (CV^{-1})
coefficient of viscosity	η	poise, dekapoise (newton second/metre^2) (kilogram/metre/second)	 Nsm^{-2} $kgm^{-1}s^{-1}$
density	ρ	kilogram/metre^3 kilogram/millilitre	kgm^{-3} $kgmL^{-1}$
displacement, distance	S	metre	m
electric charge	Q, q	coulomb	C
electric current	I, i	ampere (coulomb/second)	A Cs^{-1}
electrical energy	—	megajoule, kilowatt-hour	MJ kWh
electric intensity, field strength	E ($= -dV/dr$)	newton/coulomb volt/metre	NC^{-1} Vm^{-1}
electric p.d.	V	volt (joule/coulomb)	V (JC^{-1})
electrical power	—	watt (joule/second)	W (Js^{-1})
electromotive force (e.m.f.)	E	volt (watt/ampere)	V (WA^{-1})
electrical conductance	S	siemen, ohm^{-1}	AV^{-1}
electrical resistance	R	ohm (volt/ampere)	Ω (VA^{-1})
electric permittivity	ε	farad/metre	Fm^{-1}
frequency	f	hertz (cycles/second)	Hz (s^{-1})
force	F	newton (kilogram metre/second^2)	N ($kgms^{-2}$)
gravitational intensity, field strength	—	newton/kilogram	Nkg^{-1}
heat capacity of a body	ms	joule/degree Kelvin	JK^{-1}

PHYSICAL QUANTITY	SYMBOL	S.I. UNIT	SYMBOL
inductance	L	henry	H
		(volt second/ampere)	(VsA^{-1})
		(weber/ampere)	(WbA^{-1})
induced e.m.f.	e	volt	V
		(weber/second)	(Wbs^{-1})
magnetic field strength	H	ampere/metre	Am^{-1}
magnetic flux	Φ	weber	Wb
magnetic flux density	B	tesla	T
		(weber/metre^2)	(Wbm^{-2})
magnetic permeability	μ	henry/metre	Hm^{-1}
mass	m	kilogram	kg
mechanical power	—	watt	W
		(joule/second)	(Js^{-1})
moment of intertia	I	kilogram metre^2	kgm^2
momentum	mv	kilogram metre/second	$kgms^{-1}$
pressure	P	pascal	Pa
		(newton/metre^2)	Nm^{-2}
quantity of substance	—	mole	mol
specific heat capacity	s	joule/kilogram/ degree Kelvin	$Jkg^{-1}K^{-1}$
specific latent heats of fusion, vaporization	L	joule/kilogram	Jkg^{-1}
surface tension	T, γ	newton/metre	Nm^{-1}
torque, moment of force, moment of couple	—	newton metre	Nm
velocity gradient	dv/dr	metre/second/metre	$(ms^{-1}m^{-1})$
			s^{-1}
velocity, speed	u, v	metre/second	ms^{-1}
		kilometre/hour	kmh^{-1}
volume	V	metre^3	m^3
		millilitre	mL
wavelength	λ	metre	m
weight	W	newton,	N
		kilogram-force	kgf
work, energy	—	joule	J
		(newton metre)	(Nm)

Index

Circuit symbols

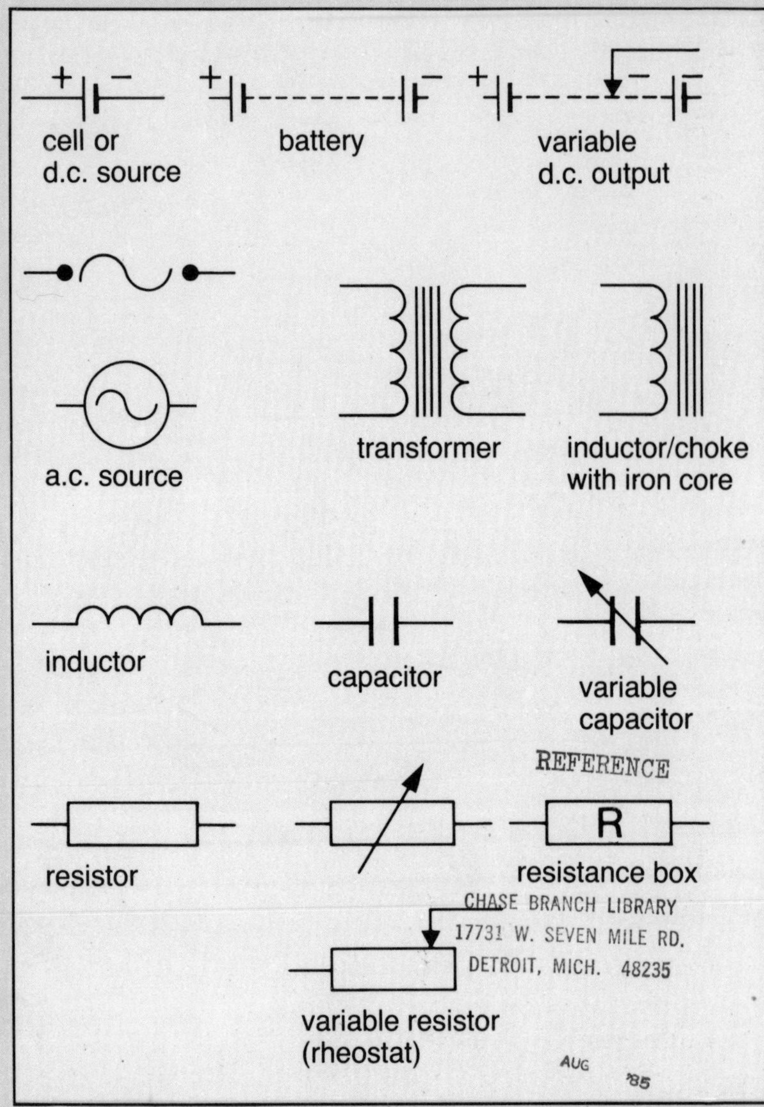

cell or
d.c. source
battery
variable
d.c. output
a.c. source
transformer
inductor/choke
with iron core
inductor
capacitor
variable
capacitor
resistor
R
resistance box
variable resistor
(rheostat)